Applied Geology

Marina De Maio • Ashwani Kumar Tiwari
Editors

Applied Geology

Approaches to Future Resource Management

Springer

Editors
Marina De Maio
DIATI
Polytechnic University of Turin
Turin, Italy

Ashwani Kumar Tiwari
DIATI
Polytechnic University of Turin
Turin, Italy

SES
Jawaharlal Nehru University
New Delhi, India

ISBN 978-3-030-43952-1 ISBN 978-3-030-43953-8 (eBook)
https://doi.org/10.1007/978-3-030-43953-8

This Springer imprint is published by the registered company Springer Nature Switzerland AG.
The registered company address is: Gewerbestrasse 11, 6330 Cham, Switzerland

Name: Marina De Maio
From: 28 October 1967 to 25 November 2019

Prof. Dr. Marina De Maio was a Professor of Applied Geology and GIS at Politecnico di Torino, Turin, Italy. She obtained her Master's degree in Geology from the University of Catania in 1993 and her PhD in Engineering Geology from Politecnico di Torino in 1998.

She was a Co-founder of ISE-NET, a spin-off of Politecnico di Torino, a Member of the Scientific Committee of the Fondazione Montagna Sicura and a Board Member of the International Association for Engineering Geology and the Environment (IAEG), the

International Association of Hydrogeologists (IAH) and the Italian Association of Applied and Environmental Geology (AIGA).

Furthermore, she was a Representative of Politecnico di Torino in the "European Technical and Scientific Committee (ETSC)" and European Water Association (EWA) and Senior Consultant for UNESCO. In the last 2 years, she was the Contact Person for the European network "Copernicus Academy" in Politecnico di Torino.

She has published many articles as Author and Coauthor in various reputed journals. Her contributions to the scientific community in the fields of geology, aquifer vulnerability, hydro-geochemistry, water contamination, climate change and GIS were very well appreciated by the researchers. She had always been strongly committed to planning in favour of the territory, especially the mountain area in relation to snow, glaciers and water resources.

In the last period, despite her illness, she still carried out important research activities and with great energy stimulated new scientific projects and training courses dedicated to the mountain in collaboration with the Università della Valle d'Aosta. She organised the 6th AIGA National Conference at Courmayeur, Italy, in 2018. She gave her full support and assistance during the review of the chapters and organised the book. Her efforts and contributions to this book will always be remembered.

She has never been discouraged, not even in the face of illness, and always supported her ideas with strength and determination, stimulating reflection on the importance of proper management.

Your colleagues will all feel the lack of your inexhaustible energy and unparalleled ability to put into practice the results of scientific activities.

We will always miss you, Prof. Marina De Maio.

From
Her family, colleagues and research group

Foreword

I take this opportunity to praise the engagement of Prof. Marina De Maio and Dr. Ashwani Kumar Tiwari in collecting articles selected among the most innovative studies presented during the 6th National AIGA (Italian Association of Applied and Environmental Geology) Congress. This book is the demonstration of the vitality of researchers belonging to various research institutions in different fields, such as hydrogeology, aquifer contamination, geomechanics, urban processes, landslide monitoring and control, climate change and natural hazards and risk management. These subjects represent typical and fundamental application areas of the technical geology.

There is no doubt that, in natural processes, past and current human activities are jointly responsible for events whose scenarios are often difficult to interpret. This is even more true today because of changes in precipitation pattern, with more intense and frequent extreme events.

The understanding of the mechanisms of the processes is therefore at the base of risk mitigation and in this latter can be identified as one of the key roles of applied geology and of researchers working for the development of knowledge.

The methodical and well-organised collection of articles, which the reader will find in this volume, will make it possible to investigate case histories of international scientific interest and to employ innovative technological applications that are increasingly essential both in data collection and in the monitoring of geological and environmental matrices. Therefore, this book provides strategies to make decisions for the sustainable management of environmental problems.

Prof. Francesco Maria Guadagno
Department of Science and Technology (DST)
University of Sannio, Italy
Former President of AIGA
(Associazione Italiana di Geologia Applicata ed Ambientale)

Preface

Recurrent natural hazards, depletion of water resources and their contamination, and rapidly rising atmospheric CO_2 levels are major concerns facing the globe, primarily due to a combination of various geogenic and anthropogenic processes. Lack of enough public awareness of environmental issues further complicates the scenario. In order to understand how pattern affects the environment, we must first be able to identify and quantify such patterns. Hence, this book aims to provide essential management tools (e.g., decision-support systems) with detailed information for a healthy and sustainable environment.

This book has 18 chapters that deal with the hydrogeological characterizations of springs, aquifer contamination, CO_2 sequestration, slope stability, climate change impact, landslide monitoring and control and role of remote sensing and GIS. It focuses on basic earth science of hazards, impacts of climate change and roles of human processes and its effects on the environment, providing a balanced approach to sustain the environment. It aids professors, researchers, students and policymakers to understand environmental earth processes and their consequences to society.

We are very grateful to the Italian Association of Applied and Environmental Geology (AIGA), Rome, Italy, and the Department of Environment, Land and Infrastructure Engineering (DIATI), Politecnico di Torino, Turin, Italy, for their support and for allowing the publication of this book. We thank all the authors who contributed their significant work in this book and all the reviewers for their valuable time and suggestions to improve this book. We are also grateful to Dr. Enrico Suozzi, DIATI, Politecnico di Torino, for his endorsement. Many thanks go to the

Springer publishing team for their dedication and support during the publication of this book. Finally, we thank the organisation committee, scientific members, sponsors and all participants for their support and encouragement during the 6th AIGA National Conference at Courmayeur, Italy.

Email: ashwani.enviro@gmail.com, ashwani.tiwari@polito.it

Turin, Italy
Turin, Italy
New Delhi, India

Marina De Maio
Ashwani Kumar Tiwari

Contents

About the Editors

Marina De Maio was a Professor of Applied Geology and GIS in the Department of Environment, Land and Infrastructure Engineering at Politecnico di Torino, Turin, Italy. She was a Representative of Politecnico di Torino in the "European Technical and Scientific Committee (ETSC)" and European Water Association (EWA) and Senior Consultant for UNESCO. She has published several articles as Author and Coauthor in various reputed national and international journals. Her research activity was mainly on geological risk assessment, aquifer vulnerability, pollution risk, water resources management, climate change, mining and GIS. She has developed some part of "SINTACS" method to evaluate aquifer venerability during her research work and has completed several international and national research projects. She has also served as Reviewer for many international and national journals.

Ashwani Kumar Tiwari is working as an Assistant Professor in the School of Environmental Sciences at Jawaharlal Nehru University, New Delhi, India. His teaching and research areas are water resources management and GIS, hydro-geo-chemistry, pollution of water resources by geogenic and anthropogenic activities, groundwater-seawater interaction and aquifer vulnerability. He was a Postdoctoral Researcher at the Department of Environment, Land and Infrastructure Engineering, Politecnico di Torino, Turin, Italy for around four years. He has completed his M.Sc., M.Phil., and Ph.D. in Environmental Science. He obtained his Ph.D. degree from Indian Institute of Technology (Indian School of Mines), Dhanbad, India, in 2016. He was awarded Erasmus Mundus and Marie Skłodowska-Curie Actions Scholarships. He has travelled to Canada, Chile, Malta, Germany, Estonia, Bulgaria and Finland for academic-/research-related pursuits. He has published several research articles in various reputed international and national journals, and his works on respective fields are very well appreciated by the many workers across the world. He has served as a Reviewer for many international journals.

Email: ashwani.enviro@gmail.com, ashwani.tiwari@polito.it

Part I
Hydrogeology and Aquifer Contamination

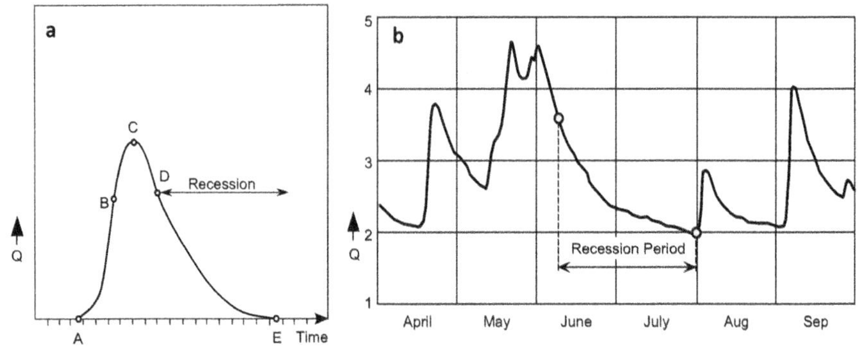

Fig. 1.1 (**a**) Components of a discharge hydrograph (Modified from Kresic and Bonacci 2010). (**b**) Example of spring discharge hydrograph (Modified from Kresic 2007)

Its shape depends on the size of the drainage area, as well as the precipitation intensity, and it is constructed according to the discharge values (Q), measured during a hydrogeological year.

It is defined by a rising limb (AB), a crest (BCD) and a falling or recession limb (DE), which corresponds to the recession period (Kresic and Bonacci 2010) (Fig. 1.1a).

Particularly, the recession limb (DE) is that part of the spring hydrograph that extends from a discharge peak to the base of the next rise (Fig. 1.1b). It contains information on storage properties and different types of media, such as porous, fractured and cracked lithology, and provides a function that quantitatively describes the temporal discharge decay and expresses the drained volume between specific time limits (Sayama et al. 2011).

The recession curve analysis has been useful in many areas for estimating the hydrological significance of the discharge function parameters and the hydrological properties of the aquifer (Amit et al. 2002).

Over the decades, many studies were made on recession curve: initial methods for the analysis of hydrograph recession periods, developed in the early 1900s, are based on Boussinesq (1904) and Maillet (1905) equations, which both give the dependence of the flow rate at specified time (Q_t) on the flow at the beginning of recession (Q_0).

In 1904, Boussinesq developed an exact analytical solution of the diffusion equation that describes flow through a porous medium, by considering the simplifying assumptions of a porous, free, homogeneous and isotropic aquifer, with the aquifer being limited by an impermeable horizontal layer at the level of the outlet (Eq. 1.1):

$$Q_t = \frac{Q_0}{(1 + \alpha[t - t_0])^2} \tag{1.1}$$

where Q_t (m^3/s) is the flow rate value at $t \neq t_0$; Q_0 is the flow rate at $t = t_0$; α is a constant depending only on the aquifer hydraulic systems (**recession coefficient**) (Eq. 1.2).

$$\alpha = \frac{\sqrt{Q_0} - \sqrt{Q_t}}{\sqrt{Q_t}t} \tag{1.2}$$

Maillet (1905) showed instead that the recession of a spring can be represented by an exponential formula, implying a linear relationship between the hydraulic head and flow rate (Eq. 1.3):

$$Q_t = Q_0 e^{-\alpha(t-t_0)} \tag{1.3}$$

where Q_t (m^3/s) is the flow rate value at $t \neq t0$; Q_0 is the flow rate at $t = t_0$; α is a constant depending only on the aquifer hydraulic systems (**recession coefficient**) (Eq. 1.4).

$$\alpha = \frac{\log Q_0 - \log Q_t}{e(t - t_0)} \tag{1.4}$$

Both in Boussinesq (1904) and Maillet (1905), the recession coefficient equation was used to determine important hydrogeological parameters as W_0, the groundwater volume stored above the spring level at the end of spring season (beginning of recession), and W_d, the groundwater volume stored at the end of autumn season (end of recession) (Table 1.1).

Table 1.1 Boussinesq (1904) and Maillet (1905) equations to determine the groundwater volume stored at the end of spring season (W_0) and the groundwater volume stored at the end of autumn season (W_d)

1904	**Boussinesq** $Wo = \frac{Q_0}{\alpha\,(1+\alpha t)^2} \times 86400$ (1.5); $W_d = \left[\frac{Q_0}{\alpha} - \frac{Q_0}{\alpha(1+\alpha t)}\right] \times 86400$ (1.6)	Where: Q_t (m^3/s) is the flow rate value at $t \neq t0$; Q_0 is the flow rate at $t = t_0$; α is a constant depending only on the aquifer hydraulic systems (recession coefficient)
1905	**Maillet** $W_0 = \frac{Q_0}{\alpha} \times 86400$ (1.7) $W_d = \frac{(Q_0 - Q_t)}{\alpha} \times 86400$ (1.8)	

About 50 years later, Castany (1967) approximated the connection between the karst spring discharge (Q) during a recession period and the hydraulic head in the aquifer (H) by the following equation (Eq. 1.9):

$$Q = \lambda H^{n} \tag{1.9}$$

where λ is a coefficient connected to conductivity and effective porosity of the aquifer; n = 1 when the discharge is linear; n > 1 when the discharge is nonlinear.

A great contribution to recession curve analysis was also given by Schoeller (1967), Drogue (1967) and Mangin (1970). As described in Civita (2008), according to Schoeller (1967) and Drogue (1967) methods, it can be useful to record daily values of discharge on a semi-log scale in order to represent the depletion curve points along a straight line: the intersection of the best fitting line for these points with the ordinate axis gives the value of initial flow at t = 0 (Q_0). By inserting this value into Maillet's equation and by calculating the depletion coefficient α, it is indeed possible to obtain the equation representing depletion of the aquifer network reserves over time.

In 1975, Mangin developed an expression that can be used to analyse spring recession curve by including an empirical homographic function that represents the first section of the recession curve in the presence of infiltration phenomena (Eq. 1.10):

$$Q_t = Q_0 e^{-\alpha(t-t_0)} + q0 + \frac{1 - n' t}{\varepsilon' t} \tag{1.10}$$

where q_0 represents the difference between the peak flow rate (Q_{max}) and the initial flow rate (Q_0); η is the infiltration velocity coefficient (d^{-1}) as it is inversely proportional to the duration of infiltration event; ε is the heterogeneity of outflow (d^{-1}) and it identifies attenuation of infiltration rate over time.

Eventually, Dewandel et al. (2003) offered a comprehensive review of approaches for analysing recession by conducting numerical simulations of shallow aquifers with an impermeable floor at the level of the outlet by comparing Boussinesq and Maillet analytical solutions.

In their results, they showed how the Boussinesq formula should be preferred to the Maillet exponential form in such hydrogeological situations: if the Boussinesq method fits the entire recession and could provide correct estimates of the aquifer's parameters, the Maillet solution largely underestimated the dynamic volume of the aquifer.

1.2.2 Analysis of Recession Curve – Time Series Analysis Methods

Assuming that the spring discharge is not affected by a rapid inflow of water into the aquifer, the recession analysis by using analytical relationship provides good insight into the aquifer structure, allowing to predict the volume of groundwater stored at the end of spring and autumn seasons.

However, ideal recession conditions with a long period of several months without precipitation are infrequent in moderate, humid climates: precipitations can disturb the recession curve and may not be removed unambiguously during analysis (Kresic and Bonacci 2010).

As a consequence, the majority of spring modelling approaches so far to better characterize spring discharge are based on various applications of time series analyses as well as general statistical and probabilistic methods.

Mathematical tools as the auto and cross-correlation functions are frequently used in the past years in hydrogeology for the analysis of a time series of values (Panagopoulos and Lambrakis 2006; Fiorillo and Doglioni 2010; Kresic and Stevanovic 2010; Kresic and Bonacci 2010).

Statistically, the autocorrelation of a random process (e.g. spring discharge) describes the correlation between values of the process (Q) at different points in time (t), as a function of the two times or the time difference; the cross-correlation represents a time-dependent relationship between an output process (e.g. daily spring discharge) and an input process (e.g. daily precipitation).

Primary hydrogeological information about the spring system can be derived by applying such analytical techniques on available spring monitoring dataset of Q, T and EC.

Through hydrograph analysis and calculated correlations between Q, T and EC dataset as a function of infiltration input (P), Galleani et al. (2011) revealed three broad aquifer behavioural categories, based on the drainage network effectiveness (i.e. Replacement, Piston and Homogenization), and developed an index to quantify spring vulnerability (Vulnerability Estimator for Spring Protection Areas).

In Banzato et al. (2011), the VESPA index method, proposed by Galleani et al. (2011), has been examined through a statistical analysis of time series with respect to the following aspects: (1) the time extension of the dataset, (2) the use of averaged data and (3) the number of missing data: the outcomes led to show that the VESPA index is a reliable way to identify higher vulnerability levels, and some uncertainty occurs in the identification of low and medium vulnerability levels.

In Lo et al. (2014), auto and cross-correlation analyses were applied to overall monitoring dataset of Q, P, T and EC of four alpine springs, supplied by porous and shallow aquifers, located in the Aosta Valley. Particularly, by evaluating the auto and cross-correlation coefficient of a time series, they have estimated the spring 'memory effect', a parameter that reflects the duration of system's reaction to input signal (Mangin 1984), understanding how the system reacts to infiltration phenomena induced by precipitation.

1.3 Study Area

The Rodoretto Valley is a left tributary of the Germanasca Valley (Italian Western Alps), located inside the Mesozoic formation of the Greenstone and Schist Complex (Piedmont Zone) (Fig. 1.2).

The geological units have been attributed to the Pennidic paleogeographic-structural domain (Sandrone et al. 1993), a multi-fold system represented by units of Mesozoic oceanic crust (Piedmont Zone) and basements in which have been distinguished the Upper Pennidic Units (Monte Rosa, Gran Paradiso, Dora Maira Massifs), the Intermediate System (Gran San Bernardo Massif) and the Lower Pennidic Units.

The Rodoretto Valley represents a glacial valley with steep rocky mountain sides and a flat valley floor.

The main elements of the landscape are small lateral and frontal moraines and many gravitational elements as minor scarps and counterscarps, linear trenches and landslide niches and landslides: these glacial and gravitational evidences, connected with deep-seated gravitational slope deformations (DSGSDs) (Agliardi et al. 2009), have been recognized on both sides of the valley (Forno et al. 2011, 2013).

The presence of a DSGSD, the widespread presence of open fractures and the consequent high degree of loosening of the bedrock, particularly evident on the upper sector of the right side of the valley (Forno et al. 2011), proved to be very favourable to the storage of large quantities of water (Fig. 1.3).

Several small springs are indeed located at the base of the right side of the Rodoretto Valley: their location appears strongly influenced by the tectonic discontinuities developed in the bedrock and by the morphostructures related to DSGSDs.

1.3.1 Spring 1 CB (Cavallo Bianco)

The spring analysed was named Spring 1 CB (Cavallo Bianco). It is located in the upper sector of Rodoretto Valley at 1954 m a.s.l., elevated about 10 m above the valley floor (Fig. 1.4).

Among the mountain spring located at the base of the slope right side of the Rodoretto Valley it been selected as its location and the outflow mode during summer season made it possible to measure the Q, T and EC parameters with a high degree of reliability and accuracy.

It outflows at the contact between the bedrock and an original slight cover sheet of glacial deposits, representing the aquifer which supply the spring (Fig. 1.5).

Part of the water, equal to 6.5 l/s, is collected by means of a reinforced concrete structure and reaches, the Rodoretto village (Comune di Prali 2009).

The Spring 1 CB was monitored first during the period between June 15th and October 28th 2016, and second between 11th June and 18th October 2017, with samples taken every 15–20 days approximately.

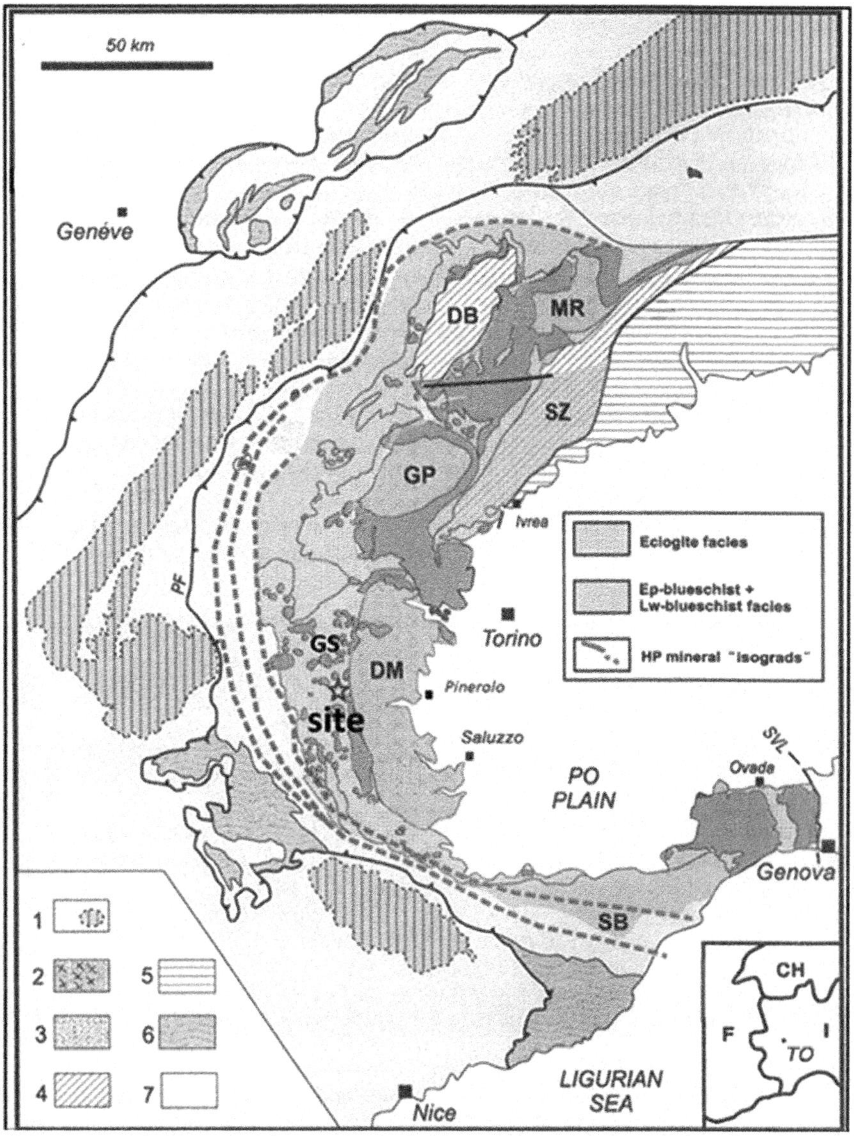

Fig. 1.2 Western Alps sketch map. (Modified from Compagnoni 2003)
1. Jura, Helvetic Domain and external Penninic Domain; 2. Internal Crystalline Massif of the Penninic Domain. Dm: Dora Maira Massif; 3. Piedmont Zone; GS: Greenstone and Schist Complex; 4. Austroalpine Domain; 5. South Alpine Domain; 6. Helminthoid Flysch Nappes; 7. Swiss Molasse

The very steep slopes of the Rodoretto Valley, engraved by a hydrographic network, favour the accumulation of huge snow masses during the winter and spring seasons, made the sector investigated inaccessible in the months from November to

Fig. 1.3 Upper sector of the right side of the Rodoretto Valley (Italian Western Alps)

Fig. 1.4 (**a** and **b**) Spring 1 CB

June (Fig. 1.6). Furthermore, traditional method for sampling groundwater physical parameters (Q, T and EC continuous datasets) by using a fixed multi-parameter probe and data logger was not allowed because it was not possible to obtain permission from the competent authority.

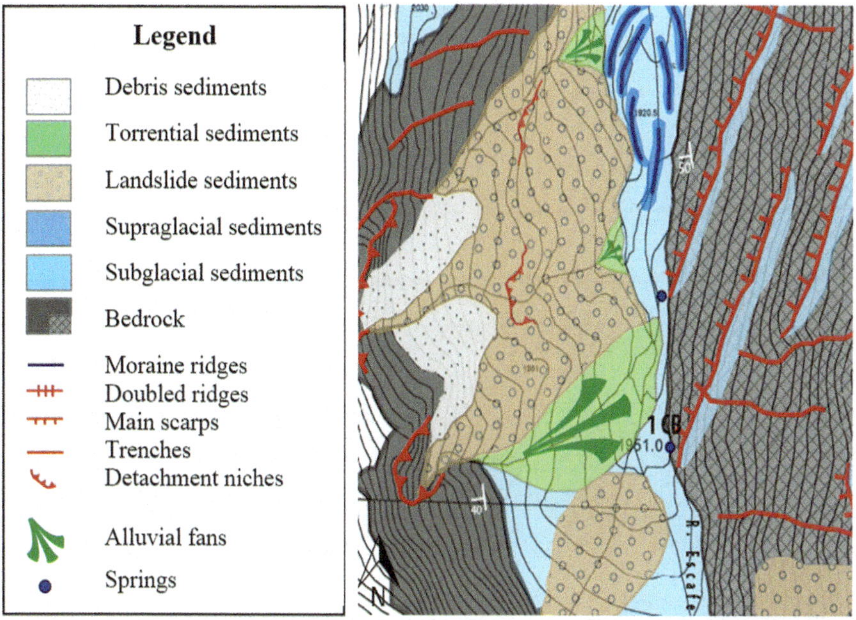

Fig. 1.5 Geological map of the Spring 1 CB area (scale 1:10000)

Fig. 1.6 (**a** and **b**) Accumulation of snow masses in the upper sector of Rodoretto Valley (11/06/2017 to 15/07/2017)

1.4 Results and Discussion

The majority of methods of spring characterization are based on various applications of time series analysis as well as general statistical methods described above in the methods paragraph. Such methods require to be applied on multi-years' time series of data on aquifer recharge (P), spring discharge (Q), temperature (T) and electrical

Table 1.2 Discharge values of Spring 1 CB during the monitoring period 2016–2017. Discharge values of spring calculated by Boussinesq (1904) and Maillet (1905) solutions

	Date	Days	Q (m^3/s)	Boussinesq (m^3/s)	Maillet (m^3/s)
2016					
1	29/06/2016	–	0.0125	–	–
2	13/07/2016	0	0.0118	0.0118	0.0118
3	28/07/2016	15	0.0109	0.0097	0.0098
4	21/08/2016	39	0.0087	0.0064	0.0066
5	03/09/2016	52	0.0075	0.0051	0.0051
6	25/09/2016	74	0.0069	0.0040	0.0040
7	28/10/2016	–	0.0070	–	–
2017					
1	30/06/2017	–	0.0130	–	–
2	15/07/2017	0	0.0124	0.0124	0.0124
3	28/07/2017	13	0.0119	0.0105	0.0107
4	23/08/2017	39	0.0088	0.0062	0.0063
5	12/09/2017	59	0.0067	0.0040	0.0041
6	28/09/2017	75	0.0066	0.0035	0.0035
7	18/10/2017	–	0.0067	–	–

conductivity (EC): this represents, therefore, a key limiting factor for many practical hydrogeological spring investigations with short execution times (Kresic 2007).

As in Rodoretto Valley, remote mountain settings generally lack easy access and continuous groundwater monitoring data collection with suitable probes or data logger is prevented by logistical problem, especially in winter and spring months.

Consequently, in many cases, the only tools available to hydrogeologists to derive hydrogeological information about mountains springs recharging systems are non-continuous and non-long-term dataset of discharge (Q), temperature (T) and electrical conductivity (EC) parameters.

Considering (1) the frequency at which the spring 1 CB discharge (Q) could be measured (see Table 1.2); (2) the non-kars condition of analysed spring 1 CB; (3) the absence of exceptional precipitation or new water rapid inflow events during the monitored period (summer and autumn seasons 2016/2017) which affect the summer spring discharge (Fig. 1.8), it has been applied Boussinesq (1904) and Maillet (1905) approach to analyse the depletion curves (un-influenced stage) with a modest degree of reliability.

1.4.1 Hydrograph Analysis

Available data related to discharge (Q) for the investigated springs are shown in Table 1.2.

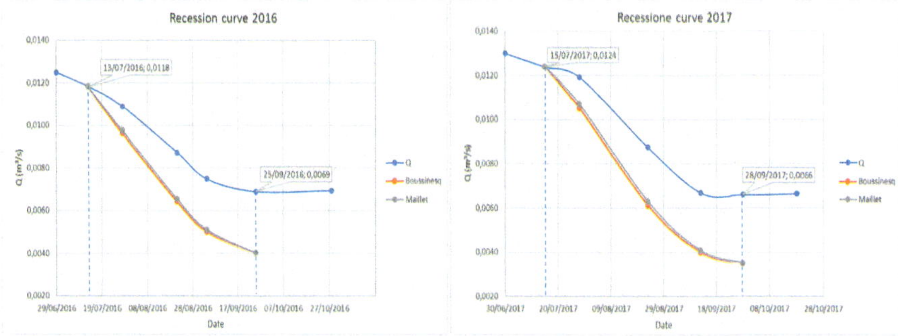

Fig. 1.7 Comparison between recession curve characteristics for Spring 1 CB (2016–2017) and estimated recession curves by Boussinesq (1904) and Maillet (1905) solutions

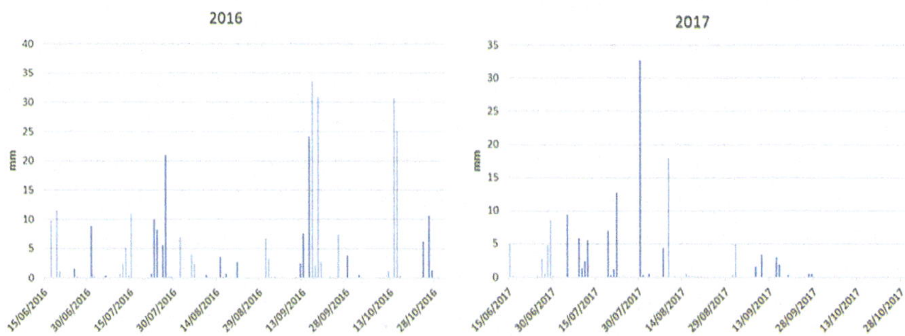

Fig. 1.8 Available rainflow data (Villa di Prali meteorological station, Arpa Piemonte 2016–2017)

The investigated periods were between (1) 13/07/2016 and 25/09/2016 and (2) between 15/07/2017 and 28/09/2017. During these periods, the recession curves do not show quick flow variations or pronounced discharge fluctuations.

The spring 1 CB shows a relatively low discharge rate with a discharge peak in early summer: the maximum value of discharge recorded in 2016 and 2017 is 0.013 m^3/s (29/06/2016 to 30/06/2017). The minimum value of discharge in both 2016 and 2017, recorded at the end of September, is 0.007 m^3/s (25/09/2016 to 28/09/2017).

For the test site, the Boussinesq (Eq. 1.1) and Maillet (Eq. 1.3) solutions were applied and the calculated discharge data were found to be consistently lower than the available discharge ones (Table 1.2; Fig. 1.7).

Due to the measurement uncertainties, both Boussinesq (Eq. 1.1) and Maillet (Eq. 1.3) solutions will cause errors in estimating the recession coefficient (α).

Nevertheless, the estimated recession coefficients allowed to quantify two important hydrogeological parameters, useful for the quantification of the resources available during the recession period: W_0 is the estimated groundwater volume stored above the spring level at the end of spring season (beginning of recession)

Table 1.3 Estimated recession coefficient (α), groundwater volume stored at the end of spring season (W_0), groundwater volume stored at the end of summer season (W_d)

	α	W_0 (m^3)	W_d (m^3)
2016			
Boussinesq (1904)	0.004	186,707,23	57,764,68
Maillet (1905)	0.007	140,295,09	58,466,17
2017			
Boussinesq (1904)	0.005	158,142,40	58,621,62
Maillet (1905)	0.008	127,414,77	59,597,23

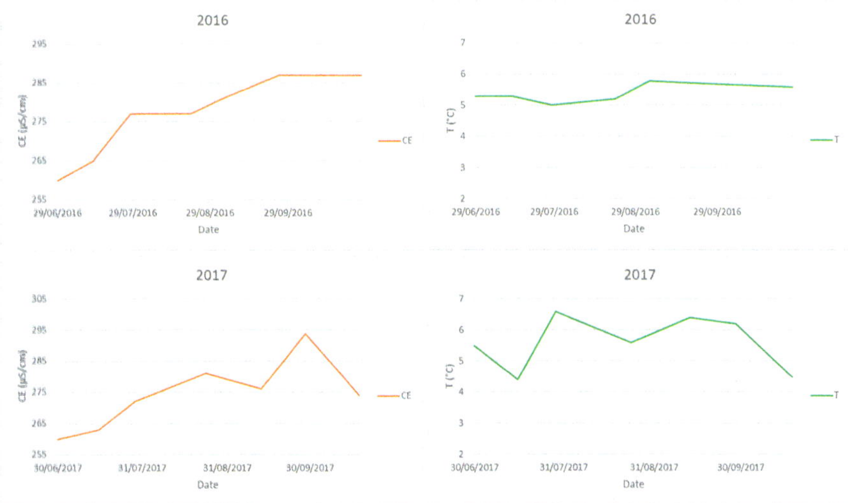

Fig. 1.9 Temporal variations of electrical conductivity (EC) and temperature (T) of the Spring 1 CB (2016–2017)

and W_d represents the estimated groundwater volume stored at the end of summer season (end of recession) (see Table 1.3).

1.4.2 *Thermograph and EC Graph*

In addition, thermograph and EC graph were plotted using available data, recorded during the monitoring period between June and October 2016 and 2017 (Fig. 1.9).

EC and Q parameter recorded very weak variations during 2016–2017 summer seasons: in the first summer months, the EC appears to feature slightly lower values (260 µS/cm) due to the dilution effect of the ionic species, favoured by the supply of glacial melt and the consequent higher values of Q.

Moreover, analysing the values relating to the water temperature, we note that these remain overall constant over the period of time between June and October 2016–2017 (average value of about 5.0 °C).

As described above, Spring 1 CB outflows at the base of a sliding scarp that displaces an original slight cover sheet of glacial deposits, representing the aquifer which supplies the spring.

The constancy in the values of CE and T parameters is in accordance with (1) the absence of freshly water recharge events induced by rainflow or snow-melting phenomena that affect in short times the spring summer discharge and physical parameters and (2) the presence of a porous glacial aquifer that supplies the spring of water with same physical characteristics.

1.5 Conclusions

As continuous groundwater monitoring in the upper sector of Rodoretto Valley (Germanasca Valley, Italian Western Alps) is hampered by logistical problem of instrumentation and data collection during winter and spring months, the only tools available to derive hydrogeological information were non-continuous and non-long-term dataset of discharge (Q), temperature (T) and electrical conductivity (EC) parameters.

A small selected mountain spring (Spring 1 CB) was investigated by applying on non-continuous monitoring discharge (Q) dataset the analytical solutions developed by Boussinesq (1904) and Maillet (1905) on spring recession curve (2016–2017).

The outcomes of the elaborations highlighted the limits of applicability of these methods in the presence of a non-continuous discharge dataset: both Boussinesq (1904) and Maillet (1905) gave a good fit of the shape of measured data-recession curve, but the estimated discharge values as a function of recession time were found to be consistently lower than the available discharge ones.

As a consequence, the estimated groundwater volume stored above the spring level at the end of spring season (W_0) and the groundwater volume stored at the end of summer season (W_d) are underestimated and can gave only hydrogeological-qualitative information.

Continuous (hourly value) and long-term (complete hydrogeological year) Q and P datasets are indeed needful not only in applications of time series analyses as well as statistical and probabilistic methods but also for the application of Boussinesq (1904) and Maillet (1905) depletion curves analysis.

Finally, the absence of a hydrodynamic impulsion response to the infiltrative processes and the recorded very weak variations of T and EC parameters suggest the presence of a low effective drainage system and an extended saturated zone which allow freshly infiltrated water-resident circulating groundwater homogenization phenomena.

If continuous (hourly value) Q, EC and T monitoring datasets have been available for a significant time interval, they would have allowed to understand in detail to what extent the presence of a DSGSD influences the position, discharge, the chemical–physical characteristics but also the infiltration dynamics and the vulnerability of analysed spring.

References

Agliardi F, Crosta G, Zanchi A, Ravazzi C (2009) Onset and timing of deep-seated gravitational slope deformations in the eastern Alps, Italy. Geomorphology 103:113–129
Amanzio G, Marchionatti F, Lavy M, Ghione R, De Maio M (2016) Springs monitoring data analysis with a frequency and time domain approach: the case study of Mascognaz spring (Aosta Valley). Geam. Geoingegneria Ambientale E Mineraria 147(1):5–12
Amit H, Lyakhovsky V, Katz A, Starinsky A, Burg A (2002) Interpretation of spring recession curves. Ground Water 40(5):543–551
Banzato C, Vigna B, Galleani L, Lo Russo S (2011) Application of the Vulnerability Estimator for Spring Protection Areas (VESPA index) in mountain quaternary aquifers. GeoHydro2011 – CANQUA-IAH, Quebec City (CA), pp 28–31
Banzato C, Butera I, Revelli R, Vigna B (2017) Reliability of the VESPA index in identifying spring vulnerability level. J Hydrol Eng 22(6):04017008
Bonacci O (1993) Karst springs hydrographs as indicators of karst aquifers. Hydrol Sci J 38 (1–2):51–62
Boussinesq J (1877) Essai sur la the'orie des eaux courantes do mouvement nonpermanent des eaux souterraines. Acad Sci Inst Fr 23:252–260
Boussinesq J (1904) Recherches théoriques sur l'écoulement des nappes d'eau infiltrées dans le sol et sur le débit des sources. J Math Pure Appl 10:5–78
Castany G (1967) Introduction a l'étude des courbes de tarissements. Chron Hydrogeol 10:23–30
Civita M (2008) An improved method for delineating source protection zones for karst. Hydrogeol J 16:855–869
Clow DW, Schrott L, Webb R, Campbell DH, Torizzo A, Dornblaser M (2003) Ground water occurrence and contributions to streamflow in an alpine catchment, Colorado Front Range. Ground Water 41(7):937–950
Compagnoni R (2003) HP metamorphic belt of the Western Alps. Episodes 26:200–204
Comune di Prali (2009) Allegato 01, Inquadramento geografico: ubicazione delle sorgenti e delle opere di captazione. Stralcio CTR foglio 172 NO Perrero, Sezione 050
Desmarais K, Rojstaczer S (2001) Inferring source water from measurements of carbonate spring response to storms. J Hydrol 260:118–134
Dewandel B, Lachassagneb P, Bakalowicz M, Wengb P, Al-Malki A (2003) Evaluation of aquifer thickness by analysing recession hydrographs. Application to the evaluation Oman ophiolite hardrock aquifer. J Hydrol 274:248–269
Dewandel B, Perrin J, Ahmed S, Aulong S, Hrkal Z, Lachassagne P, Samad M, Massuel S (2010) Development of a tool for managing groundwater resources in semi-arid hard rock regions: application to a rural watershed in South India. Hydrol Process 24:2784–2797
Drogue C (1967) Essai de détermination des composantes de l'écoulement des sources karstiques. Evaluation de la capacite de rétention par chenaux et fissures. Chron Hydrogeol 10:43–47
Fiorillo F, Doglioni A (2010) The relation between karst spring discharge and rainfall by cross-correlation analysis (Campania, southern Italy). Hydrogeol J 18(8):1881–18951
Fiorillo F, Revellino P, Ventafridda G (2012) Karst aquifer draining during dry periods. J Cave Stud 74:148–156
Forno MG, Lingua A, Lo Russo S, Taddia G (2011) Improving digital tools for quaternary field survey: a case study of the Rodoretto Valley (NW Italy). Environ Earth Sci 64:1487–1495
Forno MG, Lingua A, Lo Russo S, Taddia G (2012) Morphological features of Rodoretto Valley deep-seated gravitational slope deformations. Am J Environ Sci 8:648–660
Forno MG, Lingua A, Lo Russo S, Taddia G, Piras M (2013) GSTOP: a new tool for 3D geomorphological survey and mapping. Eur J Remote Sens 46:234–249
Galleani L, Vigna B, Banzato C, Lo Russo S (2011) Validation of a vulnerability estimator for spring protection areas: the VESPA index. J Hydrol 396:233–245

Giacopetti M, Aringoli D, Materazzi M, Pambianchi G, Posavec K (2016) Groundwater recharge estimation using spring hydrographs: the case of the Tennacola carbonate aquifer (central Apennine, Italy), Rend. Online Soc Geol It 41:61–64
Istat (2017) Statistiche Istat, la giornata mondiale dell'acqua 2017
Jakada H, Chen Z, Luo M, Zhou H, Wang Z, Habib M (2019) Watershed characterization and hydrograph recession analysis: a comparative look at a karst vs. non-karst watershed and implications for groundwater resources in Gaolan River Basin, Southern China. Water 11:743
Kresic N (2007) Hydrogeology and groundwater modeling, 2nd edn. CRC Press/Taylor and Francis, Boca Raton
Kresic N, Bonacci O (2010) Spring discharge hydrograph. In: Groundwater hydrology of springs CAP.4. Elsevier Inc., Oxford, UK
Kresic N, Stevanovic Z (2010) Groundwater hydrology of springs. Elsevier Inc., Oxford, UK
Lo Russo S, Amanzio G, Ghione R, De Maio M (2014) Recession hydrographs and time series analysis of springs monitoring data: application on porous and shallow aquifers in mountain areas (Aosta Valley). Environ Earth Sci 73:7415–7434
Luo M, Chen Z, Yin D, Jakada H, Huang H, Zhou H, Wang T (2016) Surface flood and underground flood in Xiangxi River Karst Basin: characteristics, models, and comparisons. J Earth Sci 27:15–21
Maillet E (1905) Essais dı'hydraulique souterraine et fluviale, vol 1. Herman et Cie, Paris
Mangin A (1970) Méthode d'analyse des courbes de décrue et tarissement dans les aquifères karstiques (analysis method of recession and exhaustion curves of the karst aquifers). CR Acad Sci Paris 270:1295–1297
Mangin A (1984) Pour Une meilleure connaissance des systems hydrologiques à partir des analyses correlatoire et spectrale [improving the hydrological systems knowledge using correlation and spectral analysis]. J Hydrol 67:25–43
Panagopoulos G, Lambrakis N (2006) The contribution of time series analysis to the study of the hydrodynamic characteristics of the karst systems: application on two typical karst aquifers of Greece (Trifilia, Almyros Crete). J Hydrol 329(3–4):368–376
Piras M, Taddia G, Forno MG, Gattiglio M, Aicardi I, Dabove P, Lo Russo S, Lingua A (2016) Detailed geological mapping in mountain areas using an unmanned aerial vehicle: application to the Rodoretto Valley, NW Italian Alps. Geomat Nat Haz Risk 8:137–149
Sandrone R, Cadoppi P, Sacchi R, Vialon P (1993) Pre-mesozoic geology in the Alps. In: Von Raumer JF, Neubaur F (eds) The Dora-Maira massif. Springer, Berlin, pp 45–61
Sayama T, McDonnell JJ, Dhakal A, Sullivan K (2011) How much water can a watershed store? Hydrol Process 25:3899–3908
Schoeller H (1967) Hydrodynamique dans le karst: Ecoulement et emmagasinement (Karst hydrodynamics, flow and storage). Chron Hydrogéol 10:7–21
Tobin BW, Schwartz BF (2016) Using periodic hydrologic and geochemical sampling with limited continuous monitoring to characterize remote karst aquifers in the Kaweah River Basin, California, USA Hydrol Process 30:3361–3372
United Nation (2018) The United Nations World Water Development Report 2018

Chapter 2
Evaluation and Prediction of Seepage Discharge Through Tailings Dams When Their Height Increases

Viacheslav V. Fetisov and Elena A. Menshikova

Abstract Significant seepage through tailings dams is one of the main disadvantages of upstream dams. In some cases, seepage through dams can lead to large water losses from recycling system of the dressing plant. Seepage evaluation should be carried out when designing of dams' height increase in a framework of water balance prediction. In this chapter, the authors present the results of evaluations of seepage through dams of a large tailings storage (with total area of 17.5 km^2 and with dam's average height of 62 m as of 2017) performed by both field hydrometric measurements and analytical calculations. For seepage analytical calculations, the grain size distribution data of 327 samples of wet magnetic separation (WMS) tailings were used to obtain the hydraulic conductivity (K) on the basis of empirical equations. The results of the seepage evaluation performed using both approaches were compared. It was shown that the results of the seepage evaluation obtained in the framework of hydrometric studies and analytical calculations are well comparable. Taking into consideration good comparability of the results, the Dupuit equations were used for prediction of seepage discharge through the tailings dams in case of their raising, according to design decision, to 28 m in average for the next 10 years.

Keywords Tailings dams · Seepage control · Dupuit equation · Grain size

2.1 Introduction

Ore processing at dressing plants gives the concentrates of valuable components and nonutilizable wastes – tailings consisting of separated waste rocks. The amount and properties of the tailings are determined by the nature of the minerals being processed and the used enrichment process. The volume of the tailings is normally far in excess of the liberated resource (Kossoff et al. 2014). The main method for the

V. V. Fetisov (✉) · E. A. Menshikova
Perm State National Research University, Perm, Russia
e-mail: fetisov@psu.ru

M. De Maio, A. K. Tiwari (eds.), *Applied Geology*,
https://doi.org/10.1007/978-3-030-43953-8_2

removal of wet comminuted tailings from a dressing plant is their hydraulic transportation to a natural or artificially constructed storage facility – the tailing impoundment, where solid particles sediment and accumulate, and water is clarified. The clarified water is taken in from the tailing impoundment through the spillway facility and reused as backwater at the dressing plant. Therefore, the dressing plant is being constructed simultaneously with a group of facilities and equipment for hydraulic transportation, hydraulic placement of tailings, and recycling of wastewater.

Any tailings impoundments require the construction of enclosing earth-fill dams. The design of the enclosing dam depends on the topography, the base rocks, and the required storage volume (Vick 1990; Design and evaluation ... 1994). The enclosing dam consists of primary and secondary embankments. In most cases, the primary dam is constructed from delivered low permeable soils, materials found on site, and overburden rocks, and it provides the storage capacity for the first years of the dressing plants operation. In the time following, the useful capacity of the tailing storage is maintained by the construction of secondary embankment dams.

One of the ways to maintain the useful capacity of the tailing storage of mining enterprises is to construct secondary dams by upstream method using the previously stored waste materials. On the one hand, this method of waste storage is the most economical; on the other hand, its main disadvantage is significant seepage losses of water (Vick 1990).

Many authors note that seepage through enclosing dams is one of the main factors controlling their stability and one of the main causes of emergent situation (Mittal, Morgenstern 1976; Klohn 1979; Kossoff et al. 2014; Hu et al. 2015; Aboelela 2016; Peng et al. 2016; Naeni and Akhtarpour 2018; Shen et al. 2011; Lyu et al. 2019). In addition, seepage through dams can lead to pollution of downstream groundwater system (Kossoff et al. 2014; Shen et al. 2011; Lyu et al. 2019). In some cases, seepage through embankment dams can lead to significant losses of water used in recycling system of the dressing plant (Mironenko and Rumynin 1999). Therefore, control of seepage is a major requirement in the design and operation of tailings dams.

Different approaches exist to quantify the seepage through tailings dams. In particular, calculations of seepage through dams using mathematical models based on the finite difference method (La Touche, Garrick 2012) and finite elements method (Kealy, Busch 1971; Cowherd et al. 1993; Rykaart et al. 2001; Peng et al. 2016; Chong et al. 2018; Yuan and Lei 2019) have widespread use. Along with computer modeling, the well-known analytical dependencies based on the Darcy's Law and the Dupuit equation can be successfully used to evaluate the seepage through dams composed of fairly homogeneous sandy permeable sediments. Along with field and laboratory studies, computational methods based on empirical dependencies are used to estimate the conductivity, taking into account the grain-size distribution of the tailings (Aubertin et al. 1996; Adajar and Zarco 2014).

The theoretical predictions based on analytical solutions, performed in this study, have been confirmed by actual field measurements.

The problem of seepage through tailings dams has been investigated and discussed in this chapter, primarily in the context of losses of water used in recycling system of the dressing plant.

2.2 Description of Study Area and Tailing Storage Facilities

The area under study is located on the eastern slope of the Middle Urals, in Sverdlovsk region (Russia). JSC "Evraz KGOK" (Kachkanarsky Ore Mining and Processing Enterprise) is one of the largest mining enterprises in the Urals and develops Gusevogorskoe deposit of titanium magnetite ores that contain vanadium inclusions.

The climate of the region is continental with long cold winters and short warm summers. Average annual precipitation is 540 mm according to long-term observations. Most of the precipitation (65% on average) occurs in the warm period of the year (from May to September).

In accordance with the plant's technological process of ore dressing using wet magnetic separation (WMS), iron–vanadium concentrate and waste (WMS tailings in the form of pulp with a solid content of about 10%) are produced. The WMS tailings are pumped by the hydraulic transportation system into the tailing storages. The tailing dam is used to store tailings of WMS, pulp clarifying, and wastewater reuse and recycling for needs of the plant's facilities.

The tailing dam of JSC "EVRAZ KGOK" is located in the valley at the junction of the Vyia river and its right inflow Rogalevka river at a distance of 1 km from the dressing plant. It was put into operation in 1963 (Kuznetsov et al. 2013).

The tailings impoundment includes three compartments: Rogalevsky (of Rogalevka river), Intermediate, and Vyisky (of Vyia river) (Fig. 2.1). Rogalevsky and Intermediate compartments are designed to store the tailings of wet magnetic separation and to clarify a liquid phase of the pulp, and Vyisky compartment is designed to receive clarified water from the ponds of Rogalevsky and Intermediate. The compartments are arranged in a cascade manner: Rogalevsky, Intermediate, and Vyisky.

The tailing reservoirs of Rogalevsky and Intermediate compartments are enclosed by upstream dams and by slopes of hills from northeast, east, and southeast.

The Vyisky compartment is surrounded by two dams. The first one, on the north of the compartment, impounds the valley of the Vyia river. The second one, on the west, separates the Vyisky compartment from the Lower Vyia water reservoir.

The structure of the Rogalevsky compartment includes dams: Riverside, South, Dam # 3, Dam # 4, and Dividing. The structure of the Intermediate compartment includes dams: Separating, Dam # 1, Dam # 2, Dam # 5, and East one. The position of the dams is shown in the draft, superposed with the satellite image dated 2018 (Fig. 2.1).

The total area of tailing storage inside the enclosing embankment dams of Rogalevsky and Intermediate compartments is about 17.5 km^2. Within the dressing

Fig. 2.1 Scheme of the tailings storage facilities

plant's performance, more than 800 million m^3 of tailings of wet magnetic separation were stored in the tailing impoundments (Kuznetsov et al. 2013).

The starter Riverside embankment dam was constructed upward to a height of 3–5 m using loamy soils and further was raised using tailings of WMC by upstream method. Currently, nine dams composed of sandy wet magnetic separation wastes are constructed along the tailings perimeter.

The average height of dams as at 2017 is 62 m, and the average length is 1510 m. The maximum height of embankment Separating dam reaches 100 m (Kuznetsov et al. 2013). The embankment dams are raised by moving the tailings of the beach zone using bulldozers or dumping the tailings by a dragline excavator.

The following main outflow components are identified in water balance of the dressing plant and tailing facilities: seepage through dams, seepage through the tailing storage bottom, evaporation from the water surface of tailings ponds, accumulation of water in the pore space of WMC tailings, irrecoverable water losses in technological processes, and water discharge from Vyisky compartment through a syphon spillway (when Normal Water Level is exceeded).

Wastewater seepage through the dams (#1, #2, #3, #4, #5, and East) and its discharge onto local terrain are a main outflow component of the water balance of the enterprise. This study evaluates seepage flow through the listed dams of the tailings storage. Seepage through the South, Riverside, and Separating dams comes into the reservoir of Vyisky compartment directly or through the channel of Rogalevka river, so it is included in water recycling system of the plant and is not considered in this work.

2.3 Materials and Methods

Within a framework of the plant's annual water balance estimation, hydrometric studies for quantitative assessment of seepage discharge through the embankment dams of tailings storages were performed. In this case, the volumes of seepage losses were estimated in the hydrometric control sections of large streams formed by surface runoff from downstream slopes of the each dams. Along with field measurements, analytical calculations were used to estimate the seepage discharge through the dams. When assessing the filtration capacity of the sandy soils composing the dams, we considered the grain-size distribution data of 327 samples of tailing soil taken from the beach zone of dams.

This work is determined by the need of a quantitative assessment of the water balance components with regard to the plans of the enterprise development and an increase of the dams' height by 28 m on average by 2027.

2.3.1 Hydrometric Studies

Field measurements were performed every month throughout a year from August 2016 to July 2017. Hydrometric work included observations of changes of water levels in streams formed by seepage discharge through the dams, measurement of current velocity, and calculating of discharge, measurements of depth, and drawing of the channel cross sections (Figs. 2.2 and 2.3). Measurement of current velocity on velocity verticals was carried out using the GR-21M propeller type water current meter. Measurements of depths, current velocity, and calculation of the discharge are made in accordance with the requirements of the main guidelines of Roshydromet (Guide to Hydrometeorological... 1975, 1978). The locations of the hydrometric control sections considered in this chapter are shown in Fig. 2.1.

For greater accuracy, Solinst leveloggers (Solinst Canada Ltd.) were used in several control sections, the registration data of which were taken into account

Fig. 2.2 Hydrometric measurements in control sections of the stream formed by surface runoff of water seeping through the Dam # 2: September, 2016

Fig. 2.3 Hydrometric measurements in control sections of the stream formed by surface runoff of water seeping through the Dam # 4: March, 2017

when determining the channel cross-sectional area and stream discharge. Measurements of depths were made at the sites of the leveloggers installation and the corresponding channel cross sections were drawn.

Taking into account the calculated areas of channels cross sections, we plotted the charts of dependence between cross-sectional area and water level above a levelogger and obtained linear approximation equations.

2.3.2 *Seepage Through Earth Dams*

Water losses through the dam's body and basement, the position of drawdown curve, and pressure gradient are determined on the base of seepage calculations. As a rule, the calculations of seepage through earth dams are made under the following assumptions: filtration is two-dimensional, waterproof layer is impervious and horizontal, and earth materials (soil and rock) of a dam's body are homogeneous isotropic.

The analytical calculations of seepage in earth dams under above-mentioned assumptions are based on the Darcy's Law and the Dupuit equation (Nedriga 1983; Rozanov et al. 1985).

$$q = K \cdot \frac{H_1^2 - H_2^2}{2 \cdot L} \tag{2.1}$$

In this equation, q is specific discharge per meter dam length, m^2/d; K is hydraulic conductivity of dam's earth materials, m/d; H_1 is the upstream head, m; H_2 is the downstream head, m; L is the length of the flow path, m, as shown in Fig. 2.4.

If seepage through earth dams placed on pervious foundation is considered, the water flow discharge through the dam's body and its foundation can be calculated using the following formula (Nedriga 1983; Rozanov et al. 1985):

$$q = K \cdot \frac{H_1^2 - H_2^2}{2 \cdot L} + K_0 \cdot T \cdot \frac{H_1}{L + 0.4T} \tag{2.2}$$

where q is specific discharge per meter dam length, m^2/d; K is hydraulic conductivity of dam's earth materials, m/d; K_0 is the hydraulic conductivity of pervious soils of the foundation, m/d; H_1 is the upstream head, H_2 is the downstream head, м; L is the length of the flow path, m; T is the thickness of pervious soils at the base of the dam, m, as shown in Fig. 2.5.

The total flow discharge Q (m^3/day) is defined as the product of the specific discharge q (m^2/day) and the length of the dam.

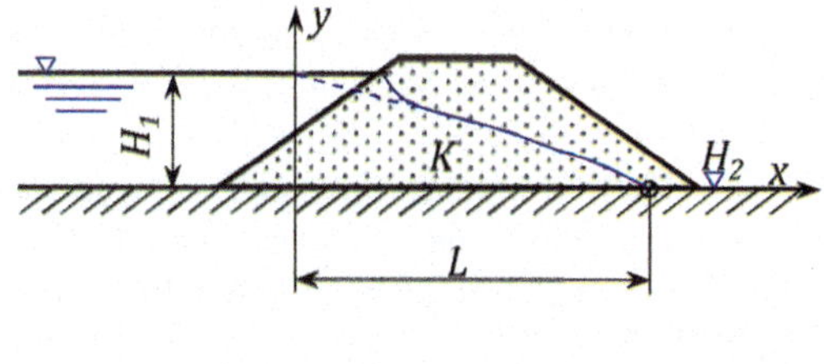

Fig. 2.4 Schematic diagram of seepage flow through earth dam placed on an impervious base

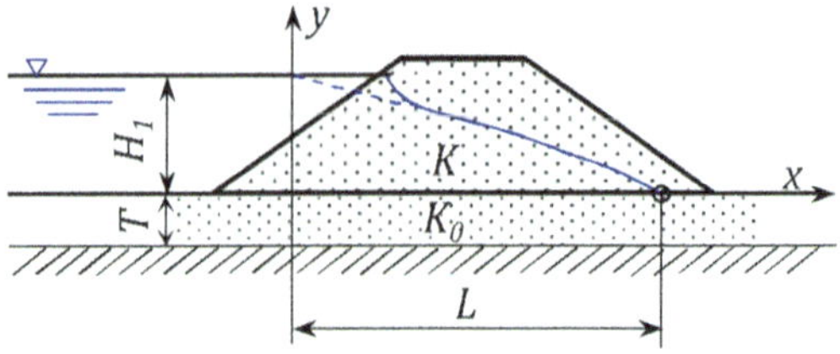

Fig. 2.5 Schematic diagram of seepage flow through earth dam placed on a pervious base

2.3.3 Seepage Characteristics of Earth Dams

For estimation of seepage in dams, the hydraulic conductivity of sand deposits was previously calculated taking into account the data of grain size distribution (composition) of tailings that make up the dams. Samples of WMS tailings were collected in 2011 from the beach zone of tailings dams. The sampling points were located on several cross-section lines of each dam at a distance of 5, 10, 25, 50, 75, 100, and 150 m from crest of dams (Summary data… 2011).

The tailings were sampled from the following dams: Dam # 1 (42 samples from 6 cross sections), East dam (47 samples from 4 cross sections), Dam # 2 (42 samples from 6 cross sections), Dam # 3 (35 samples from 5 cross sections), Dam # 4 (35 samples from 5 cross sections), South dam (42 samples from 6 cross sections), Riverside dam (21 samples from 3 cross sections), and Separating dam (63 samples from 9 cross sections).

The data of averaged grain size composition of samples in each of the dams were used in evaluation of hydraulic conductivity. Averaged grain size composition of tailings for particular dams is presented in Table 2.1 and in Fig. 2.6.

The coefficient of grain uniformity U according to Hazen (1892) is given:

$$U = \left(\frac{d_{60}}{d_{10}}\right) \tag{2.3}$$

where d_{60} and d_{10} in the formula represent the effective grain diameter in (mm) corresponding to 60 % and 10 % by weight passing through the sieves.

Porosity n according to Vukovic and Soro (1992) may be derived from the empirical relationship with the coefficient of grain uniformity U as follows:

$$n = 0.255 \cdot \left(1 + 0.83^U\right) \tag{2.4}$$

The effective diameter d_{60} and d_{10}, the coefficient of grain uniformity U, and the porosity n were obtained for all samples. The bar graphs show distribution of values of the parameters of tailings calculated for each of the samples (Figs. 2.7, 2.8, 2.9, and 2.10).

A significant amount of research is devoted to studying of relationship between grain size distribution and hydraulic conductivity K. Reviews of the empirical dependences to approximately estimate hydraulic conductivity K using grain size

Table 2.1 Averaged grain size composition and calculated parameters of WMS tailings

		Particle size, mm						$d_{g.m.}$	d_{10}	d_{60}	U	n	K_1	K_2
Dam	Samples	> 1.6	1.6–0.56	0.56–0.28	0.28–0.14	0.14–0.071	< 0.071	mm	mm	mm	–	–	m/day	m/day
Dam # 1	42	100	97.8	76.8	51.2	25.9	6.3	0.334	0.084	0.376	4.48	0.37	6.47	6.44
East dam	35	100	96.2	78.7	58.8	33.8	9.5	0.321	0.072	0.297	4.10	0.37	4.99	4.87
Dam # 2	42	100.2	98.0	73.6	44.4	19.6	3.9	0.355	0.098	0.430	4.39	0.37	8.84	8.76
Dam # 3	35	100	96.9	72.9	48.5	24.2	6.4	0.359	0.085	0.412	4.85	0.36	6.37	6.47
Dam # 4	35	100	98.6	79.9	54.4	28.2	7.3	0.318	0.080	0.341	4.27	0.37	5.97	5.88

$d_{g.m.}$ – geometric mean grain diameter
d_{10}, d_{60} – effective grain diameter
U – grain uniformity coefficient
n – porosity calculated by Eq. 2.4
K_1 – hydraulic conductivity calculated by Eq. 2.5
K_2 – hydraulic conductivity calculated by Eq. 2.6

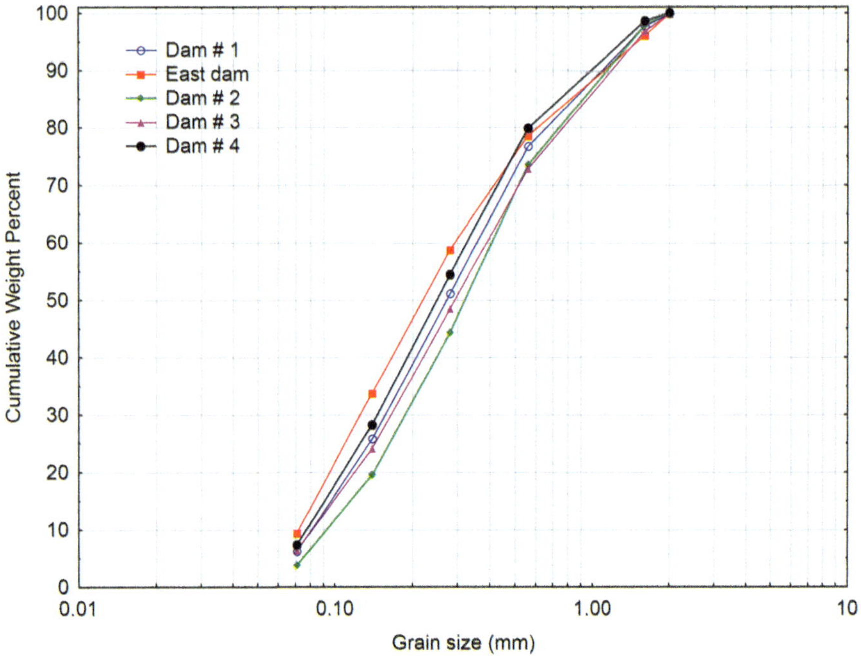

Fig. 2.6 Averaged grain size distribution curves for WMS tailings

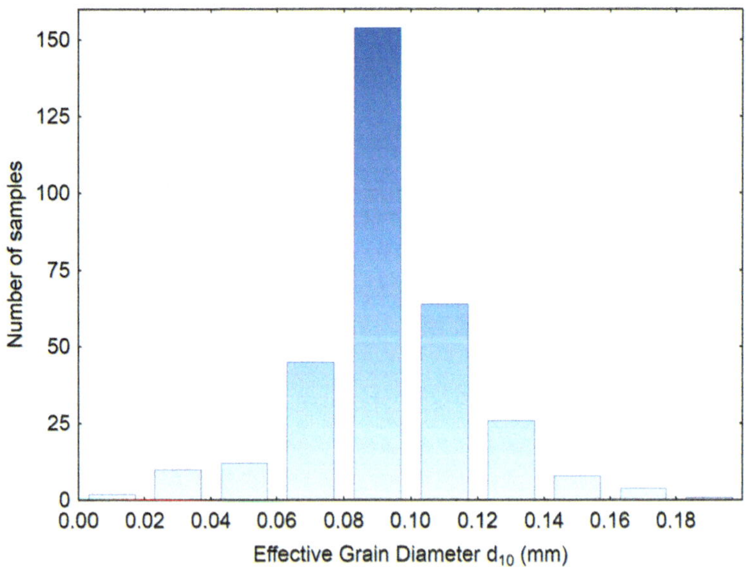

Fig. 2.7 Distribution of effective grain diameter d_{10}

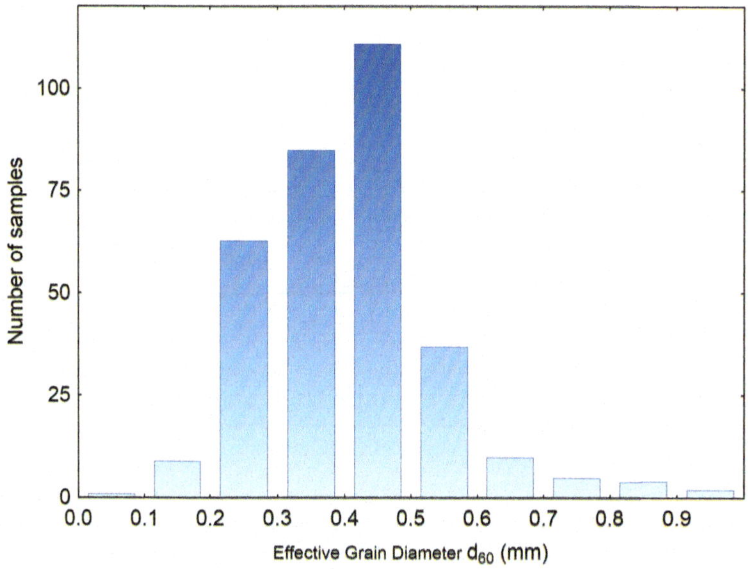

Fig. 2.8 Distribution of effective grain diameter d_{60}

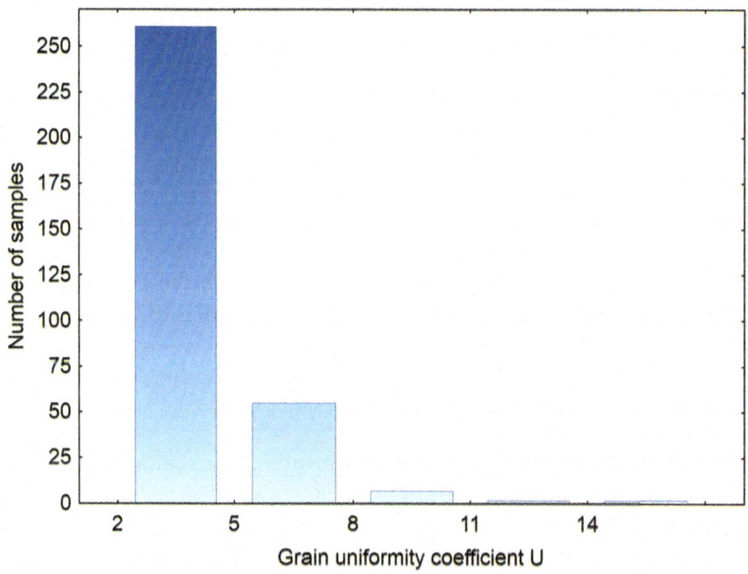

Fig. 2.9 Distribution of grain uniformity coefficient U

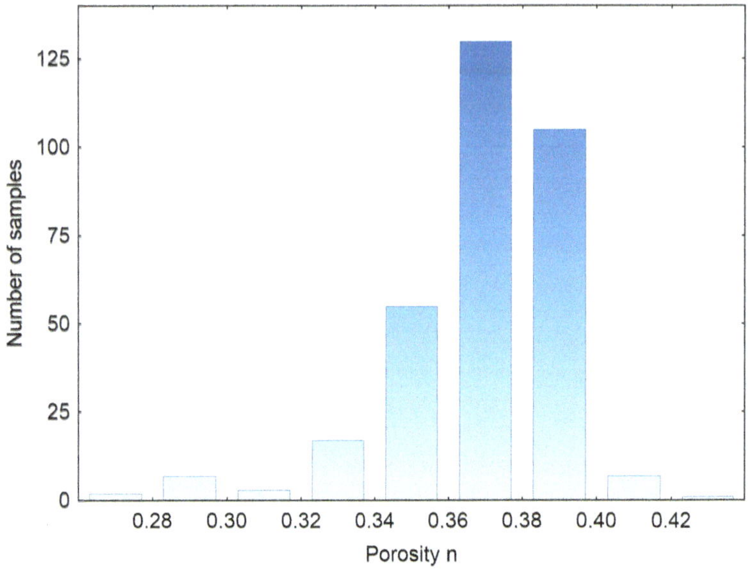

Fig. 2.10 Distribution of porosity *n*

distribution data are presented in works (Vukovic and Soro 1992; Odong 2007; Rosas et al. 2015).

As the charts (Figs. 2.7, 2.8, and 2.10) show, the effective diameter d_{10} of the tailing's grains is mainly in the range of 0.06–0.14 mm, with a prevailing value of 0.08–0.10 mm. The effective diameter of the d_{60} is mainly 0.2–0.6 mm. Grain uniformity coefficient U is in the range from 2 to 8, and mainly, it does not exceed 5.

Taking into account the limitations, two equations were selected for calculation of hydraulic conductivity K: the original Hazen equation (1892) and Beyer equation (1964).

The Hazen equation is applicable for sediments with the coefficient of grain uniformity U less than 5 and the effective grain size d_{10} between 0.1 and 3 mm:

$$K = \frac{p \cdot g}{\mu} \times 6 \times 10^{-4}[1 + 10(n - 0.26)]d_{10}^2 \tag{2.5}$$

where K is hydraulic conductivity, cm/s; p is density; g is acceleration due to gravity; μ is dynamic viscosity, g/cm·s (0.0114 at 15 ° C); n is porosity; and d_{10} is effective grain diameter.

In this chapter, the Hazen equation was used with some assumption, which is that the equation was also used if the value of effective diameter was in the range of $0.08 \leq d_{10} \leq 0.1$ – prevailing effective diameters in the samples.

The Beyer equation is applicable in condition when sediments have grain uniformity coefficient U between 1 and 20 and effective grain size d_{10} between 0.06 and 0.6 mm:

$$K = \frac{p \cdot g}{\mu} \times 6 \times 10^{-4} \times \log \frac{500}{U} d_{10}^2 \tag{2.6}$$

The parameters used for the calculations (effective diameter, grain uniformity coefficient, porosity, hydraulic conductivity) are presented in Table 2.1. As can be seen in the table, the hydraulic conductivity values *K* determined by Eqs. 2.5 and 2.6 are very close. These values of hydraulic conductivity were used when estimating seepage through the dams according to Eqs. 2.1 and 2.2.

2.4 Results and Discussion

2.4.1 Results of Hydrometric Measurements

The results of field hydrometric measurements made in control sections for estimating the seepage discharge through the earth dams are presented in Table 2.2.

The seepage discharge (m^3/s) on the date of observations and the calculated seepage volume (mln m^3) for every month are given for each of the dam. The last column of Table 2.2 contains the calculated annual seepage volume (mln m^3) for each of the dams. As can be seen, the total annual volume of seepage through dams (# 1, East dam, # 2, # 5, # 3, # 4) determined taking into account the monthly field hydrometric observations from August 2016 to July 2017 is 41,13 million m^3 per year.

2.4.2 Calculating Earth Dam Seepage

To determinate the longitudinal section area and the position of the reference level, longitudinal sections of dams were produced using elevation marks of natural ground surface at the bottom of the dams (Fig. 2.11). Data of topographic survey made before the tailing dams were constructed were used as basis for the section drawing. In further calculations, it was assumed that the reference level passes through the average elevation marks of the dams bottoms, determined on the basis of the produced longitudinal sections.

According to engineering and geological surveys (Construction... 2013), the top layer of natural soils in the bottom of tailings storage is mainly represented by loams of deluvial genesis with an average thickness of about 6 m. Their averaged hydraulic conductivity *K* is 0.008 m/day. In the case of almost impervious base of the tailings dams, Eq. 2.1 can be used to estimate seepage through dams. This equation was used in analytical calculations for the following dams: #1, East dam, # 2, # 5, # 3, constructed on impervious base.

In accordance with geological survey (Construction... 2013), the central part of the Dam # 4 is associated with a large tectonic disturbance revealed here on seismic

Table 2.2 Measurements of seepage discharge through the dams made in control sections

Dam		Measurement month												Year
		08/2016	09/2016	10/2016	11/2016	12/2016	01/2017	02/2017	03/2017	04/2017	05/2017	06/2017	07/2017	
Dam # 1	Mean, m^3/s	0.27	0.41	0.36	0.31	0.4	0.49	0.44	0.29	0.25	0.05	0.27	0.28	0.32
	mln m^3	0.71	1.05	0.95	0.81	1.07	1.31	1.06	0.78	0.66	0.12	0.71	0.75	9.99
East dam	Mean, m^3/s	0.051	0.103	0.112	0.098	0.078	0.071	0.062	0.054	0.069	0.05	0.029	0.034	0.07
	mln m^3	0.137	0.267	0.3	0.254	0.209	0.19	0.15	0.145	0.179	0.134	0.075	0.091	2.13
Dam # 2	Mean, m^3/s	0.91	0.92	0.71	0.55	0.41	0.39	0.38	0.38	0.36	0.38	0.53	0.68	0.55
	mln m^3	2.44	2.38	1.90	1.43	1.10	1.04	0.92	1.02	0.93	1.02	1.37	1.82	17.38
Dam # 5	Mean, m^3/s	0.0183	0.0225	0.0256	0.0141	0.0057	0.0025	0.0015	0.0015	0.0025	0.0036	0.0046	0.0057	0.009
	mln m^3	0.049	0.058	0.069	0.036	0.015	0.007	0.004	0.004	0.006	0.01	0.012	0.015	0.28
Dam # 3	Mean, m^3/s	0.25	0.29	0.32	0.21	0.13	0.1	0.09	0.09	0.1	0.11	0.12	0.13	0.16
	mln m^3	0.67	0.75	0.86	0.54	0.35	0.27	0.22	0.24	0.26	0.29	0.31	0.35	5.11
Dam # 4	Mean, m^3/s	0.19	0.2	0.22	0.16	0.21	0.18	0.15	0.12	0.17	0.27	0.26	0.24	0.2
	mln m^3	0.51	0.52	0.59	0.41	0.56	0.48	0.36	0.32	0.44	0.72	0.67	0.64	6.24
Mean monthly seepage discharge, m^3/s		1.69	1.94	1.74	1.34	1.23	1.23	1.12	0.93	0.96	0.86	1.22	1.37	1.30
Total volume of seepage, mln m^3		4.51	5.03	4.67	3.48	3.3	3.31	2.72	2.5	2.48	2.3	3.16	3.67	**41.13**

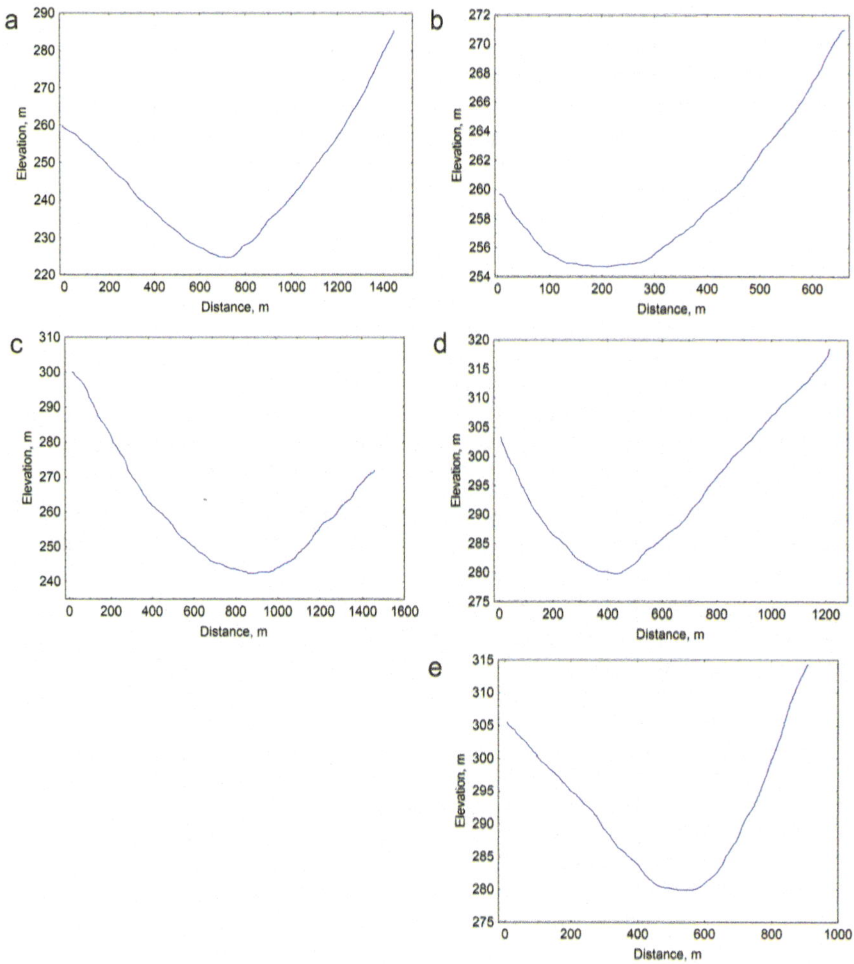

Fig. 2.11 Longitudinal profiles of natural dam base: (**a**) Dam #1, (**b**) East dam, (**c**) Dam #2, (**d**) Dam #3, (**e**) Dam #4

and electrical prospecting data. This site of the dam base comprises very pervious alluvial sandy sediments with a total thickness, exposed by drilling, of about 11 m. Therefore, the seepage discharge through the central part of Dam # 4 was calculated using Eq. 2.2 for pervious base earth dams.

The head H_1 was determined by the average annual water levels in the ponds of Rogalevsky and Intermediate compartments of the tailings storage for the considered period (August 2016–July 2017). The head H_2 lines up with zero value of the reference level (the dam base).

The length of the flow path L was determined by the contour line of the ponds in the tailings storage compartments, as well as by the position of the external dam boundary. To determinate this parameter, we used topographic survey maps of tailings storage made in 2016–2017.

The values of hydraulic conductivity K obtained at the previous step were used in calculations of seepage discharge through the dams with Eqs. 2.1 and 2.2. Table 2.3 presents parameters used in calculations and the results.

2.4.3 Comparison of the Calculation Results and the Seepage Prediction

The annual seepage discharge through the dams of tailing storage was evaluated within two independent approaches. Table 2.4 presents the volumes of seepage through dams assessed by both hydrometric measurements and analytical calculations with Eqs. 2.1 and 2.2.

As can be seen from Table 2.4, the evaluations of seepage performed by both approaches have similar results. Seepage discharge through Dam # 4 estimated by analytical calculations is 12% less than the ones performed by hydrometric studies. Here, the calculations for the central part of the dam were performed with Eq. 2.2. In this particular case, the calculation includes the thickness of alluvial sands of the dam base only opened by drilling, which is probably somewhat larger in reality.

Table 2.3 Geometric parameters of dams, parameters for calculation of seepage discharge, and the results

Parameters	*D.l.*	*D.b.*	K/K_0	*W.l.*	H_1	H_2	*L*	q/q_1	*Q*
Dam	m	m	m/day	m	m	m	m	m^2/day	mln m^3/ year
Dam # 1	1558	245	6.47	318.23	73.23	0	930	18.65	10.61
East dam	755	259	4.99	318.23	59.23	0	950	9.21	2.54
Dam # 2	1476	251	8.84	318.23	67.23	0	600	33.30	17.94
Dam # 5	409	308.5	7.00	318.23	9.73	0	250	1.33	0.20
Dam # 3	1242	286	6.37	319.77	33.77	0	320	11.35	5.15
Dam # 4	1033	292	5.97/ 11.72	319.77	27.77	0	300	7.67/ 19.43	5.47

D.l. – dam length
D.b. – average elevation marks of the bottom of the dam
K – hydraulic conductivity of tailings calculated by Eq. 2.5
K_0 – hydraulic conductivity of pervious soils of the foundation calculated by Eq. 2.5
W.l. – average annual water level in the ponds (August 2016–July 2017)
H_1 – upstream head
H_2 – downstream head
L – average length of the flow path
q – specific discharge per meter dam length calculated by Eq. 2.1
q_1 – specific discharge of alluvial sandy sediments of the foundation per meter
Q – total flow discharge

Table 2.4 Total seepage discharge through Dam # 1 and the dams on the east side of the tailing storage

Dams	Total seepage discharge per year, mln m^3										
	2017 [a]	**2017 [b]**	2018	2019	2020	2021	2022	2023	2024	2025	2026
Dam # 1	**9.99**	**10.61**	12.52	13.99	15.6	17.38	19.35	20.67	22.03	23.43	24.67
East dam	**2.13**	**2.54**	3.12	3.61	3.95	4.29	4.65	5.03	5.42	5.82	6.18
Dam # 2	**17.38**	**17.94**	19.5	20.12	20.75	21.38	22.01	23.62	25.28	27	28.52
Dam # 5	**0.28**	**0.20**	0.40	0.53	0.66	0.80	0.94	1.16	1.41	1.68	1.94
Dam # 3	**5.11**	**5.15**	6.02	6.25	6.49	6.74	7.01	7.85	8.75	9.88	10.06
Dam # 4	**6.24**	**5.47**	6.4	6.57	6.77	6.97	7.18	7.99	8.84	9.9	10.07
Total through all dams	**41.13**	**41.91**	47.96	51.07	54.22	57.56	61.14	66.32	71.73	77.71	81.44

2017 [a] – for the considered period August 2016–July 2017 made in hydrometric control sections
2017 [b] – for the considered period August 2016–July 2017 calculated analytically

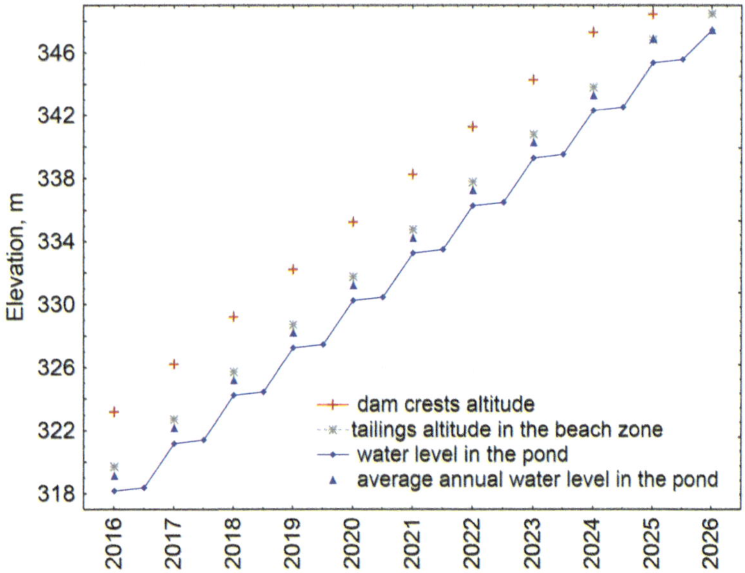

Fig. 2.12 Prediction of water level rise in the pond of Rogalevsky tailings storage compartment

Taking into consideration good comparability of the results for 2017 (Table 2.4), Eqs. 2.1 and 2.2 may be used for prediction of seepage discharge through the tailings dams for 2018–2026 (see Table 2.4).

In order to perform prediction calculations of seepage through the dams until 2026, the average annual water levels in the ponds of the Rogalevsky and Intermediate compartments for each year in the forecast period were determined in accordance with the dams rising construction plan. When estimating the average annual level, the peculiarity of its increase in ponds over a year was considered: levels grew slower in the first half of the year compared with the second one. For example, the chart of yearly water level rise in the pond of Rogalevsky compartment with calculated average annual values is shown in Fig. 2.12.

The chart in Fig. 2.13 shows the yearly prediction of the total seepage discharge through Dam # 1 as well as through the dams on the east side of the tailing storage (see also Table 2.4).

Calculations show, as the embankment dams, stored tailings altitude and accordingly water levels in ponds raised, the total seepage discharge through Dam # 1 and the dam on the east side of the tailing storage increases from 41.91 million m^3 (in 2017) to 81.44 million m^3 (in 2026). The discharge increases under the Darcy's Law: the water level in the ponds rises together with the pressure gradient and the seepage discharge values.

Calculated volumes of seepage discharge through the dams were used to evaluate the overall plant's water balance forecast for 2018–2026. The balance calculations have showed that irrecoverable seepage losses from the dams will lead to scarce of water in the plant's recycling system and to its increase with each subsequent year.

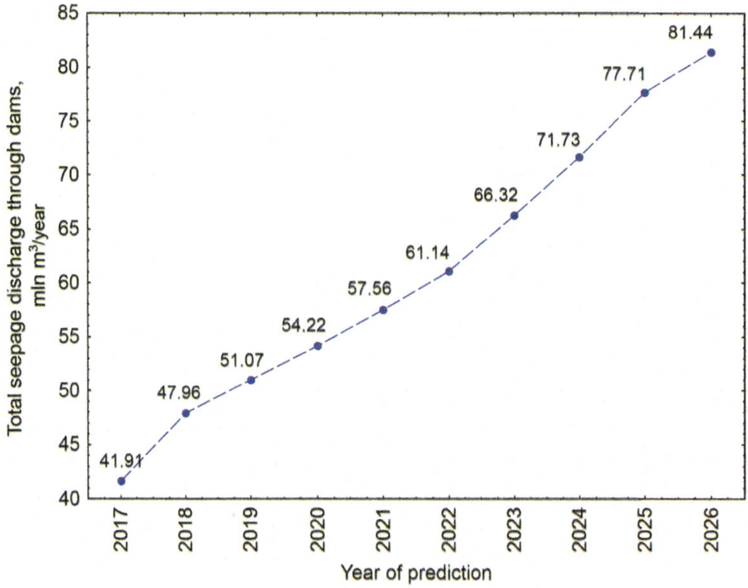

Fig. 2.13 Total seepage discharge through Dam # 1 and the dams on the east side of the tailing storage by year for a period from 2017 to 2026

Addition of other inflow components does not remedy the situation and only contributes to its partial improvement. The water intake from the Lower Vyia water reservoir, that is the main additional source of water in the recycling system, is limited by the requirements of Reservoirs Use Regulations. Thus, in case of the dams rising, the necessary condition for maintenance of the dressing plant's designed capacity is the construction and commissioning of drainage facilities for catchment and returning of water seeping through the tailings dam and running down the slopes.

2.5 Conclusions

Seepage through dams formed by the tailings of wet magnetic separation of titanium–magnetite ores was studied within the framework of annual cycle of monthly hydrometrical observations. The results were compared with the results of the seepage analytical calculations based on the Darcy's Law and the Dupuit equations.

To calculate the hydraulic conductivity, 327 samples taken from the tailings beach zone were studied. According to the results of particle size analysis, the tailings are mostly fine sands with a minor content of medium sands and coarse sand. The porosity of tails determined by the formula Vukovic and Soro (1992) varies mainly from 0.34 to 0.40. Since the effective diameter d_{10} of tailings is mainly in the range of 0.08–0.10 mm, and a grain uniformity coefficient U does not exceed 5, the empirical equations of Hazen and Beyer were used to determine hydraulic conductivity K of sandy tailings deposits. Because the tailings are mostly presented by the fine sand fraction, and the parameters of porosity (in the Hazen formula) and the uniformity coefficient (in the Bauer equation) are correlated, very similar values of the hydraulic conductivity K were obtained. Hydraulic conductivity K obtained by this way was used for analytical calculations of seepage through dams.

Considering that, the values of the seepage discharge through the dams evaluated by two independent approaches are close, valid analytical calculations were performed for seepage predictions taking into account the increase of the tailings dams' height. On the other hand, the close values of seepage through dams (measured hydrometrically and calculated analytically) allow us to say that the hydraulic conductivity has been determined correctly, and the Hazen and Beyer empirical equations are well suited for this type of sediments.

Thus, if seepage properties of tailings are well studied and volumes of seepage discharges through dams are confirmed by field measurements, the analytical dependencies can be used for seepage evaluation and prediction.

References

Aboelela M (2016) Control of seepage through earth dams based on pervious foundation using toe drainage systems. J Water Resour Prot 8:1158–1174. https://doi.org/10.4236/jwarp.2016.812090

Adajar MAQ, Zarco MAH (2014) An empirical model for predicting hydraulic conductivity of mine tailings. Int J Geomate 7(2):1054–1061

Aubertin M, Bussiere B, Chapuis RP (1996) Hydraulic conductivity of homogenized tailings from hard rock mines. Can Geotech J 33(3):470–482. https://doi.org/10.1139/t96-068

Beyer W (1964) Zur Bestimmung der Wasserdurchlässigkeit von Kiesen und Sanden aus der Kornverteilungskurve. Wasserwirtschaft-Wassertechnik 14(6):165–168

Chong L, Zhen-zhong S, Lei G et al (2018) The seepage and stability performance assessment of a new drainage system to increase the height of a tailings dam. Appl Sci 8(1840). https://doi.org/10.3390/app8101840

Construction of a new compartment of the EVRAZ KGOK tailings, first stage (2013) Report on engineering and geological surveys. PSU, Perm

Cowherd DC, Miller KC, Perlea VG (1993) Seepage through mine tailings dams. Paper presented at the International Conference on Case Histories in Geotechnical Engineering

Design and Evaluation of Tailings Dams (1994) U.S. Environmental Protection Agency, Office of Solid Waste, Washington

Guide to Hydrometeorological Stations and Posts (1975) Issue 2. Part 1, 2. Gidrometeoizdat, Leningrad

Guide to Hydrometeorological Stations and Posts (1978) Issue 6. Part 1, 2. Gidrometeoizdat, Leningrad

Hazen A (1892) Some physical properties of sands and gravels, with special reference to their use in filtration. Massachusetts State Board of Health, vol. 24th annual report, pp 539–556

Hu S, Chen Y, Liu W et al (2015) Effect of seepage control on stability of a tailings dam during its staged construction with a stepwise-coupled hydro-mechanical model. Int J Min Reclam Environ 29(2):125–140. https://doi.org/10.1080/17480930.2013.870693

Kealy CD, Busch RA (1971) Determining seepage characteristics of mill-tailing dams by the finite-element method. US Bur Mines Rep Invest 7477

Klohn EJ (1979) Seepage control for tailings dams. Paper presented at the first international mine drainage symposium. Miller Freeman Publications, San Francisco/Denver

Kossoff D, Dubbin WE, Alfredsson M et al (2014) Mine tailings dams: characteristics, failure, environmental impacts, and remediation. Appl Geochem 51. https://doi.org/10.1016/j.apgeochem.2014.09.010

Kuznetsov AG, Lytin OV, Karelin AE (2013) Tailing dump of "EVRAZ KGOK" JSC and its development prospects. Gornyi Zhurnal 9(1):17–19

La Touche GD, Garrick H (2012) Hydraulic performance of liners in tailings management and heap leach facilities. Paper presented at the annual conference of International Mine Water Association

Lyu Z, Chai J, Xu Z et al (2019) A comprehensive review on reasons for tailings dam failures based on case history. Adv Civil Eng.. Article ID 4159306. https://doi.org/10.1155/2019/4159306

Mironenko VA, Rumynin VG (1999) Problems of hydrogeoecology, vol 3(1). State University of Mining, Moscow

Mittal H, Morgenstern N (1976) Seepage control in tailings dams. Can Geotech J 13:277–293. https://doi.org/10.1139/t76-030

Naeini M, Akhtarpour A (2018) A numerical investigation on hydro-mechanical behaviour of a high centreline tailings dam. J South Afr Inst Civ Eng 60(3). https://doi.org/10.17159/2309-8775/2018/v60n3a5

Nedriga VP (ed) (1983) Gidrotekhnicheskie sooruzheniya (Hydrotechnical constructions). Stroiizdat, Moscow

Odong J (2007) Evaluation of empirical formulae for determination of hydraulic conductivity based on grain-size analysis. J Am Sci 3(3):54–60

Peng C, Chen S, Chen W (2016) The finite element analysis of tailings dam seepage. Paper presented at the 2nd International Conference on Architectural, Civil and Hydraulics Engineering (ICACHE 2016). https://doi.org/10.2991/icache-16.2016.5

Rosas J, Jadoon KZ, Missimer TM (2015) New empirical relationship between grain size distribution and hydraulic conductivity for ephemeral streambed sediments. Environ Earth Sci 73:1303. https://doi.org/10.1007/s12665-014-3484-2

Rozanov NP, Bochkarev JW, Lapshenkov VS (1985) Gidrotekhnicheskie sooruzheniya (Hydrotechnical constructions). Agropromizdat, Moscow

Rykaart M, Fredlund M, Stianson J (2001) Solving tailings impoundment water balance problems with 3-D seepage software. Geotech News 19:50–54

Shen L, Luo S, Zeng X et al (2011) Review on anti-seepage technology development of tailings pond in China. Proc Eng 26. https://doi.org/10.1016/j.proeng.2011.11.2370

Summary data on the determination of technological parameters in the samples (2011) EVRAZ KGOK, Kachkanar

Vick SG (1990) Planning, design and analysis of tailings. BiTech Publishers Ltd, Dams

Vukovic M, Soro A (1992) Determination of hydraulic conductivity of porous media from grain-size composition. Water Resources Publications, Littleton

Yuan L, Lei J (2019) The analysis of the seepage characteristics of tailing dams based on FLAC3D numerical simulation. Open Civil Eng J 9:400–407. https://doi.org/10.2174/1874149501509010400

Chapter 3
Sediment Yield in Mountain Basins, Analysis, and Management: The SMART-SED Project

Davide Brambilla, Monica Papini, Vladislav Ivov Ivanov, Luca Bonaventura, Andrea Abbate, and Laura Longoni

Abstract Sediment yield from mountain basins and solid transport in rivers are widely studied and still represent a major issue when dealing with hydrogeological hazard. The correct determination of flooding scenarios involving huge amounts of debris also has implications for cities and human infrastructure safety. However, studies focused on catchment scale modeling tend to decouple the hydraulic processes from the sediment yield processes. Indeed, a large amount of hydraulics research literature has focused on hydro-morphological river models in which the sediment yield must be provided only as a boundary condition. This approach has clear limits, and decoupling such processes could lead to a weak understanding of the complexity of interactions within the watershed. To overcome such limitations, a new approach is proposed. The project we present aims to develop a complete model able to simulate sediment yield: from slope erosion down to in flow transport. The use of innovative mathematical approaches seeks to improve accuracy and performance over classical models and to find a right balance between computational cost and detailed description of physical processes. The model relies mainly on preexisting geographic databases to retrieve data. A set of test synthetic cases are also presented in the final part of the work.

Keywords Sediment transport · Hydrogeological risk · Distributed modeling · Mountain streams · Flooding

---

D. Brambilla (✉) · M. Papini · V. I. Ivanov · A. Abbate · L. Longoni
Department of Civil and Environmental Engineering, Politecnico di Milano, Milan, Italy
e-mail: davide.brambilla@polimi.it

L. Bonaventura
MOX-Department of Mathematics, Politecnico di Milano, Milan, Italy

M. De Maio, A. K. Tiwari (eds.), *Applied Geology*,
https://doi.org/10.1007/978-3-030-43953-8_3

3.1 Introduction

Smart cities improve quality of life by implementing sustainable strategies via advanced technology and innovation. However, for many future smart cities located in flood- or landslide-prone areas, natural hazards cannot be overlooked and developing resilience and mitigation measures in response to natural disasters remains crucial for the safety of citizens and infrastructure. Cities located at the downstream end of mountain catchments are exposed to specific flood risks, in which sediment transport plays a significant role. During the past years, these natural calamities have increased in frequency, possibly due to climate change (Milly et al. 2002; Hirabayashi et al. 2013). Future smart cities must be ready to face natural hazards by implementing smart strategies. SMARTSED (Sustainable MAnagement of sediment transpoRT in responSE to climate change conditions) aims to improve substantially the tools available for management of hydrogeological risk downstream of mountain catchments (Allamano et al. 2009).

The urgent need for a territorial policy that considers sediment transport has been discussed for decades (Hartmann and Driessen 2017). Several recent studies have demonstrated the significance of transported sediment in fluvial floods (Lane et al. 2007). The European Floods Directive (Smith 2015) recommends that flood risk maps based on calamitous scenarios should include information about "areas where floods with a high content of transported sediments and debris floods can occur."

Mountain basins can produce huge quantities of sediments due to slope erosion, especially in Alpine and pre-Alpine catchments (Ballio et al. 2010; Brambilla et al. 2011b; Brambilla et al. 2011a; Longoni et al. 2016a). Moreover, the solid presence in river can substantially modify the hydraulic behavior of rivers and channels and thus increase flood hazard (Radice et al. 2012; Radice et al. 2016). However, studies focused on catchment scale modeling tend to decouple the hydraulic processes from sediment yield processes. Indeed, a large amount of hydraulics research literature has focused on hydro-morphological river models based on 1D or 2D shallow-water equations, in which the sediment yield must be provided only as a boundary condition. Concerning sediment discharge estimates, several models for sediment generation and transport have been proposed, including lumped parameters, distributed parameters, semiempirical, and physically based models (de Vente et al. 2013; Bennett et al. 2014; Kim and Ivanov 2014). SWAT, EUROSEM, KINEROS2, and CAESAR are among the most widely used models (Coulthard et al. 2002; Goodrich et al. 2002; Neitsch et al. 2011). Most of the available models focus on specific aspects, preventing a complete understanding of the processes at basin scale. Some of these models do not provide spatially distributed information, but only a prediction of the integrated quantities over the entire catchment or apply inefficient numerical solvers. Indeed, most available models use simple explicit time discretization schemes, which make them highly inefficient in case small lakes or larger water bodies are present within the simulation domains. Additionally, most of the available models require a priori identification of the extent of the riverbeds, thus being unable to handle the simulation of flood waves without ad hoc modifications.

Even commonly used models such as KINEROS2 employ approximate equation sets for surface water routing, which are not valid in general, especially in the strong transients associated with major flooding events that are responsible for a large portion of the total liquid and solid runoff. Finally, even in relatively recent models like CAESAR, the conservation of fluid and solid mass is not always guaranteed by the numerical methods employed for spatial discretization. SWAT (Soil and Water Assessment Tool) emerges as the standard method irrespective of the dimension of the considered catchments (Abbaspour et al. 2015), but it employs many parameters, which requires costly and time-consuming high-resolution monitoring of the basin.

Starting from this wide range of needs and aiming to develop an innovative model that combines efficiency and effective, rather than formal, accuracy, the SMART-SED project is presented. This chapter states the objectives of the project that has a twofold rationale: creating the model itself and compiling a comprehensive database of solid transport data for a reference basin. Data will allow for a deeper understanding of the processes and model tuning and validation. Later, the logical and mathematical model structure is described, and some simple test cases to prove model basis are described. The project is still ongoing, and shown results are a picture of the development at publishing date.

The model SMART-SED is being developed inside a wider project funded by Fondazione Cariplo, an Italian charity society, and Comune di Lecco, local government of the city, the project started in 2016 and will end in 2020.

3.2 SMART-SED Objectives and Model Conceptual Scheme

Developing SMART-SED, the authors posed a series of objectives the project should meet to mark a step forward in comparison to existing similar experiences. These objectives led to the development of a model that is both effective and yet simple enough to be widely applied. The development of the model could not be divided from a deeper detailed knowledge of the physical process, thus the need for a deep surveying of sediment transport in a test basin arose. Another request was the possibility to largely rely on open access data, to allow professionals and public agencies to apply the suggested methods and models to large areas with limited data retrieve effort.

Main objective of the SMART-SED model is to be able to model the hydraulic and hydrogeological processes over the entire catchment by a coherent approach and with efficient numerical solvers. The model can automatically identify storage and erosion zones all over the basin (hillslopes and watercourses) and simultaneously compute water discharge along river/lakes/hydraulic infrastructure. Simplifications have been necessary to make the model applicable, as it is not possible to solve each known process in the basin with a deterministic set of equation. Nevertheless, the simplification is guided by a scale principle: it is necessary to simulate each process with an impact at basin scale, simplifying local processes which effect is diluted on large areas. Indeed, sediment transport in mountain catchments is the combination of the direct erosive activity of short-term events like floods and of the longer-term impact of milder events. Therefore, SMART-SED is able to account for both time scales and model all the hydrogeological processes considered through the fundamental conservation laws of mass and momentum (Rosso 1994). From a conceptual point of view, the SMART-SED model aims to overcome the traditional static a priori division in slope and drainage cells as they can change in time due to weather conditions. From the analysis of the state of the art, the model scheme proposed in Bemporad et al. (1997) appears to be the most promising starting point for the new SMART-SED approach, thanks to its acceptable balance between input data requirements, complexity of the model, and results. However, several improvements have been introduced both in the conceptual scheme and in equation solver.

SMART-SED model is a distributed model, and thus, the computational domain, a basin, is divided into square pixels. Each pixel is configured as a biosphere portion that encloses different layers, from top to bottom: (1) Atmosphere, (2) Interception zone, and (3) Soil zone, which are internally divided into (3.1) Capillary soil, (3.2) Gravitational soil, and (3.3) Deep soil. The areas of the Interception zone in which the water depth reaches a critical threshold are automatically identified as a separate Drainage zone layer. Each pixel exchanges water and soil fluxes horizontally with neighbors and internally among the different layers in the vertical direction. The layer subdivision is purely due to modeling purposes and needs as such borders are not so sharp in nature. Scheme of layers and fluxes is reported in Fig. 3.1.

The Atmosphere A is modeled as a reservoir with almost infinite capacity, whose outflows are represented by precipitation "P," characterized by its own intensity, duration, and distribution. The latter can be rain or snow depending on the temperature. The soil can recharge the atmosphere through evapotranspiration "E" which is modeled using the formulations present in literature, such as the Thornthwaite (1948) formula.

The Interception Zone I is inserted in the model as it allows the interpretation of surface runoff processes ("f_e" fluxes) and infiltration into the ground "Q_i." It represents the terrain–atmosphere interface in which the accumulation of snow occurs at sufficiently low temperatures. Modeling snow presence and accumulation\depletion "Q_n" is remarkably important, as it contributes to increase the amount of water which infiltrates into the ground or flows as runoff. The model does not allow generating runoff until gravitational layer is fully saturated, following a standard Dunnian processing (Dunne et al. 1991). This is a valid hypothesis for coarse soils, with high infiltration rates. When a pixel is fully saturated, any other incoming contribution will leave the gravitational soil through exfiltration "E_x." In order to make SMART-SED applicable to a wide range of soil kind, even with low infiltration rates, the application of Hortonians flows (Horton 1939) is foreseen in future development. In this way, runoff will be generated if the intensity of rain is greater than soil infiltration capacity, even if Gravitational soil is not saturated.

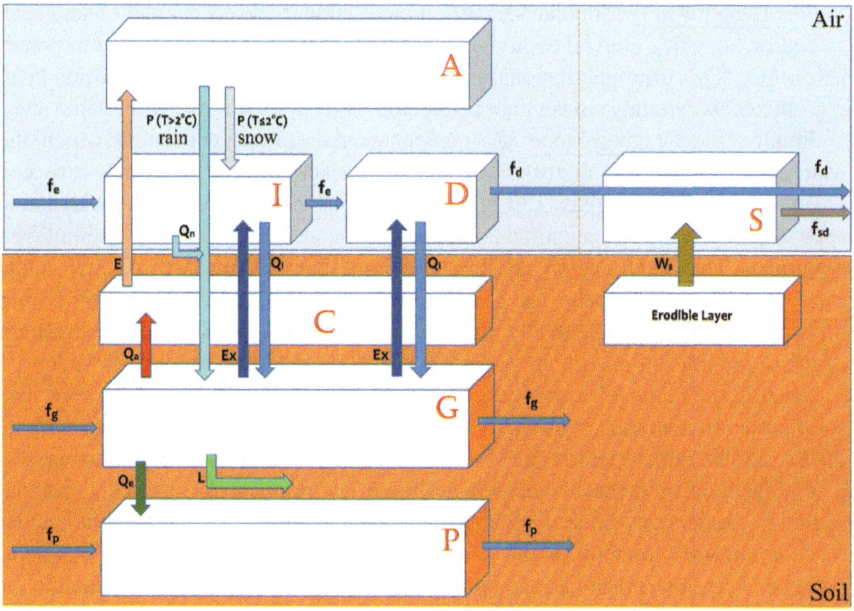

Fig. 3.1 Biosphere discretization and fluxes scheme in SMART-SED model, on the left column the hydrological model and on the right column the erosion/solid model

In the Drainage zone (D), the full de Saint-Venant system is employed to model water flow, to be able to account also for watercourses and larger water bodies such as lakes and to model accurately strong transients.

Water could move horizontally in the Interception zone from cell to nearby ones or could move vertically infiltrating in the Soil zone; infiltration depends on volume of water building up in the Interception zone.

The Soil zone is subdivided into three layers:

The Capillary Layer (C) is the uppermost, in which capillarity allows water to occupy portions of soil above groundwater level. The most common condition is partial saturation; adsorbed water flux "Q_a" is practically still and is drained from the gravitational layer and not recharged from rainfalls. Water stored is lost through evapotranspiration "E," due to the proximity to atmosphere and to the presence of the plants' roots.

The Gravitational Layer (G) is the main layer in which the motion of water is linked to gravity, and horizontal fluxes "f_g" toward nearby cells are possible. This layer loses water also toward the upper capillary layer (absorption flux "Q_a") and the lower layer (deep percolation flux "Q_e"). Water is fed to gravitational layer from the atmosphere through the infiltration of rainfall "Q_i" and snow "Q_n." Horizontal fluxes "f_g" are modeled with simple linear reservoirs equation, and flow direction is modeled from surface steepness. An improvement of this part of the model is foreseen, but applying a physical-based simulation of underground circulation will lead to unbearable numerical complexity and run time.

The Deep Layer (P) simulates the water circulation in the deeper part of soil, such as bedrock or thick alluvial\colluvial debris. Water reaches this layer due to percolation flux "Q_p" from gravitational layer. The dynamics of water flow in this layer ("f_p" fluxes) is certainly slower than in upper layers, but it is driven by the same laws.

Finally, the Drainage zone (D) constitutes the surface portion in which the watercourses flow ("f_d" fluxes). This zone is variable in time since it is activated when water level in a given cell reaches a critical threshold value. This dynamic approach avoids a priori identification of watercourses which is generally employed by models. This identification, fixed in time, introduces an arbitrary assumption that can lead to unexpected behavior. Drainage zone collects all the waters coming from the superficial outflow. Generally, the drainage areas are concentrated in correspondence of the lines of maximum slope that run along the slope or long the valley.

SMART-SED model explicitly considers soil erosion and solid transport. The solid production module predicts soil erosion on pixel surface. Formulation exploits the Erosion Potential Method (EPM) (Gavrilovic 1988) model of spread erosion. It is a semiquantitative model developed in alpine environment for estimating annual sediment production inside a watershed. The EPM equation expresses the annual mean volume of sediment "G" in function of annual sediment production W_s due to superficial erosion and annual sediment redisposition R. This equation is suitable for seasonal estimation, but in SMART SED model, it has been downscaled in order to improve its use for a daily sediment balance estimation. SMART-SED model calculates W_s using this approach, while sediment transport is explicitly accounted for with transport formulas and not evaluated with EPM R value. The term Ws is the defined as follow:

$$\mathrm{Ws} = \pi\, \mathrm{H}\, \tau_{\mathrm{g}}\, \mathrm{Z}^{2/3}\, \mathrm{A}$$

This formulation takes in account the dependence of the sediment production in function of rainfall rate (H), the temperature (τ_g), the watershed basin area (A), and the erosion coefficient (Z).

The complete mathematical model can be expressed by the following set of conservation laws. For the Drainage zone (D), the de Saint-Venant equations are assumed,

$$\frac{\partial u}{\partial t} = -g\frac{\partial \eta_D}{\partial x} - u\frac{\partial u}{\partial x} - v\frac{\partial u}{\partial y} - \gamma u$$

$$\frac{\partial v}{\partial t} = -g\frac{\partial \eta_D}{\partial y} - u\frac{\partial v}{\partial x} - v\frac{\partial v}{\partial y} - \gamma v$$

$$\frac{\partial \eta_D}{\partial t} = -\frac{\partial (u h_D)}{\partial x} - \frac{\partial (v h_D)}{\partial y} + P - Q_I$$

where u, v denote the velocity components,η_D is the height of the water free surface above a reference level, $h_D = \eta_D - b$ is the total water depth, b is the terrain elevation, γ is a friction coefficient derived from an appropriate friction model, P is the precipitation rate, and Q_I is the infiltration rate. For the whole domain, the conservation equations for the equivalent water depths of the various layer are given by the following:

$$\frac{\partial h_A}{\partial t} = -P + E$$

$$\frac{\partial h_C}{\partial t} = Q_a - E$$

$$\frac{\partial h_S}{\partial t} = -\frac{\partial(f_{Sd,x})}{\partial x} - \frac{\partial(f_{Sd,y})}{\partial y} + W_{s.}$$

$$\frac{\partial h_G}{\partial t} = -\frac{\partial(f_{G,x})}{\partial x} - \frac{\partial(f_{G,y})}{\partial y} + Q_I - Q_e$$

$$\frac{\partial h_P}{\partial t} = -\frac{\partial(f_{P,x})}{\partial x} - \frac{\partial(f_{P,y})}{\partial y} + Q_{e.}$$

The mathematical model outlined above is approximated numerically by a hybrid finite volume and finite difference technique. The spatial discretizations of all the layer depth equations are performed by a first-order conservative upwind approach, which allows to guarantee conservation of the water mass at the discrete level. For the Drainage zone layer, the de Saint-Venant equations are solved numerically by the widely applied and robust semi-implicit, Eulerian-Lagrangian discretization approach proposed in Casulli and Cheng (1992). This discretization uses a staggered Cartesian mesh built directly on the pixels of the digital elevation model and allows to employ relatively large time steps also in the presence of lakes and rivers. The implicit time discretization requires the solution of a linear system at each time step, which is well conditioned and sparse and can be solved efficiently by the conjugate gradient method. All the layer depth equations are instead discretized explicitly in time by the forward Euler method, which in these cases does not entail excessive time step restrictions.

Positivity of the water layer and consistency between the mass conservative discretization of sediment transport and that of the Drainage zone layer are guaranteed considering the remarks in Gross et al. (2002). In a future development, the approach recently proposed in Casulli (2019) will also be implemented to consider marginally resolved or under resolved watercourses. It is to be remarked that, even though formally of first order only, the effective accuracy of the method employed is adequate for situations with great uncertainties in the available data and strongly irregular data patterns.

Input data for SMART-SED are constituted by raster maps of terrain properties (pedological class, granulometry, porosity), use of soil, DTMs, and weather data. These data can be retrieved from public database for most of developed countries' territory. SMART-SED is designed to work mainly in the natural environment that characterizes mountain basins; nevertheless, it will be possible to account for main structures existing in the watershed, such as dams, retention ponds, or reservoirs. These structures modify the terrain surface and will be simulated altering the elevation of DEM or creating artificial boundaries (i.e., increasing maximum storage in Interception zone prior of flow starting) for cells interested by the structures themselves. Similarly, if wide areas of the modeling domain are interested by soil and water conservation measures, their effect will be implicitly considered choosing the correct parameters to calculate soil erosion and water flow velocity. At this stage, the authors did not feel the urge to develop a dedicated module to handle this situation, but it would be added, if needed, during future developments of the code.

3.3 Synthetic Test Cases and Capability Demonstration

In this section, some synthetic test cases regarding the application of the SMART-SED model are presented. The model is tested in a range of cases in which some simplifications of the geometry are adopted. In particular, synthetic digital elevation models are generated in order to better evaluate the behavior of the model in different terrain conditions.

Hillslopes are described by geometries such as a paraboloid or inclined planes that can simulate real situations which may be present in a complex terrain. The paraboloid can represent a sink, which is quite common in the upstream portions of a valley, where glacial circle lakes are settled. Instead, the inclined planes represent an idealized valley shape, which generally characterizes a mountain river path. The two cases are chosen in order to evaluate the ability of the model to deal with sinks and high gradients typical of the mountain regions.

The sink test, see Fig. 3.2, is set up in a simple way, evaluating only the superficial water fluxes inside the Drainage zone module, disregarding both the terrain modules and sediment module. The underlying assumption is that the terrain is completely impervious, water can flow only through the surface, and precipitation is maintained constant at rate of 10 mm/h for the entire simulation and is uniformly distributed in the domain.

The goal of this simulation is to test the ability of the discretized de Saint-Venant equations to reproduce the lake formation inside the paraboloid geometry. This approach is quite novel, since in most hydrological studies the digital elevation model is usually preprocessed and all sinks are filled. At the basin scale, this option is necessary in order to permit the water river flux to reach the closing section of the basin, but this choice is not correct from the sediment transport point of view. Due to the importance of the simulation of sediment transport in our model, sinks and terrain depressions should be included, so that the complete set of the de Saint-Venant equations is required for a better reproduction of surface water fluxes.

In Fig. 3.2, the results obtained in this test are reported. The rising lake level due to the water superficial runoff is well reproduced and corresponds to the lowest part of the paraboloid surface.

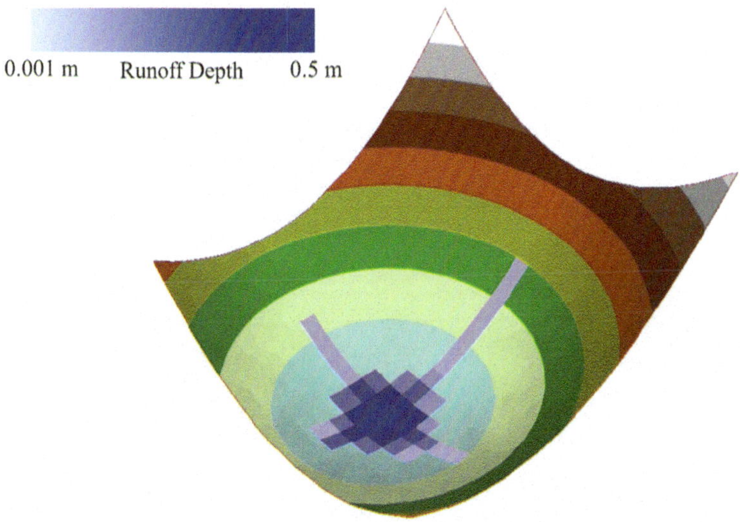

Fig. 3.2 Paraboloid sink with water accumulation. A preferential flow path is present at the valley bottom

Table 3.1 Simulation parameters

Simulation parameters	Simulation N°1	Simulation N°2
Rain rate [mm/h]	20	5
Temperature [°C]	20	5
Gavrilovic Z coefficient [/]	0.55	0.55
Test duration [h]	5	5
Rainfall event duration [h]	2.5	2.5
Model time step [n°]	1000	1000

The second test considered an idealized valley with triangular cross section. In this case, the whole model is considered, including also the terrain module and the sediment module that were excluded in the previous test case. Two different simulations have been carried out:

- The first simulates a quite intense and localized rainfall event with an intensity of 20 mm/h.
- The second simulates a smooth precipitation, uniform over the entire domain with an intensity of 5 mm/h.

The key simulation data are summarized in Table 3.1. The goal of these two simulations is to test the ability of the model to work under different meteorological conditions. Moreover, in the first, the temperature of the lowest cell of the digital elevation model is equal to 20 °C, and in the second, it is imposed to be 5 °C.

The results obtained are in good qualitative agreement with the physics of the phenomena involved and included in the SMART SED model.

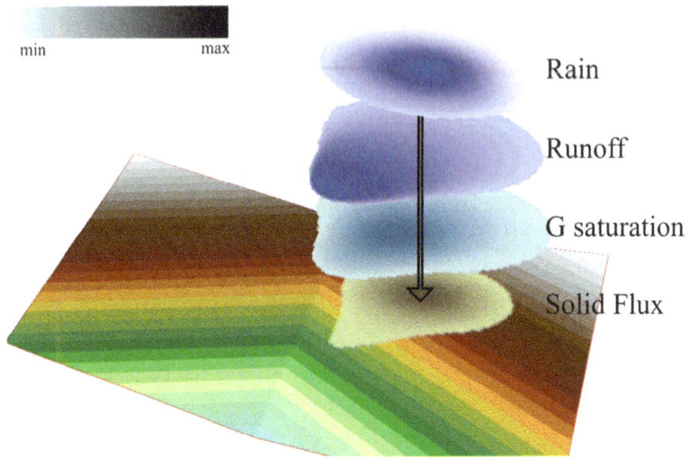

Fig. 3.3 Fictional valley case: Rain, Runoff, Gravitational layer saturation, and Solid flux intensity are represented in color scale. Color saturation is proportional to quantity of described variable, normalized with respect to maximum value. Rain is constant in time; other quantities are represented at $t_1 = 2$ h

For simulation 1, shown in Fig. 3.3, we can see the repartition of the flows inside the terrain layers and the runoff formation in the part of the valley where the precipitation is located. In addition, solid transport along the slope is present in conjunction with runoff. At the end of the simulation, the superficial fluxes are collected in the lower part of the valley and the total water and sediment balance inside the domain is preserved. It is possible to notice how runoff moves faster than gravitational soil saturation, as expected, while solid flux is slower than runoff peak.

In simulation 2, lower temperature has two effects inside the domain. The snow is accumulated in the Interception zone when temperature drops under 0 °C. Here, the absence of runoff sets to zero the Gavrilovic sediment production factor, and in this part of the basin, the sediment transport is absent. This is a typical situation for basins in the alpine area, where the snow cover prevents erosion processes over the whole basin in the winter period.

3.4 Sediment Transport Database

SMART-SED project foresees an integration of modeling and monitoring. Monitoring solid transport is fundamental to tune the model but also to improve our quantitative knowledge of these processes. Generally, not many detailed field measurements of solid fluxes are available. Two different approaches have been outlined within the project for sediment transport estimation – an Eulerian and a Lagrangian one. All the measures have been carried out on a test basin enclosing the city of Lecco (Lombardy, Northern Italy). These direct measures are usually coupled with topographical evaluation of bed shape variations (Longoni et al. 2016b), which are being carried out on the test case, so that complete results are not available yet (Fig. 3.4).

The town (where about 50,000 people live) is crossed by three streams (the Bione, Caldone, and Gerenzone) that have the typical characteristics of torrents in a pre-Alpine area. Tests are carried out on the Caldone river. The hydro-graphic basin of this watercourse is 24 km^2, with an altitude stretching from 197 m a.m.s.l. to 2118 m a.m.s.l. at the top of Grigna Meridionale. Geologically, the basin is characterized by rocky outcrops in the higher part (mainly limestone and clastic rock), while downstream toward the city, the river flows through a floodplain. The average precipitation over the city of Lecco is about 1400 mm/y. The Caldone river flows from Mount Resegone and, just before entering the city, receives the water from the Grigna torrent. From that point on, the river flows through the town, mostly within artificial banks. Waters are withdrawn by industries in the surroundings (mostly for machinery cooling) and by residential buildings. On the other hand, the stream receives a significant amount of water from the sewer network that drains the (mostly impermeable) town area. In its last kilometer before the outlet into the Lario lake, the Caldone flows within a culvert that passes below the town center. The choice of Caldone river was due to a combination between a short hydrologic time of response, high slope, intense sediment transport, and flow within a densely urban area.

The Eulerian approach is based on measures in a fixed section of the river, while the Lagrangian approach has the objective to monitor the displacements of single pebbles. The two approaches, coupled, can depict in an effective way the main features of solid transport at river scale: 1. total volume, 2. critical diameter, and

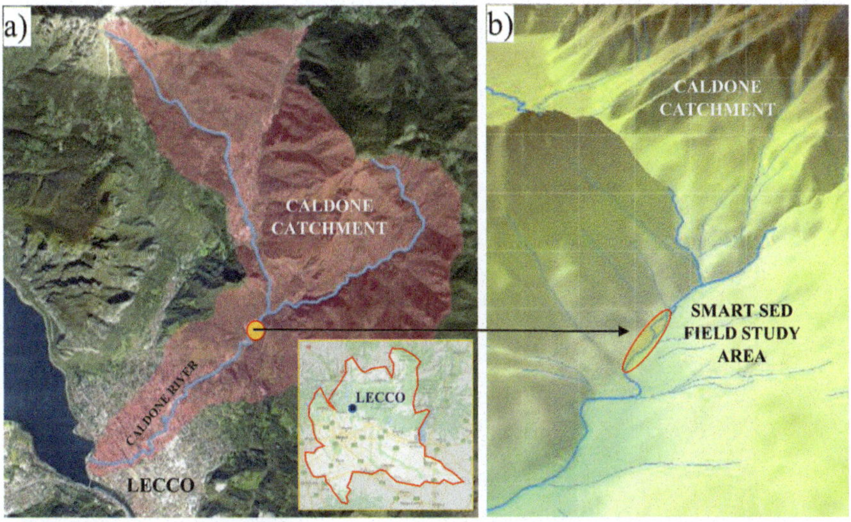

Fig. 3.4 Caldone river basin. The basin is located in northern Italy and it has been selected for its typical features. (**a**) The basin shape and the position of Lecco city. (**b**) A 3D view of the area where most of the field work is focused, just upstream of the city

3. mean pebbles' displacement. The first objectives are fulfilled with the Eulerian approach, the latter with the Lagrangian one. These techniques have been presented and widely discussed in previous works by the same authors (Ivanov et al. 2017; Longoni et al. 2017; Brambilla et al. 2018).

The Eulerian approach for the total volume measurement is based on drone ultrasonic bathymetry and critical diameter indirect measure. A simple drone was developed in-house within the SMART-SED project. The drone is a catamaran of almost 50 × 50 cm whose main scope is to stabilize the sonar probe. The sonar probe itself is a low-cost fishing sonar which transmits data via Wi-Fi connection to a smartphone. The measure range extends from 0.7 m to 70 m, the measure resolution is 0.01 m, and tested accuracy falls under 0.05 m in the range 0.7–2 m (comparable to sedimentation pool deepness). Sonic wave draws a conic beam with a top angle of 15°; the acquired measure is the mean value of hit area (Fig. 3.5). The shallow and dry part of the pool is monitored via traditional topographic approach. The data are joined and a DTM of the pool is created, successive DTMs are compared, and accumulated material is measured. Two campaigns have been carried out so far, autumn 2016 and 2017, and accumulated volume is 1260 m^3. More frequent campaign is foreseen in spring 2019 to evaluate single-event mobilized volume.

The critical diameter was measured as explained in Scheingross et al. (2013), exploiting a semi-Eulerian method (see Fig. 3.6, a and b). Square areas of 0.25 m^2 were painted in red, without disturbing the material. These areas were selected on bars, banks, and area that are dry during normal flow stages and flooded during moderate to intense events. Comparison of pre- and postevent states is done via image analysis, pictures are taken before and after any event, and the granulometric distribution is analyzed with Basegrain (Detert and Weitbrecht 2012). Eighteen tests of this kind have been done between August 2016 and November 2017. The results are summarized in Table 3.2 and a discussion of the results is presented in Brambilla et al. (2018). This database has been used to validate classic hydraulic modeling in the investigated reaches of the Caldone River.

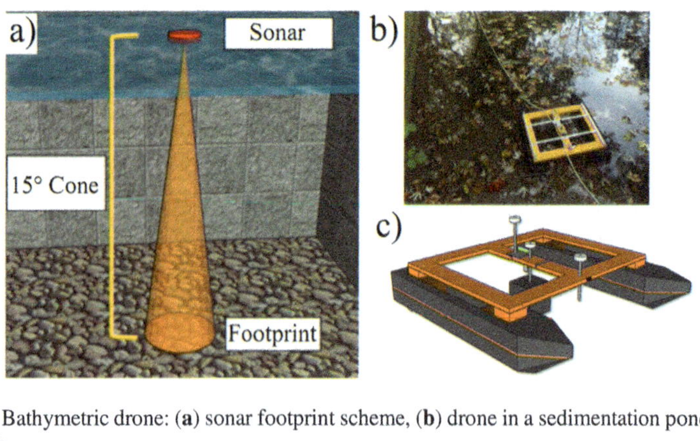

Fig. 3.5 Bathymetric drone: (**a**) sonar footprint scheme, (**b**) drone in a sedimentation pond, and (**c**) 3D sketch

Fig. 3.6 Eulerian and Lagrangian pebbles to monitor sediment transport. (**a**) Eulerian approach, painted area before event (**b**) and after event. (**c**) Lagrangian approach, SMART-PS after deployment. (Modified after Brambilla et al. 2018)

Table 3.2 Critical diameters (Dc) comparison table; flow rate measured in the same reach

Event date	Morphology type	Flow rate [m^3/s]	D_c measured [mm]
05/08/2016	Bank	5.88	>90.5
15/09/2016	Bank	1.12	45.3–64
15/09/2016	Bar	1.12	64–90.5
14/10/2016	Bank	0.67	45.3–64
14/10/2016	Bar	0.67	45.3–64
14/10/2016	Bar	0.67	32–45.3
14/10/2016	Bar	0.67	64–90.5
26/03/2017	Bar	1.4	45.3–64
26/03/2017	Bank	1.4	45.3–64
06/06/2017	Bar	5.88	>90.5
06/06/2017	Bar	5.88	45.3–64
18/08/2017	Bar	1.55	>90.5
18/08/2017	Bank	1.55	>64
18/08/2017	Bank	1.55	>100
18/08/2017	Bank	1.55	>100
15/11/2017	Bar	3.19	22.6–32
15/11/2017	Bank	3.19	<10

Modified after Brambilla et al. (2018)

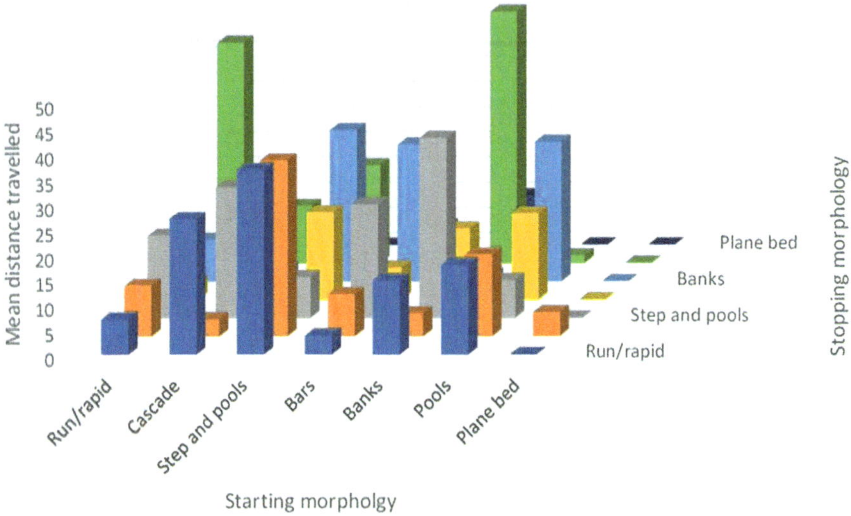

Fig. 3.7 Mean distance traveled by SMART-PS in all the recorded events, classified for starting and stopping kind of morphology

As for the Lagrangian approach, the migration of single grains along the stream was monitored by means of RFID (Radio Frequency IDentification) tags implanted in pebbles (Hassan and Ergenzinger 2003; Chapuis et al. 2015) (see Fig. 3.6 c). Starting from late spring 2016, almost 1000 pebbles have been dispersed in a chosen reach of Caldone test river in three different campaigns held at the beginning of each rain season (see Fig. 3.4). The test reach is 200 meters long; it is a natural bed with moderate steepness and encompasses areas with different morphologies. The tracers, named SMART-PS, are retrieved from the Caldone river itself, and Low Frequency Passive Transducers are glued inside each of them. Any SMART-PS has a unique identification code, and its measures and weight are recorded in a database. After each major rain event, a team was in charge to look for SMART-PS in the studied reach with a self-built antenna. Displacements are recorded along with the morphology of starting and stopping area and reported in Fig. 3.7.

The database is still being updated as new events occur in the test basin. The data constitute a valuable picture of sediment movements and transport in a mountain basin and will be used to calibrate the model when it will be ready to work on real data with adequate resolution.

3.5 Discussion and Conclusion

This chapter presents part of the planned work in the framework of the SMART-SED project. The objective of the project is to improve sediment transport management in mountain rivers to decrease hazard posed to cities and infrastructures. Our team is currently working to achieve this objective developing a numerical model that will be integrated in a freeware, open-source software model; this model will be distributed among local government authorities and professionals. The urge for models able to combine effectiveness and ease to be used is felt by stakeholders and professionals, and it was one of the reasons to propose the SMART-SED project. Moreover, the model will rely on free data, made available publicly by research or governmental institutions, allowing to use them without long and costly survey campaigns, accepting a little trade-off on accuracy. In the model planning phase, the need of reliable field data for validation arose; indeed, the databases available in literature are usually built from seasonal data while event scale data were needed. A pluriannual campaign for solid transport detailed monitoring has been designed and started in 2016, and it is still going on, providing reliable and valuable data for model tuning. SMART-SED project will last until end of 2020 and will have as final outcome the publication of SMART-SED model.

References

Abbaspour KC, Rouholahnejad E, Vaghefi S, Srinivasan R, Yang H, Kløve B (2015) A continental-scale hydrology and water quality model for Europe: calibration and uncertainty of a high-resolution large-scale SWAT model. J Hydrol 524:733–752

Allamano P, Claps P, Laio F (2009) Global warming increases flood risk in mountainous areas. Geophys Res Lett 36. https://doi.org/10.1029/2009GL041395

Ballio F, Brambilla D, Giorgetti E, Longoni L, Papini M, Radice A (2010) Evaluation of sediment yield from valley slopes: a case study. WIT Trans Eng Sci 67:149–160

Bemporad GA, Alterach J, Amighetti FF, Peviani M, Saccardo I (1997) A distributed approach for sediment yield evaluation in alpine regions. J Hydrol 197:370–392

Bennett GL, Molnar P, McArdell BW, Burlando P (2014) A probabilistic sediment cascade model of sediment transfer in the Illgraben. Water Resour Res 50:1225–1244. https://doi.org/10.1002/2013WR013806

Brambilla D, Longoni L, Mazza F, Papini M (2011a) Sediment yield from mountain slopes: a GIS based automation of the classic Gavrilovic method. River Basin Manage VI 146:301

Brambilla D, Longoni L, Papini M, Giorgetti E, Radice A (2011b) On analysis of sediment sources toward proper characterization of hydro-geological hazard for mountain environments. Int J Saf Secur Eng 1:423–437

Brambilla D, Papini M, Longoni L (2018) Temporal and spatial variability of sediment transport in a Mountain River: a preliminary investigation of the Caldone River, Italy. Geosciences 8:163. https://doi.org/10.3390/geosciences8050163

Casulli V (2019) Computational grid, subgrid, and pixels: computational grid, subgrid, and pixels. Int J Numer Methods Fluids 90:140. https://doi.org/10.1002/fld.4715

Casulli V, Cheng RT (1992) Semi-implicit finite difference methods for three-dimensional shallow water flow. Int J Numer Methods Fluids 15:629–648. https://doi.org/10.1002/fld.1650150602

Chapuis M, Dufour S, Provansal M, Couvert B, de Linares M (2015) Coupling channel evolution monitoring and RFID tracking in a large, wandering, gravel-bed river: insights into sediment routing on geomorphic continuity through a riffle–pool sequence. Geomorphology 231:258–269. https://doi.org/10.1016/j.geomorph.2014.12.013

Coulthard TJ, Macklin MG, Kirkby MJ (2002) A cellular model of Holocene upland river basin and alluvial fan evolution. Earth Surf Process Landf 27:269–288

de Vente J, Poesen J, Verstraeten G, Govers G, Vanmaercke M, Van Rompaey A, Arabkhedri M, Boix-Fayos C (2013) Predicting soil erosion and sediment yield at regional scales: where do we stand? Earth Sci Rev 127:16–29. https://doi.org/10.1016/j.earscirev.2013.08.014

Detert M, Weitbrecht V (2012) Automatic object detection to analyze the geometry of gravel grains–a free stand-alone tool. In: River flow. Taylor & Francis Group, London, pp 595–600

Dunne T, Zhang W, Aubry BF (1991) Effects of rainfall, vegetation, and microtopography on infiltration and runoff. Water Resour Res 27:2271–2285. https://doi.org/10.1029/91WR01585

Gavrilovic Z (1988) Use of an empirical method (erosion potential method) for calculating sediment production and transportation in unstudied or torrential streams. In: International conference on river regime. Hydraulics Research Limited, Wallingford, pp 411–422. 5 fig, 4 tab, 8 ref

Goodrich DC, Unkrich CL, Smith RE, Woolhiser DA (2002) KINEROS2 – a distributed kinematic runoff and erosion model. Hydrologic modeling for the 21st Century agenda: Second federal interagency hydrologic modeling conference, July 28–August 1, 2002, Riviera Hotel, Las Vegas, Nevada

Gross ES, Bonaventura L, Rosatti G (2002) Consistency with continuity in conservative advection schemes for free-surface models. Int J Numer Methods Fluids 38:307–327. https://doi.org/10.1002/fld.222

Hartmann T, Driessen P (2017) The flood risk management plan: towards spatial water governance: towards spatial water governance. J Flood Risk Manage 10:145–154. https://doi.org/10.1111/jfr3.12077

Hassan MA, Ergenzinger P (2003) Use of tracers in fluvial geomorphology. In: Tools in fluvial geomorphology. Wiley, Chichester, pp 397–423

Hirabayashi Y, Mahendran R, Koirala S, Konoshima L, Yamazaki D, Watanabe S, Kim H, Kanae S (2013) Global flood risk under climate change. Nat Clim Chang 3:816–821. https://doi.org/10.1038/nclimate1911

Horton RE (1939) Analysis of runoff-plat experiments with varying infiltration-capacity. Trans Am Geophys Union 20:693. https://doi.org/10.1029/TR020i004p00693

Ivanov VI, Brambilla D, Longoni L, Arosio D, Papini M (2017) RFID-aided sediment transport monitoring—laboratory and preliminary field test results. In: Mikoš M, Casagli N, Yin Y, Sassa K (eds) Advancing culture of living with landslides. Springer International Publishing, Cham, pp 623–630

Kim J, Ivanov VY (2014) On the nonuniqueness of sediment yield at the catchment scale: the effects of soil antecedent conditions and surface shield. Water Resour Res 50:1025–1045. https://doi.org/10.1002/2013WR014580

Lane SN, Tayefi V, Reid SC, Yu D, Hardy RJ (2007) Interactions between sediment delivery, channel change, climate change and flood risk in a temperate upland environment. Earth Surf Process Landf 32:429–446. https://doi.org/10.1002/esp.1404

Longoni L, Ivanov VI, Brambilla D, Radice A, Papini M (2016a) Analysis of the temporal and spatial scales of soil erosion and transport in a mountain basin. Ital J Eng Geol Environ 16:17–30

Longoni L, Papini M, Brambilla D, Barazzetti L, Roncoroni F, Scaioni M, Ivanov VI (2016b) Monitoring riverbank Erosion in mountain catchments using terrestrial laser scanning. Remote Sens 8:241

Longoni L, Arosio D, Brambilla D, Ivanov VI, Papini M (2017) Monitoring bedload sediment transport in a pre-alpine river: an experimental method. Rend Online Soc Geol Ital 43:57–63

Milly PCD, Wetherald RT, Dunne KA, Delworth TL (2002) Increasing risk of great floods in a changing climate. Nature 415:514. https://doi.org/10.1038/415514a

Neitsch SL, Arnold JG, Kiniry JR, Williams JR (2011) Soil and water assessment tool theoretical documentation version 2009. Water Resources Institute, Texas

Radice A, Giorgetti E, Brambilla D, Longoni L, Papini M (2012) On integrated sediment transport modelling for flash events in mountain environments. Acta Geophys 60:191–213

Radice A, Longoni L, Papini M, Brambilla D, Ivanov VI (2016) Generation of a design flood-event scenario for a mountain river with intense sediment transport. Water (Switzerland) 8. https://doi.org/10.3390/w8120597

Rosso R (1994) An introduction to spatially distributed modelling of basin response. Adv Distrib Hydrol 3:30

Scheingross JS, Winchell EW, Lamb MP, Dietrich WE (2013) Influence of bed patchiness, slope, grain hiding, and form drag on gravel mobilization in very steep streams: gravel mobilization in steep channels. J Geophys Res Earth 118:982–1001. https://doi.org/10.1002/jgrf.20067

Smith R (2015) Directive 2008/94/EC of the European Parliament and of the Council of 22 October 2008. In: Core EU legislation. Macmillan Education UK, London, pp 423–426

Thornthwaite CW (1948) An approach toward a rational classification of climate. Geogr Rev 38:55. https://doi.org/10.2307/210739

Chapter 4
Natural Groundwater Background Levels of Nitrate and Landfill Effects (Apulia, Southern Italy)

Livia Emanuela Zuffianò, Pier Paolo Limoni, Giorgio De Giorgio, and Maurizio Polemio

Abstract High nitrate concentration of fresh groundwater of a coastal karstic aquifer, not far from the Apulian regional capital (Southern Italy), caused public concern and alarm from the authorities. The attention of local communities focused on the effect of a group of landfills, the use of which started from 1975, using improved technological and safety devices to reduce groundwater pollution risks.

This chapter describes a simple approach based on the spatial and temporal variability of groundwater nitrate concentration (GNC) able to answer to public concern, circumscribing the main source of contamination, discussing of natural background values and spatial/temporal variability of GNC. For this purpose, a wide spatial and temporal range of data were discussed from 2006 to 2018. A regional survey of 2006–2007 was used to define background values or almost natural values of groundwater nitrate concentration (GNC). In the landfill sector, a monthly time series from 2007 to 2013 showed the GNC seasonality and the relationships with rainfall, and the periodical surveys of 2014–2018 showed the geochemical groundwater characteristics and the spatial variability of GNC, in both cases at the detailed scale of the study area.

The results show the source is diffuse and should be related to the intensive agriculture activities in the area. The results correspond perfectly with the results obtained using a multimethodological complex and expensive approach based on the use of several parameters, including nitrogen isotopes.

Keywords Nitrate · Natural background nitrate · Nitrate pollution · Landfill karstic coastal aquifer · Apulia · Italy

L. E. Zuffianò (✉) · P. P. Limoni · G. De Giorgio · M. Polemio
CNR-IRPI, National Research Council – Research Institute for Hydrogeological Protection, Bari, Italy
e-mail: l.zuffiano@ba.irpi.cnr.it

M. De Maio, A. K. Tiwari (eds.), *Applied Geology*,
https://doi.org/10.1007/978-3-030-43953-8_4

4.1 Introduction

Natural soil and rock minerals containing nitrogen are generally very soluble and, therefore, rare (Appelo and Postma 1996). In contrast, compounds containing nitrogen can be found in organic matter, such as in amino acids of proteins. The natural nitrate concentration of groundwater is worldwide low; concentrations greater than a few milligrams per litre can be generally due to anthropogenic activities (WHO 2011).

High nitrate concentration of groundwater is a worldwide concern (Strebel et al. 1989; Burow et al. 2010; Thorburn et al. 2003). Potential anthropogenic sources are usually septic waste systems, inorganic fertilizers, animal manure and leachate of landfill. The correlation between agriculture activities and nitrate groundwater content is considered high in the literature (Heaton 1986; Sheikhy Narany et al. 2017).

The evolution from traditional agricultural practices to the intensive cultivation forms and the development of the chemical products has determined the spread of nitrogenous fertilizers.

As leachate contains heavy metals and nitrogenous solutes, the surrounding landfill areas have a greater probability of groundwater pollution due to the risk of leachate leakage (Mor et al. 2006). High nitrate concentration was highlighted as anomalously high, alarming the local communities and leaving them worried about the landfill activity, especially after 2011, when the latest landfill became operative.

Natural background levels of nitrate in the groundwater should be commonly used to serve as references when assessing the contamination status of groundwater units (Lucon et al. 2018; Parrone et al. 2019). This chapter depicts a simple spatial temporal method, based on the discussion of groundwater nitrate concentration (GNC), assessing if groundwater nitrate pollution by landfill leachate mixing exists and, in the negative case, suggesting alternative anthropogenic sources of nitrate. The approach is validated selecting a site for which a previous study was done. It used complex and expensive chemical and isotope parameters of groundwater and leachate, analysis of land use and the assessment of the role of nitrogen deriving from inorganic fertilizers, and mineralogical analysis of groundwater suspended particles and showed that the role of landfills in the area on the groundwater quality was null or insignificant (Cossu et al. 2018). High GNC, somewhere higher than the standard threshold, was explained as the effect of inorganic fertilizers.

4.2 Geological and Hydrogeological Settings

Four main hydrogeological structures (HSs), Gargano, Tavoliere, Murgia and Salento, can be distinguished in Apulia (Fig. 4.1) (Polemio 2016). Apart from Tavoliere, which corresponds to a shallow porous aquifer, the remaining HSs constitute wide coastal karstic aquifers, offering high-quality groundwater resources:

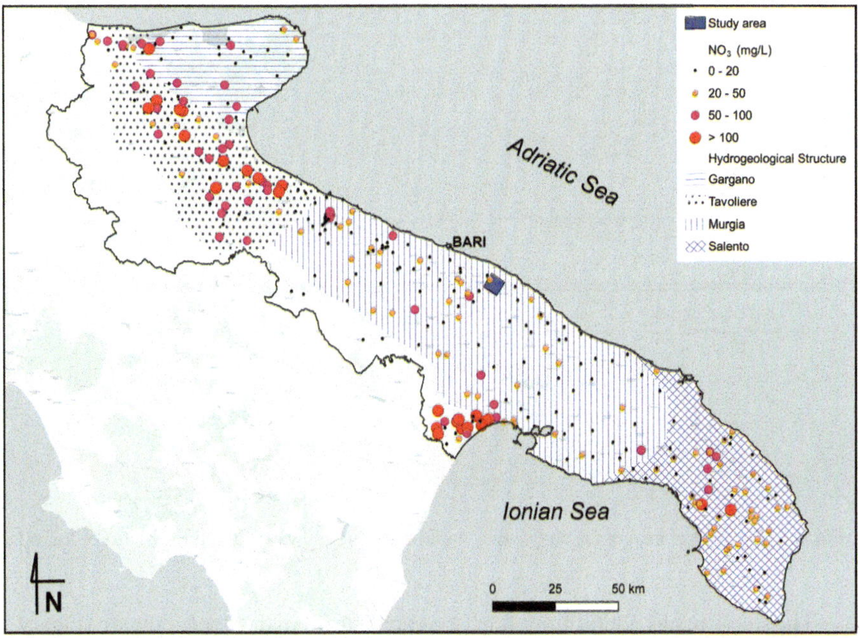

Fig. 4.1 Boundaries of main hydrogeological structures of Apulia and groundwater nitrate concentration of 2006–2007, from the regional nitrate survey of the Apulia region (Ancona et al. 2010)

Gargano, Murgia and Salento. HSs show common characteristics: they consist of Mesozoic calcareous and/or calcareous-dolomitic rocks; the permeability, which is heterogeneous and anisotropic due to the karst and the fracturing, is from medium to high.

Focusing on the Murgia HS, it is the central part of the Apulian carbonate platform, in which can be distinguished the Mesozoic limestone formation called 'Calcari di Bari', roughly 2 km thick, which outcrops in the area (Fig. 4.2). The Mesozoic limestone formation was involved by paleo-tectonic and neo-tectonic movements (Pieri et al. 1997). It shows surface and deep karstic features, that is, poljes, swallow holes, dolines and dry valleys. These rocks outcrop widespread also if are locally overlapped by subsequent formations, constituted by calcarenite, sandy clay and alluvial deposits. The groundwater saturated flow involves predominantly the limestone formation, which is part of a deep karstic aquifer, involving the whole Murgia HS (Cotecchia et al. 2005). The recharge area includes inland portions of the Murgia HS (Polemio 2005); the outflow happens along the Adriatic coast and the Ionian coast (Zuffianò et al. 2016). Serious seawater intrusion effects are known for this coastal aquifer (Polemio 2016). Several degradation effects of groundwater quality and quantity of this aquifer were observed (Polemio and Limoni 2006).

The salinity threshold of pure fresh groundwater of the aquifer was assessed equal to 0.5 g/L, value mainly due to water–carbonate rock interaction (Polemio et al. 2009).

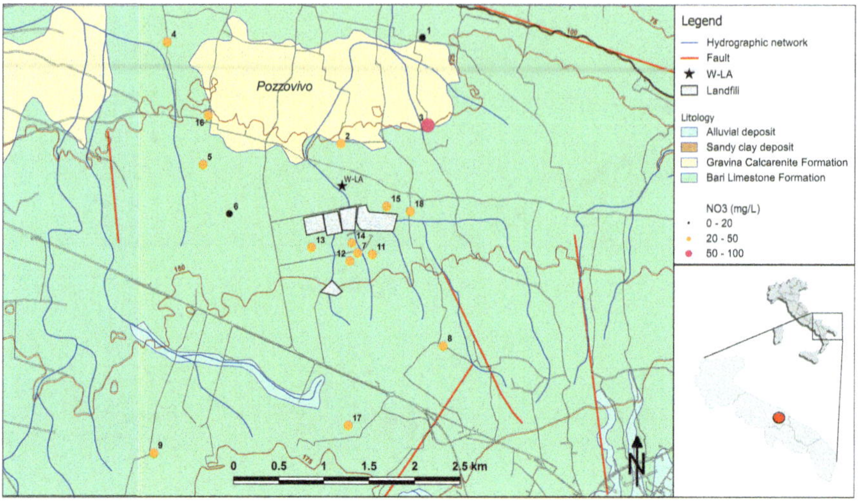

Fig. 4.2 Geological map of the study area, landfills and groundwater sampling points (W-La, 1-18)

The study area is not far from Bari, the capital town in the Apulia region (Figs. 4.1 and 4.2), in the Conversano Municipality. The land surface is a vast rural area in which agriculture is largely prevalent. Vineyards, orchards, olive trees and arable crops cultivations are very diffuse. The study area includes five solid urban waste landfills, which are located roughly at 5 km northwest from the town of Conversano, where the closest rainfall gauge of the regional network is located (Fig. 4.2). As the rest of the Murgia HS, the study area is dominated by limestone outcrop (Fig. 4.2). The local groundwater flow is towards the Adriatic Sea (6 km away, north–north-east). The depth of piezometric level from ground surface, 120–137 m, is so high to complicate any type of survey. Local groundwater overexploitation for irrigation purposes seems worsening seawater intrusion effects.

4.3 Materials and Methods

Groundwater nitrate concentration (GNC) data of 361 sampling points were considered for the period 2006–2007, involving all Apulian HSs (Fig. 4.1). The sampling points were selected for the Regional Nitrate Plan, including wells of the Regional Groundwater Monitoring Network of the Apulia region (Ancona et al. 2010). Period and wideness of this surveyed area were selected to define background or almost natural GNC values respect to fearing news published after 2011.

Moving to the detailed scale of the study area, the well W-LA was used to study the GNC monthly time series from 2007 to 2013, showing the GNC seasonality and the relationships with monthly rainfall, using the closest gauge of the monitoring

Table 4.1 Descriptive statistics for physical–chemical parameters of groundwater

	EC (mS/cm)	T (°C)	pH (–)	DO mg/L	Eh (mV)
Min.	0.68	12.07	6.73	1.39	−52.00
Average	1.18	16.75	7.14	4.35	147.18
Max	3.17	19.48	7.58	8.25	346.00
Std. dev	0.52	1.03	0.20	1.73	79.12

EC electrical conductivity at 25 °C, *T* temperature, *DO* dissolved oxygen, *Eh* redox potential

network of the Apulia region (5 km from landfills, very close to the SE corner of Fig. 4.2).

A total of 18 groundwater samples (Fig. 4.2) were collected between November 2014 and January 2018 during four surveys. Before sampling, each well was purged, and groundwater samples were collected using a flow cell, connected to the pumping system, and a multiparametric probe. Each groundwater sample was taken only after constant values of electrical conductivity (EC), pH, temperature (T), dissolved oxygen (DO) and redox potential (Eh) were established after purging (Zuffianò et al. 2018). The descriptive statistic for physical–chemical parameters is shown in Table 4.1.

On-site measurements of EC, pH, T, DO and Eh were conducted during sampling. The water samples were filtered (pore size 0.45 μm) using a membrane of cellulose acetate and then acidified adding HNO_3 to a pH < 2 for cation analysis, avoiding negative precipitation or absorption effects. The sample for anion analysis was unacidified. Moreover, alkalinity (as HCO_3^-) was determined by titration with HCl.

The chemical study focused on GNC and some ions (Na^+, K^+, Ca^{2+}, Mg^{2+}, Cl^-, NO_3^-, SO_4^{2-}) to define the geochemical features, operating by means of ICP-OES spectrometry and ion chromatography. The maximum accepted error of charge balance was mandatory within 5% for each analysed sample. Figure 4.4 shows the analysis results and the water classification.

The discussion was supported by classical statistical, geostatistical and time series tools, assessing main statistical parameters and frequency distribution of spatial GNC at aquifer or regional scale, cross-correlation and trend.

4.4 Results and Discussion

All HSs show GNC values greater than 20 mg/L (Fig. 4.2). A global rough estimation of the upper limit for pure or natural groundwater could be considered 1 g/L (Dubrovsky et al. 2010). In the United States, the natural GNC was assessed less than 4–9 mg/L, depending on lithological situation under aerobic conditions (WHO 2011). For both the criteria, the whole Apulian region is characterized by too high GNC values, too high to be natural. The widespread quality degradation is clear due to a diffuse source: the use of fertilizers. High GNCs correspond to cultivation of crops for which a significant nitrogen surplus is commonly used (Serio et al. 2018). Serio et al. (2018) report the negative effects of some cultivations such as of hard wheat in the Tavoliere HS, of fig trees in the Bari hinterland (Murgia HS), of sugar beets together with tobacco in some areas of the Salento HS, all areas in which, also due to the availability of irrigation facilities, intensive agricultural activities still exist. If the subset of data of the Murgia HS is considered, the range is 0.4–94.9 mg/L. The 2° quartile, equal to the median value, is 11.7 mg/L, while the 1° quartile is 3.0 mg/L: a preliminary assessment of the natural GNC of the Murgia HS could be considered not very much higher than 3.0 mg/L and in any case lower than 11.7 mg/L.

The well W-LA is downward if the landfill district is considered. The GNC monthly time series of the well W-LA permits to discus NO_3^- variability, cross-checking this with monthly rainfall variability (Fig. 4.3). Data are available from January 2007 to March 2013 with some gaps.

The GNC range is from 18 (June 2007) to 192 (November 2012) mg/L, with mean value equal to 67 mg/L, showing that there is an almost continuous anthropogenic modification of GNC in this well. The very high GNC monthly variability shows a self-evident seasonality: if the mean monthly GNC is considered, the minimum values are in summer (from June to August the range is 33–45 g/L) and the peak values are from November (156 mg/L) and January (the range is 90–152 mg/L), the wet season in the area. After deseasonalizing, the trend GNC was analysed using the Mann-Kendall test, showing a significant decreasing trend (at 5%). This type of temporal pattern does not seem physically justifiable with the release of a leachate plume from landfill. The rainfall time series shows a seasonal pattern almost like that of GNC. The cross-correlation with rainfall shows that the peak cross-correlation, which is statistically significant, is for lag 0 (cross-correlation coefficient equal to 0.248). The rainfall excess is able to cause relevant infiltration, especially in this karstic aquifer: the result is an increase of the GNC.

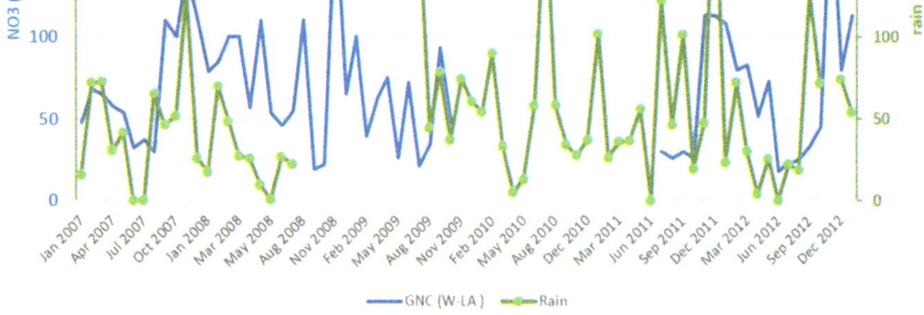

Fig. 4.3 Monthly GNC (groundwater nitrate concentration of well W-LA) and rainfall time series for the period 2007–2013

Four surveys were realized on site from 2014 to 2018 (Fig. 4.2). Statistics of on-site parameters of groundwater samples are reported in Table 4.1.

The EC range is from 0.68 to 3.17 mS/cm, 1.18 mS/cm as mean value. The peak EC was measured at the measuring point 1 (Fig. 4.2), corresponding to the highest well depth (300 m), which is reasonably the most exposed to the risks of upconing of brackish groundwater. pH is from 6.73 to 7.58, almost common values for this type of groundwater.

Piper diagram (Fig. 4.4) was used to show the geochemical facies and water types. Ca-HCO_3 and Ca-Mg-Cl are considered to be the main water types in the whole area surrounding the landfills: the former type is pure fresh groundwater (Polemio et al. 2009), whereas the latter type is due to the mixing caused by seawater intrusion (Custodio and Llamas 1996).

The samples 17 can represent the pure fresh groundwater in the study area: this sample together with the sea water sample should be considered the end members of the mixing (Fig. 4.4), focusing on the effect of seawater intrusion. Fig. 4.4 confirms graphically that the well-known mixture of pure fresh groundwater and saline water due to seawater intrusion is the phenomenon to be considered dominant in the study area (Polemio et al. 2009).

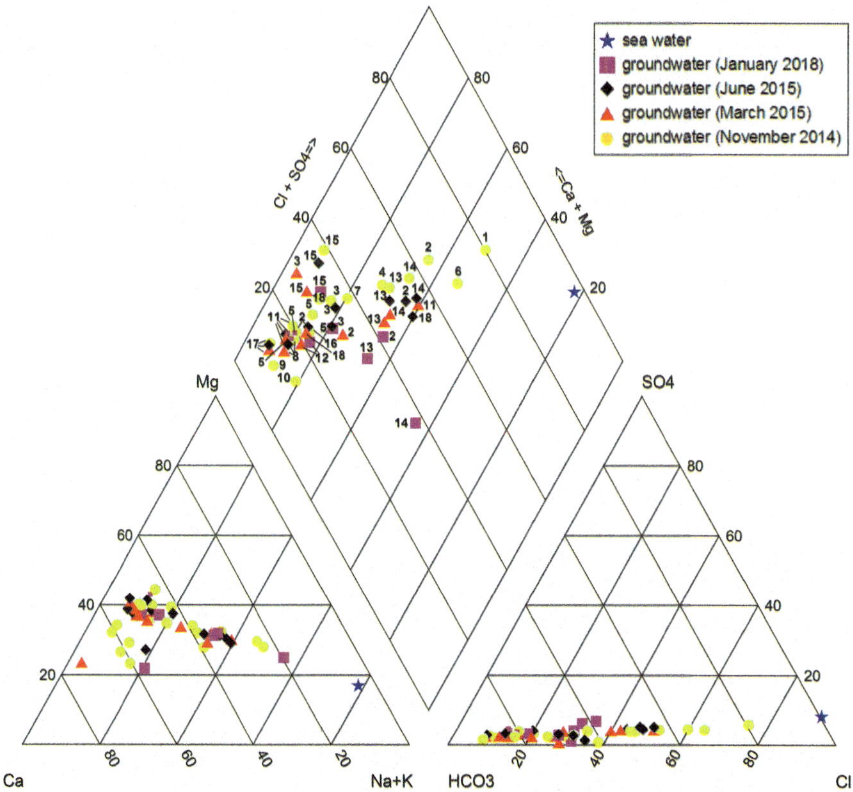

Fig. 4.4 Piper diagram of sampled groundwater

The GNC ranged from the minimum of 13.40 mg/L to the peak value of 59.90 mg/L (Fig. 4.2), with a unique sample slightly exceeding 50 mg/L, considered a worldwide upper value for good-quality waters. The spatial distribution is not affected by the position respect to landfills and confirms a diffuse effect of anthropogenic degradation of groundwater quality.

All results show that there is no evidence of negative effects of landfills on the GNC of the area. This result is with the previous results obtained with an innovative multi-methodological approach, including the use of more isotopes, as a nitrogen isotope (Cossu et al. 2018) showing the key role of the inorganic fertilizer utilization to justify the GNC in the study area.

4.5 Conclusions

Wide portions of main hydrogeological structures of Apulian region are characterized by high and unnatural GNC. High GNCs are due to anthropic effects, especially for the excessive use of fertiliser in agricultural.

The spatial temporal approach based on the discussion of groundwater nitrate concentration (GNC) was able to clarify the lack of groundwater pollution due to leachate leakage and mixing; it allowed to formulate realistic hypotheses, confirmed by the use of other methodologies or other studies, at different scales, on the main source of groundwater nitrate.

Acknowledgements This research was supported by 'Società Progetto Gestione Bari Cinque Srl'.

Author Contributions Authors' contribution should be considered equal.

References

Ancona V, Uricchio V, Lopez N, Caniglia G, Ripa S (2010) Attività di monitoraggio dei nutrienti nelle acque e nei suoli pugliesi a supporto delle strategie gestionali per la riduzione e razionalizzazione del consumo dei fertilizzanti. Ital J Agrometeorol 3:61–66

Appelo CAJ, Postma D (1996) Geochemistry, groundwater and pollution, 4th corrected print. Balkema/Brookfield, Rotterdam

Burow KR, Nolan BT, Rupert MG, Dubrovsky NM (2010) Nitrate in groundwater of the United States, 1991–2003. Environ Sci Technol 44:4988–4997. https://doi.org/10.1021/es100546y

Cossu R, Zuffianò LE, Limoni PP, De Giorgio G, Pizzardini P, Miano T, Mondelli D, Garavaglia R, Carella C, Polemio M (2018) How can the role of leachate on nitrate concentration and groundwater quality be clarified? An approach for landfills in operation (southern Italy). Waste Manag 77:156–165. https://doi.org/10.1016/j.wasman.2018.05.014

Cotecchia V, Grassi D, Polemio M (2005) Carbonate aquifers in Apulia and seawater intrusion. G Geol Appl 1:219–231. https://doi.org/10.1474/GGA.2005-01.0-22.0022

Custodio E, Llamas MR (1996) Hydrologia subterranea. Omega, Barcelona

Dubrovsky NM, Burow KR, Clark GM, Gronberg JM, Hamilton PA, Hitt KJ, Mueller DK, Munn MD, Nolan BT, Puckett LJ, Ruper MG, Short TM, Spahr NE, Sprague LA, Wilber WG (2010) The quality of our Nation's waters—nutrients in the Nation's streams and groundwater, 1992–2004: (No. 1350.). US Geological Survey Circular, Reston

Heaton THE (1986) Isotopic studies of nitrogen pollution in the hydrosphere and atmosphere: a review. Chem Geol Isot Geosci Sect 59:87–102. https://doi.org/10.1016/0168-9622(86)90059-X

Lucon TN, Costa AT, Galvão P, Leite MGP (2018) Natural background levels and seasonal influence on groundwater chemistry of the Upper São Francisco karst region, MG. Brazil Braz J Geol 48:867–879. https://doi.org/10.1590/2317-4889201820180071

Mor S, Ravindra K, Dahiya RP, Chandra A (2006) Leachate characterization and assessment of groundwater pollution near municipal solid waste landfill site. Environ Monit Assess 118:435–456. https://doi.org/10.1007/s10661-006-1505-7

Parrone D, Ghergo S, Preziosi E (2019) A multi-method approach for the assessment of natural background levels in groundwater. Sci Total Environ 659:884–894. https://doi.org/10.1016/j.scitotenv.2018.12.350

Pieri P, Festa V, Moretti M, Tropeano M (1997) Quaternary tectonic activity of the Murge area (Apulian foreland – southern Italy). Ann Geofis 40(5):1395–1404

Polemio M (2005) Seawater intrusion and groundwater quality in the southern Italy region of Apulia: a multi-methodological approach to the protection. In: Maraga F, Arattano M (eds) Progress in surface and subsurface water studies at the plot and Small Basin scale, technical documents in hydrology, UNESCO, IHP, Paris, pp 171–178

Polemio M (2016) Monitoring and Management of Karstic Coastal Groundwater in a changing environment (southern Italy): a review of a regional experience. Water 8(4):1–16. https://doi.org/10.3390/w8040148

Polemio M, Limoni PP (2006) Groundwater pollution and risks for the coastal environment (southeastern Italy). In: Sivapalan M, Wagener T, Uhlenbrook S, Zehe E, Lakshmi V, Liang X, Tachikawa Y, Kumar P (eds) Predictions in ungauged basins: promise and progress. IAHS Publications, Wallingford, pp 477–486

Polemio M, Dragone V, Limoni PP (2009) Monitoring and methods to analyse the groundwater quality degradation risk in coastal karstic aquifers (Apulia, southern Italy). Environ Earth Sci Former Environ Geol 58:299–312. https://doi.org/10.1007/s00254-008-1582-8

Serio F, Miglietta PP, Lamastra L, Ficocelli S, Intini F, De Leo F, De Donno A (2018) Groundwater nitrate contamination and agricultural land use: a grey water footprint perspective in southern Apulia region (Italy). Sci Total Environ 645:1425–1431. https://doi.org/10.1016/j.scitotenv.2018.07.241

Sheikhy Narany T, Aris AZ, Sefie A, Keesstra S (2017) Detecting and predicting the impact of land use changes on groundwater quality, a case study in northern Kelantan. Malaysia Sci Total Environ 599–600:844–853. https://doi.org/10.1016/j.scitotenv.2017.04.171

Strebel O, Duynisveld WHM, Böttcher J (1989) Nitrate pollution of groundwater in Western Europe. Agric Ecosyst Environ 26:189–214. https://doi.org/10.1016/0167-8809(89)90013-3

Thorburn PJ, Biggs JS, Weier KL, Keating BA (2003) Nitrate in groundwaters of intensive agricultural areas in coastal northeastern Australia. Agric Ecosyst Environ 94:49–58. https://doi.org/10.1016/S0167-8809(02)00018-X

WHO (2011) Nitrate and nitrite in drinking-water, Background document for development of WHO Guidelines for drinking-water quality

Zuffianò LE, Basso A, Casarano D, Dragone V, Limoni PP, Romanazzi A, Santaloia F, Polemio M (2016) Coastal hydrogeological system of Mar Piccolo (Taranto, Italy). Environ Sci Pollut Res 23:12502–12514. https://doi.org/10.1007/s11356-015-4932-6

Zuffianò LE, Limoni PP, De Giorgio G, Polemio M (2018) Data to clarify the landfill role in the case of groundwater quality degradation (southern Italy). Data Brief 20:1489–1499. https://doi.org/10.1016/j.dib.2018.08.201

Part II
Geology and Urban Areas

Chapter 5
Sinkholes in the Friuli Venezia Giulia Region Focus on the Evaporites

Chiara Calligaris, Luca Zini, Stefania Nisio, and Chiara Piano

Abstract Sinkholes are well-known phenomena in the Italian peninsula, and for this reason, since 2002, ISPRA (Istituto Superiore per la Protezione e la Ricerca Ambientale) started a national project aiming at inventory them. From the beginning, it emerged that Friuli Venezia Giulia Autonomous Region (FVGAR) was one of the most sinkhole-prone areas. Since 2014, the Geological Department of the FVGAR founded specific projects aimed at the identification, classification, and characterization of sinkholes present on the regional territory. These phenomena are widespread in the northern part where deeply karstified carbonates and evaporites are widely present; thousands are the solution sinkhole type phenomena. The present research focused indeed only on the subsidence sinkholes, which up to date are 1199. From the data analyses, it emerges that the sagging sinkholes are not present, while cover suffosion sinkholes and collapse sinkholes are the most frequent. In this chapter are presented two meaningful case studies, representative of the phenomena due to their magnitude and the linked risk.

Keywords Karst · Evaporites · Gypsum · Sinkhole · Friuli Venezia Giulia Autonomous Region

C. Calligaris (✉) · L. Zini
Dipartimento di Matematica e Geoscienze, Università degli Studi di Trieste, Trieste, Italia
e-mail: calligar@units.it

S. Nisio
Istituto Superiore per la Protezione e la Ricerca Ambientale (ISPRA), Rome, Italy

C. Piano
Direzione centrale difesa dell'ambiente, energia e sviluppo sostenibile, Servizio Geologico FVG, Trieste, Italy

M. De Maio, A. K. Tiwari (eds.), *Applied Geology*,
https://doi.org/10.1007/978-3-030-43953-8_5

5.1 Introduction

The term sinkhole is used by the engineering geologists to indicate a natural circular depression linked to a sometimes complex, subterranean drainage system. These features can reach depths of tens of meters and diameters of more than hundreds. The researchers described them using the dominant processes behind their development and/or the material involved (Beck 2004; Waltham et al. 2005; Gutiérrez et al. 2008, 2014). The process may include sudden catastrophic collapses, more frequent in evaporites than in carbonate rocks (Cucchi and Forti 1993; Cucchi et al. 1994; Klimchouk et al. 1996; Dreybrodt and Eisenlohr 2000; Jeschke et al. 2001; Waltham et al. 2005; Ford and Williams 2007; Zini et al. 2015a; De Waele et al. 2017). Due to their rapidity of occurrence, sinkhole's nature and distribution represent a challenge to be met in order to guarantee a more conscious use of the territory. In this framework, at the national level in Italy, in 2002, there arose the necessity to implement an inventory dedicated to this type of phenomena. The inventory collected all the available information related to subsidence sinkholes (sensu Gutiérrez et al. 2014). The number of inventoried phenomena was far from the real number of existing sinkholes, so each region was left to contribute to the project, helping in the database implementation. The initial picture at the national level constituted more than 1500 sinkholes (Nisio 2003) (Fig. 5.1). From the database analyses, it emerged that the inventoried sinkholes are mainly located in plain areas and reached variable dimensions from few meters up to 200 m in diameter and generally shallow depths, which occasionally could reach 50 m. More than 50% of the investigated phenomena are occasionally water filled small lakes or ponds. Referring to the typology and distribution, it is noted that sinkholes are strongly linked to the different geological, structural, and geomorphological conditions of the Italian territory (Nisio 2008a, 2010).

In Sicily, their presence is mainly linked to the Messinian evaporitic bedrock whose dissolution creates an upward erosion of the cover sediments. Sicilian sinkholes are mainly located inside the island in close basins in the Miocene foredeep.

Cover collapse sinkholes in Calabria are correlated to salt dissolution, which is mantled by marine clays, flysch, and metamorphic or igneous sequences. Consequently, the reported cases are just isolated sinkholes in coastal or small valleys triggered sometimes by high-magnitude earthquakes. Sinkholes in Apulia are mainly of karst origin and are located on the coastal areas often affecting silty-sandy marine sediments of reduced thickness overlying mainly carbonatic sequences.

In the investigated areas of Central Italy, the most frequent phenomena are the Deep Piping Sinkhole (DPS) linked to the presence of raising fluids from depths (enriched in CO_2 and H_2S as free and dissolved gases) and the cover collapse sinkholes (Nisio and Salvati 2004), which are mainly located along the Tyrrenian margin and in some intermontane valleys of the Apennine ridge. Cover collapse sinkholes have also been detected in Sardinia and in the northern Apennine (Caramanna et al. 2008, Nisio 2008b). In the Emilia Romagna Region, the most

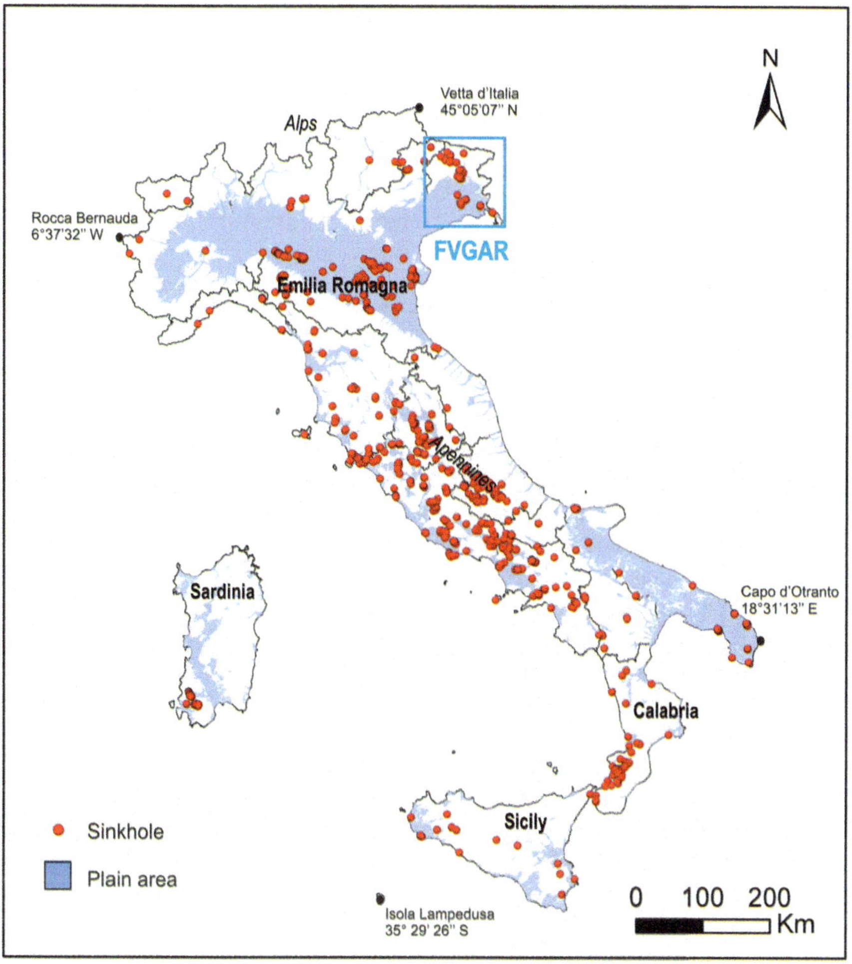

Fig. 5.1 Sinkhole's distribution in the Italian peninsula. The blue rectangle highlights the FVGAR

significant phenomena are the suffusion sinkholes sometimes linked to the gas uprising along the main tectonic structures. In all the Alpine area, few are the sinkhole's identifications, mainly linked to the evaporites (De Waele et al. 2017).

One of the regions which already at the starting of the inventory had more prone sinkhole areas was the Friuli Venezia Giulia (FVGAR). In its territory, since 2014, a new project has been started aimed at the identification, characterization, and inventory of the subsidence sinkholes. Almost the whole inventoried phenomena are due to karst processes, and the area interested by the most hazardous phenomena is linked to the presence of Permian and Triassic evaporites. The inventory realized has been done following the classification proposed by Nisio et al. (2007), but Gutiérrez et al. (2008, 2014) published a new genetic classification in order to compare what is present in the study territory with the international literature, and their classification was also adopted.

In the present chapter, we propose a description of the karst areas identified in the carbonates and evaporites of the FVGAR, a summary of the sinkholes inventoried, an insight into the classification used and a focus on two impressive risky case studies: the area of Quinis-Enemonzo and the Ovaro village.

5.2 Environmental Settings and Karst Features in the FVGAR

The FVGAR is a small territory (in the NE corner of the Italian peninsula, Fig. 5.1) of about 8000 km^2 where it is possible to find karst areas from the Paleozoic to Cenozoic (Carulli 2006; Venturini 2009). In the north, at the Austrian border, the older rocks outcrop (Ordovician–Triassic). Moving south, a wide belt which includes the Carnian and Julian Alps and Prealps is characterized by the presence of Triassic and Jurassic outcrops, sometimes deeply karstified. The Cretaceous and Paleogene outcrops are identifiable only south of the reliefs, just upstream the lowlands.

The regional territory was exposed to the Alpine orogeny (Miocene sup.–Present). The collision between the African and European plates crushed either side of the Apula microplate; this tectonic evolution is characterized by two polyphasic contractional events: the Insubric event (Chattian sup.–Langhian) and the Neoalpine event (Serravallian–Present) (Zanferrari et al. 2013). The latter caused a crustal shortening due to N–S compression and the formation of the major thrusts, which are E–W oriented. The mountain areas present in the northern part of the FVGAR are mainly due to these compressions. The easternmost part of the FVGAR is interested by the southwest verging Dinaric structures (Busetti et al. 2010; Carulli 2006) with a NW–SE orientation.

In the FVGAR, 5000 km^2 are mountainous or hilly areas, 1900 km^2 of which are characterized by outcropping carbonate rocks (Calligaris et al. 2016a; Calligaris et al. 2017a; Calligaris et al. 2018; Cucchi and Zini 2009; Cucchi et al. 2010; Fabbri et al. 2013; Zini et al. 2013; Zini et al. 2015b). These lithologies are the core of notable karst landforms developing, thanks to the high rainfall rate (Cicogna 2008) and to a prolonged exposition to meteoric agents. So, karst processes with a prevailing acidification (karst) are present in almost 40% of the regional territory.

In the NW part of the FVGAR, several areas, characterized by a karst in which prevail the solubilization (hyperkarst), develop in the evaporites with an outcropping percentage close to 1%, of which 25% are Permian and 75% are Triassic in age. The subcropping percentage is of interest because in these zones, several are the recognized sinkhole phenomena having important dimensions. Most parts of the evaporites are mantled by alluvial, detritical, and glacial deposits or are overlain by non-karst rocks.

5.2.1 Karst Areas in the Carbonates of the FVGAR

Throughout the regional territory, the karst development is demonstrated by the presence of more than 7000 caves discovered and explored and by widespread epigean karst forms, such as dolines, sinkholes, karrenfeld, grikes, and grize, which often reach truly remarkable dimensions. The karstic territories also show very different characteristics which derive from different exposure conditions, different evolutionary history, and the presence of lithologies with different characteristics of karstificability. From north to south, the main karst areas of the FVGAR are presented (Fig. 5.2).

- **Paleocarnian Range**: This area is related to the karst morphologies developed on the Devonian carbonatic facies, on the transitional facies, and on the facies typical of shallow waters. The Mt. Cavallo di Pontebba can be identified as representative of this type of alpine karst.
- **Carnian Alps**: The mountains are located south of the Paleocarnian Range, reaching the northern part of the Tagliamento River. The massif characterizing this area is the Mt. Amariana (1906 m a.s.l.), whose core is made of light gray micritic limestones Megalodont-bearing m-thick beds Upper Triassic in age.
- **Julian Alps**: In these areas, it is possible to recognize alpine karst features at elevations of about 1800–2000 m a.s.l., formed mainly on Rhaetian limestones in platform facies typical of the Dachstein Fm. These areas present an articulated superficial morphology, result of several erosive and glacio-karstic phases set on

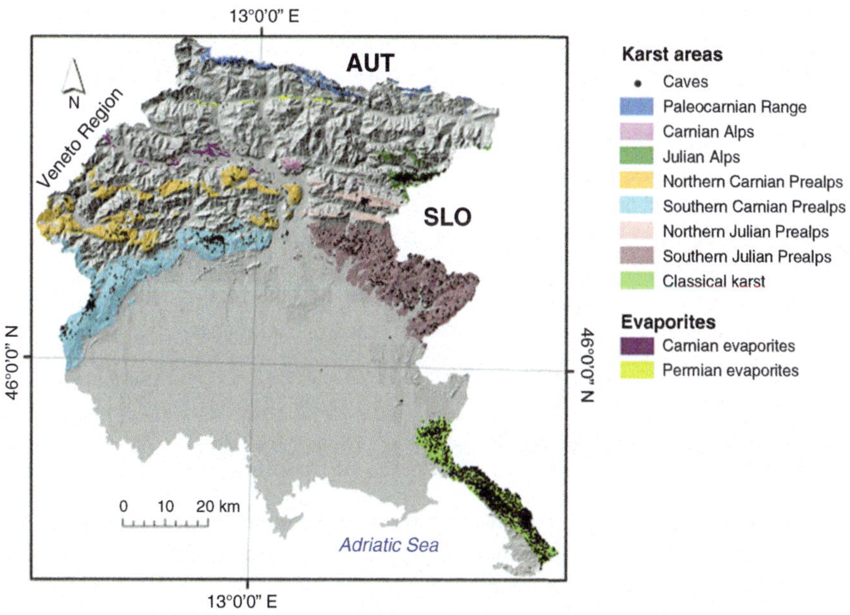

Fig. 5.2 Karst areas in the FVGAR. Black dots represent the known caves

important tectonic lines. On these, deep hypogean karst features formed, as Mt. Canin, with its complex intertwinement of pits, tunnels, gorges, conduits, with over 800 caves (Zini et al. 2015c).

- **Northern Carnian Prealps**: The karst phenomena develop mainly over the tops, in the highlands, in correspondence to the Triassic and Jurassic carbonatic formations. The karst features mostly progress along the strata, where sinkholes are usually aligned with karrenfelds having the latter, several and deep crevasses.
- **Southern Carnian Prealps**: The surface is characterized by the outcropping of rounded massive blocks with holes, karren, and grikes, similar to a ruiniform karst. The presence of sinkholes is typical; they are usually aligned along the main fractures and faults. On the sinkhole's bottom can be found pits as the Abisso del Col della Rizza, 800 m deep and 4 km long, which is the deepest cave in the Cansiglio area. To the presence of karst features are also linked the important springs, which give rise to the Livenza River.
- **Northern Julian Prealps**: They are characterized by the Mt. Musi range, E–W oriented, where stand out several different peaks among 1800 and 1869 m a.s.l. The area is the rainiest one not only at the regional level, but also at the national level, with more than 3000 mm/y of rain. The southern side of the range is made of Dolomia Principale Fm. overlaid by the Dachstein limestones Fm. on which strata, manly on the northern slopes, important caves formed as well as karst features characteristics of high altitudes.
- **Southern Julian Prealps**: They are different from other areas because they develop in Flysch, an alternation of marl and sandstone layers inside which massive holistolitic megabanks are present. Karst features as caves develop in the carbonatic megabanks, especially at the contact with the marls and the sandstones. Typical karst areas can be identified in correspondence to Bernadia Mt. and Mia Mt. where it is possible to find caves, sinkholes, and typical karst springs.
- **The Classical Karst**: It entirely develops on Cretaceous and Paleogene limestones (Jurkovšek et al. 2016). On the surface and subsurface, important karst morphotypes are present, frequent and different in type, witnessing a mature karst where more than 3500 caves have been inventoried. More than 6000 sinkholes/dolines cover an area of 200 km^2, and the karrenfelds have a total development of a few tens of km^2. The cave density reaches a value of about 30 caves/km^2, and sinkholes reach a value of about 48 sinkholes/km^2 (Zini et al. 2015b).

5.2.2 Karst Areas in the Evaporites of the FVGAR

From the tectonic point of view, the evaporitic sequences are often a level that acts as a large-scale tectonic lubricant, allowing first the separation from the Permian or Permotriassic basement to the overlying Triassic–Jurassic units, then their translation into ramps and flat even for thousands of meters. It follows that the evaporites, in the FVGAR, do not outcrop areally, but are often present at the base of the reliefs or in the valley bottoms. Here, they become, due to their high solubility and erodibility, a predisposing factor for innumerable instability phenomena (landslides and sinkholes). The solution of evaporitic clasts present in the loose alluvial or moraine deposits and the solution of evaporitic bedrock mantled by loose permeable deposits give rise to phenomena including cover collapse sinkholes (Zini et al. 2015a) and cover suffosion sinkholes. In other cases, the high erodibility and the very high solubility can lead, where the geological and geomorphological characteristics allow it, to a continuous and fast cut at the toe of the reliefs, generating widespread and continuous landslides from the overlying slopes (I.F.F.I. Italian Inventory, http://www.isprambiente.gov.it/it/progetti/suolo-e-territorio-1/iffi-inventario-dei-fenomeni-franosi-in-italia).

- **Karst in the Permian Evaporites**: Hyperkarst features from prevalent solubilization can be identified in the evaporites of the Bellerophon Fm. (Upper Permian), characterized by a transgressive sequence where gypsum is placed at the bottom (Venturini 2009). Gypsum is usually microcrystalline, but can also be saccharoidal with thin interbedded layers of blackish limestones and dolomites. Moving upward into the sequence, gypsum is gradually replaced by limestones-dolomites-chalky vacuolar breccias and by dolomitic limestones that form the intermediate member of the "vacuolar dolomites," later replaced by the frankly limestone upper member.
- **Karst in the Carnian Evaporates:** Hyperkarst features can be identified in the Raibl Fm. (Venturini 2009), defined as Travenanzes Fm. (Upper Carnian), a transgressive sequence characterized by a red pelites member, a gypsum and gray dolostones member, and a dolostones and marls member (Zanferrari et al. 2013). The saccharoidal gypsum, gray and pink-colored, is rich in clayey impurities and it can be recognized in mm-thick rhythmic laminae, typical of evaporitic lagoon deposits.

5.3 Sinkholes in the FVGAR

Sinkholes are high-risk phenomena due to their low predictability and rapid evolution, sometimes characterized by sudden collapses, which can also involve urban areas, infrastructures, agricultural areas, sometimes with a cost in human lives, but often causing considerable damages and constituting a social problem.

From a legal point of view, almost all the Hydrogeological Asset Plans (Piano di Assetto Idrogeologico – P.A.I.) adopted by the Italian Basin Authorities do not take into consideration, if not within the landslide hazard inventory, the sinkhole hazard, and only some Italian regions have enacted specific rules requiring detailed studies in areas susceptible to sinkholes. In the FVGAR, the sinkhole's inventory has been realized following the national methodological protocol, as (1) data collection; (2) detailed geological, geomorphological, and hydrogeological surveys; (3) photo-interpretation; (4) data informatization; (5) study of genetic processes related to the geological–structural, geomorphological, and hydrogeological contexts, which lead to the sinkholes formation.

In the region, it is possible to recognize the following:

Sinkholes in the Alluvial Plain Areas They form in the alluvial plains and do not have any connection with the karst processes; they are exclusively evorsion sinkholes. They usually occur on levees and are not easy to be mapped; it is seen that the recovery teams operate continuously filling them restoring the original topography.

Sinkhole in the Karst Flysch Areas The widespread presence of these phenomena is the morphological expression of the karstic corrosion acting on calcarenites and calcilutites characterizing the flysch of the karst areas in the southern Julian Prealps. They are mostly present with sudden collapses occurring along preferential water flow lines (Fig. 5.3a).

Sinkholes in the Carbonates Solution sinkhole is a typical epigean karst feature, which characterizes all the carbonatic karst areas. Classical Karst area where there have been already inventoried almost 6000 sinkholes with a diameter of less than 100 m, and more than 180 phenomena with a diameter of more than 100 m (Fig. 5.3b). The research here presented focused primarily on those type of sinkholes that can potentially represent a risk due to their sudden catastrophic evolution. For this reason, solution sinkholes were not considered and the attention was focused on the subsidence sinkholes which represent a small percentage in the Classical Karst.

Sinkholes in the Evaporites These phenomena are due to hyper-karst processes developing mainly in the chalks and in the vacuolar dolomites as well as in all the secondary deposits, which are the result of the evaporites dismantling. This is why

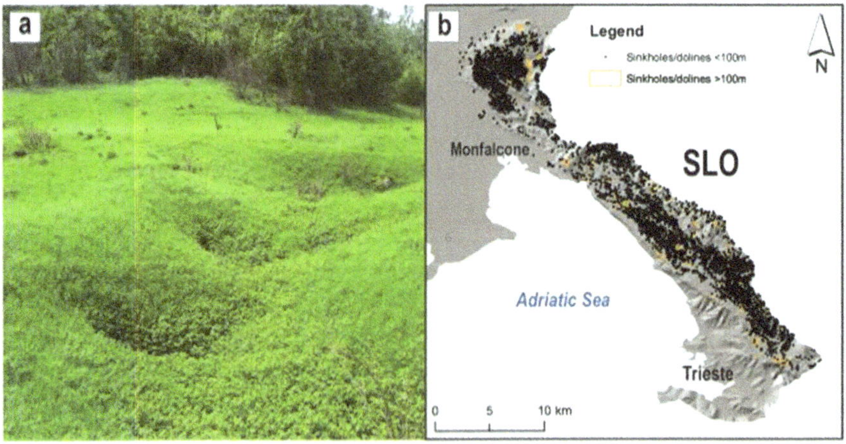

Fig. 5.3 (**a**) A series of sinkholes in the Flysch lithology induced by the caves developed in the megabanks (Taipana Municipality, UD); (**b**) Sinkholes/dolines in the Classical Karst (Italian side). In black, the phenomena with a diameter lower than 100 m, and in orange, the ones with a diameter over 100 m are indicated

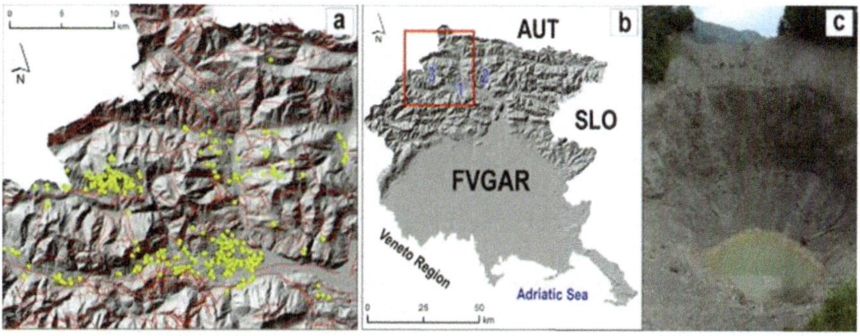

Fig. 5.4 (**a**) In red the tectonic lines, and in yellow, the sinkholes that occurred in the evaporites are indicated; (**b**) FVGAR, the red rectangle identifies the zoom in (**a**); 1 corresponds to Enemonzo, 2 to Ovaro, and 3 to Sauris; (**c**) a bedrock collapse sinkhole in the evaporites (Sauris Municipality, UD) having a diameter of more than 100 m and a depth of 35 m

the areas in which sinkholes occur are quite much larger than the outcrops. The solubilization of the gypsums bedrock and the enlargement of the fractures present in it due to the hypogean waters (phreatic and pseudo-artesian ones) in the cover, in the evaporites, in the interface between the evaporites, and in the more or less permeable caprocks cause voids that can generate sudden collapses, giving rise to sinkhole. These events are clearly identifiable in correspondence to some of the main tectonic structures (Cucchi and Piano 2002; Burelli et al. 2004) (Fig. 5.4a).

In these areas, thanks to the favorable lithostratigraphical conditions, also DPS can be identified. In Quinis, a hamlet of Enemonzo (UD) presents the only possible DPS of the FVGAR. The area is historically known for the presence of sinkholes (Cosano 1948; Gortani 1965; Marinelli 1897) whose evolution is also related to gas-rich groundwater upwelling (identified in correspondence to one drilled piezometer in the center of the investigated area) and water table oscillations capable of a mechanical upward erosion of the unconsolidated or less cohesive strata overlying the bedrock (Zini et al. 2015a).

5.4 Sinkhole's Geodatabase: Data Summary

For the entire regional territory, 1199 sinkholes phenomena were inventoried. The types of sources from which data were collected are different: 441 phenomena were identified for the first time, thanks to the geomorphometrical analyses realized on the DTM at 1 m available for the whole regional territory; 312 phenomena were inventoried for the first time, thanks to the field surveys realized; and 446 phenomena were already present in specific reports, in Municipality Plans or in inventories.

According to Gutiérrez et al. (2014) classification, 139 are bedrock collapse sinkholes, 77 are caprock collapse sinkholes, 178 are cover collapse sinkholes,

Table 5.1 Comparison between two classifications: (A) Gutiérrez et al. (2014); (B) Nisio and Salvati (2004)

A	B
Anthropogenic sinkhole	Anthropogenic sinkhole
Solution sinkhole	Solution sinkhole/doline
Bedrock collapse sinkhole	Cave collapse sinkhole
Caprock collapse sinkhole	Cave collapse sinkhole
Cover collapse sinkhole	Cover collapse sinkhole
Cover suffosion sinkhole	Suffosion sinkhole
	Cover subsidence sinkhole
	Eversion sinkhole
Cover collapse and suffosion sinkhole	Deep piping sinkhole (DPS)

360 cover suffusion sinkholes, 1 is cover collapse and suffusion sinkhole, and 444 are undefined. According to the Italian classification (Nisio and Salvati 2004), 216 are cave collapse sinkholes, 178 cover collapse sinkholes, 344 cover suffusion sinkholes, 16 evorsion sinkholes, and 1 deep piping sinkhole (Table 5.1).

Obviously, this is not a synoptic picture of the entire FVGAR. In interpreting the collected data, it is important to consider that:

- What has been done is the first sinkhole's inventory, which is not yet complete because the authors focused especially on the areas characterized by the presence of hyperkarst phenomena. In the remaining areas, surveys are still ongoing.
- Being an inventory related only to natural origin phenomena, the anthropogenic sinkholes were not considered at all.
- Among the evorsion sinkholes in the plain areas, identified along the embankment (the so-called "fontanazzi"), only the most emblematic have been mapped.
- Almost one-third of the phenomena have not yet been defined in type.

Among all the inventoried phenomena, 245 are defined as active, about 20.4% of the total amount; the percentage of inactive phenomena is around 9%, the dormant phenomena are approximately 30%, and the rest have an unknown state of activity. On the field, sometimes, it is very difficult to assign the correct state of activity because some of the phenomena have been partially filled by brushwood and grass or because it is not possible to go back to this information from the consulted documents present in the different archives. Among the active sinkholes, most part of them are cover suffusion and cover collapses, only one-third are represented by caprock and bedrock collapses.

Regarding the dimensions, most part of the phenomena are circular or sub-circular in shape with diameters ranging between 1 and 289 m. The depth is in the range between 10 cm and 35 m.

5.5 Sinkhole's Hazard in the Evaporites, Two Meaningful Examples: Enemonzo and Ovaro Municipalities

In the FVGAR, several are the small realities affected by sinkholes. Most of them are sited in the valley floors in the northern part of the region in the mountain areas. The most striking examples are related to Quinis (Enemonzo municipality) and Ovaro areas. In both places, infrastructures and buildings have suffered and are still suffering important damages within the years up to the necessity to relocate the injured houses.

In the Quinis area, the first scientific document on sinkholes is dated back to the 1960s when Gortani (1965) described an important event that occurred close to the Tagliamento River (Fig. 5.5d).

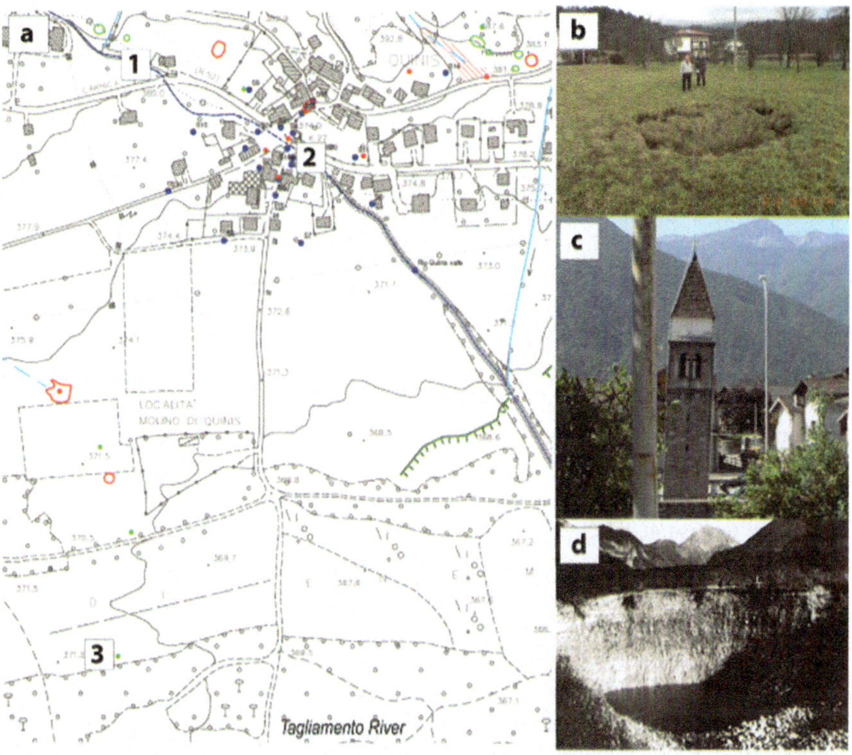

Fig. 5.5 (**a**) The area of Quinis (Enemonzo Municipality), to red perimeters correspond the active sinkholes, to green ones, the dormant phenomena; red and green dots are related to active and dormant sinkholes for which it is not possible to reconstruct a perimeter; blue dots represent the piezometers present in the area; oblique lines are placed in correspondence to depressed areas; (**b**) Marked with number 1 on the map, a sinkhole reactivated in 2013; (**c**) the bending bell tower in the center of Quinis, which survived to the church breaking down, is numbered 2 on the map; (**d**) the historical sinkhole occurred in the sixties and described by Gortani in 1965, it occurred close to the Tagliamento River (marked in the map as number 3)

After that, there has been a recruiting of the reported events up to present when 6 houses are characterized by important damages and for some of them, the inhabitants are going to be relocated. The predisposing factors must be sought in the geological and structural settings. The Quaternary deposits, Late Pleistocene–Holocene in age, mantling the 99% of the evaporites, consist of glacial till, alluvial, and colluvial deposits. These materials are extremely heterogeneous due to the complex and articulated depositional patterns conditioned in turn by several factors including the tectonic setting, the alternation of glacial and interglacial periods, the recent depositional events of the Tagliamento River (recent alluvial deposits) and the overlapping alluvial fan present in the northern Quinis. These loose deposits have variable thicknesses in the range 0–50 m: close to 0 m in the northern part of the area and more than 50 m close to the south moving toward the Tagliamento riverbed. The evaporitic bedrock is highly karstified and geothenically poor, especially the weathered part which is in contact with the mantling deposits. To these intrinsic characteristics, it is necessary to add the role of fluctuating groundwaters. The bedrock morphology with its depressions due to the different karstification degree of the evaporites conditions the groundwater flow, which is also influenced by the extreme heterogeneity of the Quaternary deposits. The presence of clayey and clayey-silt layers with limited lateral extent generates local perched aquifers. Seasonal water-level fluctuations are large, with oscillations between 6 and 32 m (Calligaris et al. 2019; Zini et al. 2015a), causing an acceleration in the evaporite's solution as well as in the sinkhole's evolution. During high flow conditions, the recharge due to the leakages of the Tagliamento River is important and the phreatic levels close to the riverbed and in the plain inhabited areas are comparable. The water table fluctuations rise up with velocities overpassing 0.4 m/h (recorded in the Quinis area) (Zini et al. 2015a).

The overall hazard of the area is high. In the paper of Calligaris et al. (2017b), the susceptibility map highlighted the most sinkhole-prone zones of the Enemonzo municipality and incorporates the parameters that better represent the conditioning factors that generate and/or accelerate the sinkhole processes. In the map, Quins has been well represented including the areas with the actually injured houses (Fig. 5.6).

Ovaro is sited over a slope where the moraine deposit mantles the evaporitic bedrock, Permian in age. The Quaternary deposits consist of moraines and alluvial sediments. The Upper Pleistocene moraine give rise to a gentle morphology, which is often hummocky. The bedrock is present at very low depths, as witnessed by a 3-m-high outcrop of gypsum near the main street of Ovaro (Fig. 5.7b). The study area is crossed by a regional structural discontinuity known as Sauris-Ovaro-Arta Terme thrust, E–W-oriented (Venturini 2009), which contributes to the geotechnical weakness of the involved materials. This peculiar geological setting favors the dissolution and/or erosion of the evaporitic rocks.

Before the 1950s in the area, several have been the occurred sinkholes. After that period, the situation remained quite stable up to 1990s when new events happened. An in-depth study realized in 1997 using mainly geophysical approaches (Cucchi

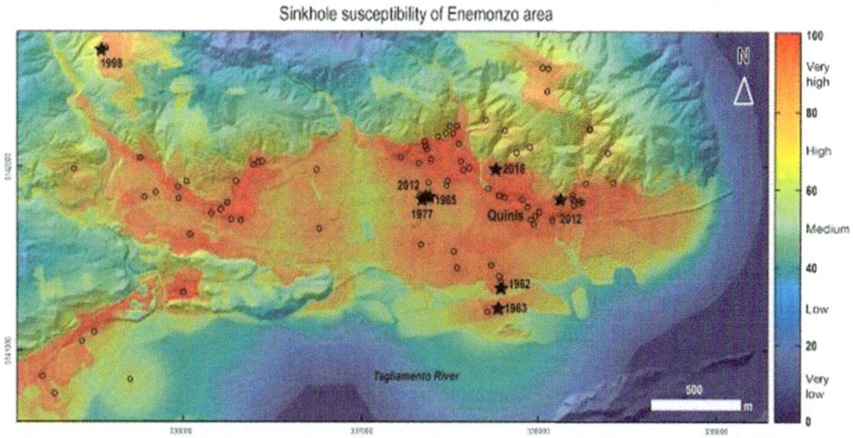

Fig. 5.6 Sinkhole susceptibility map of the Enemonzo area (Calligaris et al. 2017b)

and Giorgetti 1997) witnessed the high karstification degree of the bedrock and the presence of flowing subsurface waters, which accelerate the sinkhole's occurrence. In 2014 and at the end of October 2018, there has been a reactivation of old known events present in correspondence to the village center, close to the houses, affecting the buildings (Fig. 5.7c).

5.6 Conclusions

The presence of sinkholes is a clear intrinsic hydrogeological vulnerability of the FVGAR and their inventory can only increase the awareness of the need to pay particular attention to this instability in the territorial management and planning. Up to now, the problem perception was essentially identified by the knowledge of the areas historically affected by these types of event; with the inventory presented in this chapter, the wide spread of this phenomenon in the whole regional territory has emerged (Fig. 5.8).

Although in the region the emergent priority of evaporitic sinkholes remains due to the paroxysm and the velocity of the dynamic karstic solutional processes, the study has highlighted the spread of sinkholes even in not typically karst areas such as those in the Flysch of the Southern Julian Prealps, where up to now, the only known karst morphologies were the "confined caves."

The phenomenologies present in the FVGAR cannot therefore be framed in a single geological and/or geomorphological context, nor are they characterized by the same triggering mechanism and evolutionary process. For this reason, in light of hazard definition, it will be necessary to start from a local quantification and then

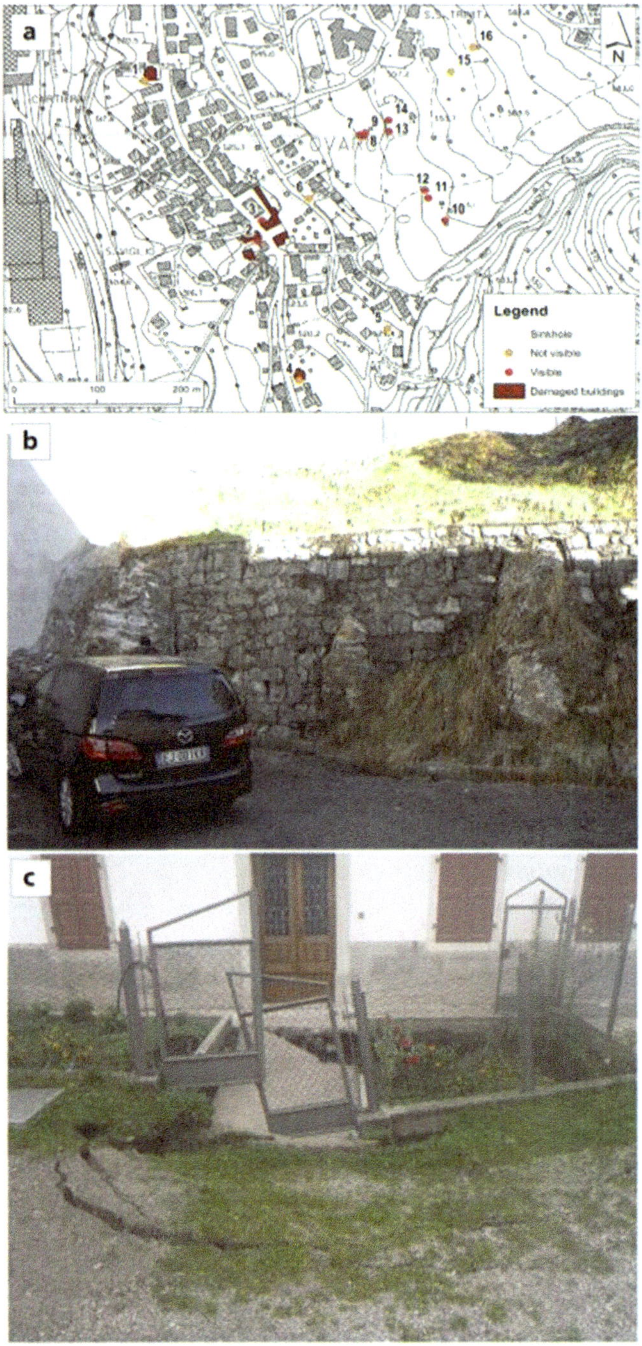

Fig. 5.7 (**a**) Overview of the injured buildings (brownish color): red dots represent the visible sinkholes, yellow ones, the not anymore visible phenomena (modified after Calligaris et al. 2016b); (**b**) one of the few gypsum outcrops in the study area; (**c**) the last event occurred in order of time: October 2018, it is a reactivation of the phenomena identified with number 2 in the **a** map (*ph. Courtesy: Ing. Burba, Civil Defence of FVGAR*)

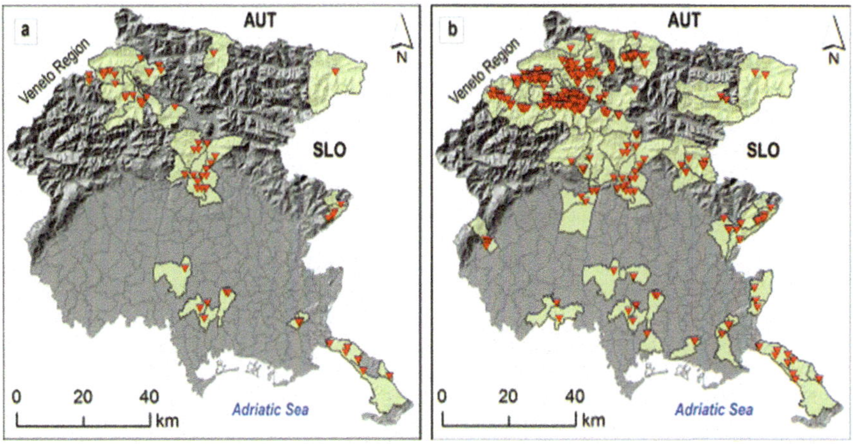

Fig. 5.8 Comparison between municipalities with sinkhole vulnerability (pale yellow) before (**a**) and after (**b**) the realized inventory. Red triangles represent sinkholes. (**a**) 27 municipalities involved and 73 sinkholes inventoried; (**b**) 67 municipalities involved and 1199 sinkholes inventoried

identify, if possible, the common characters of the areas susceptible to sinkhole. In concomitance, it will be necessary to respond to the need for recommendations and/or guidelines in order to develop, also at the local level, the adaptive capacity of territorial planning in sinkhole areas and thus reduce costs for the society, avoiding the constant historical recourses that often occur in these areas. Just not to forget, we wanted to include here, some observations present in an old thesis 1947–1948 written by P.A.B. Cosano and supervised by prof. Ardito Desio:

> From Enemonzo I can reach the hamlet of Quinis, a village of a few hundred inhabitants built on a very unhappy site: in fact, numerous cracks visible on the buildings denote the instability of the area; indeed, on the national road, a house is unsafe. There is obviously a large instability in the area and the nature of the land, offers the most favorable conditions for slippages with dark marls and clays Carnian in age alternating with lenses of white and pink saccharoidal chalk easily removable by the waters. Part of the inhabited area, destroyed during the last world war, could have been rebuilt in a safer area (for example, in the plain area, in the right side of Rio Quinis), instead the houses were rebuilt in the same dangerous zone.

Therefore, the inventory presented here is only the first step to pursue the development of strategies and action plans to increase the resilience to sinkhole, i.e., the transformation of the concept of emergency prevention into ordinary planning (at all levels: municipal and regional) of the territory. The consequence of all this is the "inaction costs" decrease, i.e., the reasonable savings given by the planning interventions (such as mitigation works and urban precautions) in avoiding the costs of damages (especially in terms of human lives) and the consequent interventions in an emergency.

Acknowledgments The Authors would like to acknowledge the field geologists Stefano Devoto and Alessandro Pavan, Prof. Franco Cucchi for his valuable suggestions and the critical review of the text. The initiative has been realized in collaboration between the University of Trieste - DMG and the Friuli Venezia Giulia Autonomous Region (FVGAR) Servizio Geologico under the R.L. 23/2013, art. 4 in the framework of four different agreements:

1. *Realizzazione del censimento regionale dei sinkhole e relativo GIS (Geographic Information System) con corso formativo e predisposizione delle linee guida di rilevamento ed informatizzazione nonché glossario dei termini utilizzati (Prot.no. 801 del 28/10/2013, CUP: J92F16001310002)*;
2. *Accordo attuativo di collaborazione per la definizione e quantificazione della pericolosità dei sinkhole nei litotipi evaporitici Carniani dell'alta valle del Tagliamento (Prot.no. 877 del 06/10/2016);*
3. *Accordo attuativo di collaborazione per la definizione e quantificazione della pericolosità dei sinkhole negli eventi collegati ai litotipi evaporitici Permiani (Prot.no. 1229 del 30/11/2017)*;
4. *Accordo attuativo di collaborazione per la definizione e quantificazione della pericolosità dei sinkhole negli eventi collegati ai litotipi carbonatici del Carso Classico (Prot.no. 1407 del 12/12/2018, CUP: J99C18000160002).*

References

Beck BF (2004) Soil piping and sinkhole failures. In: White WB (ed) Encyclopedia of caves. Elsevier, New York, pp 523–528

Burelli G, Cleva S, Cucchi F, Oberti Di Valnera S (2004) Stato di evoluzione di sinkholes in alcune aree montane del Friuli Venezia Giulia. In: Atti 1° Seminario "Stato dell'arte sullo studio dei fenomeni di sinkholes". Roma, Maggio 2004, APAT (ed), pp 159–170

Busetti M, Volpi V, Nicolich R, Barison E, Romeo R, Baradello L, Brancatelli G, Giustiniani M, Marchi M, Zanolla C, Wardell N, Nieto D, Ramella R (2010) Dinaric tectonic features in the Gulf of Trieste (northern Adriatic). Boll Geofis Teor Appl 51:117–128

Calligaris C, Boschin W, Cucchi F, Zini L (2016a) The karst hydrostructure of the Verzegnis group (NE Italy). Carbonates Evaporites 31:407. https://doi.org/10.1007/s13146-016-0320-7

Calligaris C, Devoto S, Zini L, Cucchi F (2016b) An integrated approach for investigations of ground-subsidence phenomena in the Ovaro village (NE Italy). In: Renard P, Bertrand C (eds) EuroKarst 2016, Neuchâtel. Advances in karst science, vol 8. Springer, Cham, pp 71–77. https://doi.org/10.1007/978-3-319-45465-8_8

Calligaris C, Devoto S, Zini L (2017a) Evaporite sinkholes of the Friuli Venezia Giulia region (NE Italy). J Map 13(2):406–414. https://doi.org/10.1080/17445647.2017.1316321

Calligaris C, Devoto S, Galve JP, Zini L, Pérez-Peña JV (2017b) Integration of multi-criteria and nearest neighbour analysis with kernel density functions for improving sinkhole susceptibility models: the case study of Enemonzo (NE Italy). Int J Speleol 46(2):191–204. https://doi.org/10.5038/1827-806X.46.2.2099

Calligaris C, Mezga K, Slejko FF, Urbanc J, Zini L (2018) Groundwater characterization by means of conservative ($\delta^{18}O$ and $\delta\ ^{2}H$) and non-conservative ($^{87}Sr/^{86}Sr$) isotopic values: the classical karst region aquifer case (Italy–Slovenia). Geosciences (Switzerland) 8(9):321. https://doi.org/10.3390/geosciences8090321

Calligaris C, Ghezzi L, Petrini R, Lenaz D, Zini L (2019) Evaporite dissolution rate through an on-site experiment into piezometric tubes applied to the real case-study of Quinis (NE Italy). Geosciences (Switzerland) 9(7):298. https://doi.org/10.3390/geosciences9070298

Caramanna G, Ciotoli G, Nisio S (2008) A review of natural sinkhole phenomena in Italian plain areas. Nat Hazards 45:145. https://doi.org/10.1007/s11069-007-9165-7

Carulli GB (2006) Carta geologica del Friuli Venezia Giulia, Regione Autonoma Friuli Venezia Giulia, Direzione Centrale Ambiente e Lavori Pubblici. Servizio Geologico, Trieste

Cicogna A (2008) Atlante climatico del Friuli Venezia Giulia. 1.Precipitazioni. Agenzia Regionale per la Protezione dell'Ambiente del Friuli Venezia Giulia. p 42

Cosano P (1948) Rilievo Geologico dei dintorni di Enemonzo (Carnia). Tesi di Laurea – Relatore Prof. A Desio

Cucchi F, Forti P (1993) Dissoluzione sottocutanea nei gessi: analisi e considerazioni. In: Proceedings of Atti Congresso Naz Speleol Udine 1990 1, pp 89–100

Cucchi F, Giorgetti F (1997) Relazione sulle risultanze delle campagne geofisiche e dei rilevamenti geologici eseguiti sul terreno. Final report. Department of Geological, Environmental and Marine Sciences, University of Trieste, p 23

Cucchi F, Piano C (2002) Ipercarsismo superficiale e sepolto nelle evaporiti del Friuli-Venezia-Giulia. In: Le voragini catastrofiche, un nuovo problema per la Toscana. Edizioni Regione Toscana, Grosseto, pp 35–41

Cucchi F, Zini L (2009) Paesaggi carsici nel Friuli Venezia Giulia. DiSGAM Università di Trieste—R.A. Friuli Venezia Giulia, Tipografia Stella Arti Grafiche, Trieste

Cucchi F, Forti P, Ulcigrai F (1994) Znizevanje krskega povrsja zaradi korozije (Valori di abbassamento per dissoluzione di superfici carsiche). Acta Carsologica 23:55–61

Cucchi F, Finocchiaro F, Zini L (2010) Karst geosites in NE Italy. Environ Earth Sci:393–398. https://doi.org/10.1007/978-3-642-12486-0_61

De Waele J, Piccini L, Columbu A, Madonia G, Vattano M, Calligaris C, D'Angeli IM, Parise M, Chiesi M, Sivelli M, Vigna B, Zini L, Chiarini V, Sauro F, Drysdale R, Forti P (2017) Evaporite karst in Italy: a review. Int J Speleol 46(2):137–168.. Tampa, FL (USA). https://doi.org/10.5038/1827-806X.46.2.2107

Dreybrodt W, Eisenlohr L (2000) Limestones dissolution rates in karst environments. In: Klimchouk AB, Ford DC, Palmer AN, Dreybrodt W (eds) Speleogenesis: evolutions in karst aquifers. National Speleological Society, Huntsville, pp 136–148

Fabbri P, Ortombina M, Piccinini L, Zampieri D, Zini L (2013) Hydrogeological spring characterization in the vajont area. Ital J Eng Geol Environ 6:541–553. https://doi.org/10.4408/IJEGE.2013-06.B-52

Ford D, Williams P (2007) Karst hydrogeology and geomorphology. Wiley, West Sussex

Gortani M (1965) Le doline alluvionali. Natura e Montagna 3:120–128

Gutiérrez F, Cooper AH, Johnson KS (2008) Identification, prediction and mitigation of sinkhole hazards in evaporite karst areas. Environ Geol 53:1007. https://doi.org/10.1007/s00254-007-0728-4

Gutiérrez F, Parise M, De Waele J, Jourde H (2014) A review on natural and human-induced geohazards and impacts in karst. Earth-Sci Rev 138:61–88. https://doi.org/10.1016/j.earscirev.2014.08.002

Jeschke A, Vosbeck K, Dreybrodt W (2001) Surface controlled dissolution rates of gypsum in aqueous solutions exhibit nonlinear dissolution kinetics. Geochim Cosmochim Acta 65:27–34. https://doi.org/10.1016/S0016-7037(00)00510-X

Jurkovšek B, Biolchi S, Furlani S, Kolar-Jurkovšek T, Zini L, Jež J, Tunis G, Bavec M, Cucchi F (2016) Geology of the classical karst region (SW Slovenia–NE Italy). J Map 12:352. https://doi.org/10.1080/17445647.2016.1215941

Klimchouk A, Lowe D, Cooper A, Sauro U (1996) Gypsum karst of the world. Int J Speleol 25:1–307

Marinelli O (1897) Fenomeni di tipo carsico nei terrazzi alluvionali della Valle del Tagliamento. Mem Soc Geogr Ital 8(2):415–419

Nisio S (2003) I fenomeni di sprofondamento: stato delle conoscenze ed alcuni esempi in Italia centrale. Il Quaternario Ital J Quat Sci 16:121–132

Nisio S (2008a) I sinkholes nelle altre regioni. Mem Descr Carta Geol D'it 85:419–426

Nisio S (2008b) I fenomeni naturali di sinkhole nelle aree di pianura italiane. Mem Descr Carta Geol D'it 85:475

Nisio S (2010) I Sinkholes. Gli sprofondamenti catastrofici nell'ambiente naturale ed in quello antropizzato. 2nd International workshop, ISPRA Roma 3–4 Dicembre 2009. ISBN: 978-88-448-0400-8, 1008pp

Nisio S, Salvati R (2004) Fenomeni di sprofondamento catastrofico. Proposta di classificazione applicata alla casistica italiana. In: Atti Conv. "Stato dell'arte sullo studio dei fenomeni di sinkholes e ruolo delle amministrazioni statali e locali nel governo del territorio Roma 20-21 maggio 2004", pp 573–584

Nisio S, Caramanna G, Ciotoli G (2007) Sinkholes hazard in Italy: first results on the inventory and analysis of some case studies. In: Parise, Gunn (eds) Natural and anthropogenic hazards in karst areas: recognition, analysis and mitigation, Geological Society, vol 279. Special Publications, London, pp 23–45

Venturini C (2009) Note illustrative della Carta Geologica d'Italia in scala 1:50.000, Foglio 031 Ampezzo. Serv Geol D'it, p 232

Waltham T, Bell F, Culshaw M (2005) Sinkholes and subsidence, karst and cavernous rocks in engineering and construction. Springer, Berlin/Heidelberg/New York

Zanferrari A, Masetti D, Monegato G, Poli ME (2013) Geological map and explanatory notes of the Italian geological map at the scale 1:50.000: sheet 049 "Gemona del Friuli". ISPRA – SERV GEOL D'IT – Regione Autonoma Friuli Venezia Giulia, 262 pp. http://www.isprambiente.gov.it/Media/carg/friuli.html

Zini L, Calligaris C, Zavagno E (2013) Classical karst hydrodynamics: a sheared aquifer within Italy and Slovenia. In: Evolving water resources systems: understanding, predicting and managing water-society interactions (IAHS Publication 364), pp 499–504. https://doi.org/10.5194/piahs-364-499-2014, 2014

Zini L, Calligaris C, Forte E, Petronio L, Zavagno E, Boccali C, Cucchi F (2015a) A multidisciplinary approach in sinkhole analysis: the Quinis village case study (NE-Italy). Eng Geol 197:132–144. https://doi.org/10.1016/j.enggeo.2015.07.004

Zini L, Calligaris C, Cucchi F (2015b) The challenge of tunneling through Mediterranean karst aquifers: the case study of Trieste (Italy). Environ Earth Sci 74:281. https://doi.org/10.1007/s12665-015-4165-5

Zini L, Casagrande G, Calligaris C, Cucchi F, Manca P, Treu F, Zavagno E, Biolchi S (2015c) The karst hydrostructure of the Mount Canin (Julian Alps, Italy and Slovenia). Environ Earth Sci 1:219–226. https://doi.org/10.1007/978-3-642-17435-3_24

Chapter 6
Collapses in Calcarenitic Deposits Along the Sides of the Ginosa Ravine in South Italy

Angelo Doglioni and Vincenzo Simeone

Abstract A collapse of large stretch of a calcareous sandstone cliff occurred in the night between December 2nd and 3rd, 2017, on the upper levels of the right side of the deep narrow valley called Gravina di Ginosa, Apulia, southeast Italy. This event happened during a nonparticularly rainy period and, apparently, without a clear external triggering component. Even if the involved rock mass is not particularly extended, the collapse had a strong impact on the local community, since it involved part of the archaic rupestrian settlements of the ancient town of Ginosa. Several rupestrian grottoes were damaged, as well as the collapsed mass damaged some ancient inhabited buildings, at the lower levels of the valley. Here, this event is described and analyzed with remote sensing approaches and laboratory tests on rock samples. Moreover, the particular geomorphological and geological features of the area are emphasized, together with the effects on the stability related to the digging of artificial caves on the calcarenitic flank of the valley.

Keywords Calcarenites · Collapse · Laboratory tests · Ravine · Rupestrian grottoes · Tectonic stress · Terrestrial laser scanner · UAV photogrammetry

6.1 Introduction

The southern boundary of the carbonatic platform in Apulia, southeast Italy, is characterized by deep narrow valleys, which can be classified as ravines, locally named Gravine. These are characterized by relatively soft and poorly cemented Plio-Pleistocenic calcareous sandstone, named calcarenites (Cotecchia and Grassi 1975; Cherubini et al. 1996; Lionetti and Simeone 2015), outcropping in the upper layers of their flanks. The lower layer, that is, the bottom of these ravines, is constituted of Cretaceous limestones (Doglioni and Simeone 2014), which are in direct contact

A. Doglioni (✉) · V. Simeone
Department of Civil, Environmental and Structural Engineering and Chemistry, Technical University of Bari, Bari, Italy
e-mail: angelo.doglioni@poliba.it

M. De Maio, A. K. Tiwari (eds.), *Applied Geology*,
https://doi.org/10.1007/978-3-030-43953-8_6

with the upper calcarenites (Cotecchia and Grassi 1975; Simeone et al. 2019). These ravines are usually oriented along NNW–SSE direction, mainly parallel to the Apennine, being characterized by steep and straight paths, sometimes interrupted by local tortuous paths, apparently similar to meanders, oriented along the WSW–ENE direction. The depth of these ravines ranges between few tens of meters up to about 200 m. Water flows at their bottoms only when severe rainfall events occur, activating the runoff through the stream network (Doglioni et al. 2012). Usually, they are completely dry; therefore, ravines can be classified as ephemeral streams (Doglioni et al. 2012, 2015). The genesis of these deep valleys is controversial, since they were traditionally considered as consequence of stream cutting erosion. However, due to their slopes and their straight paths, besides their depth, it may be more likely that these valleys were generated as deep cracks during the uplift of the Apulian platform, due to tectonics of Apennine orogenesis during the Pleistocene (Doglioni et al. 1994). These cracks likely started working as drain paths for runoff, being the relatively easiest flow path (Doglioni et al. 2012).

The flanks of these ravines were colonized since the prehistoric age (Lionetti and Simeone 2015; Simeone et al. 2019), being prone to be carved and dug thus providing safe and relatively comfortable dwellings for people. The relatively low permeability of calcarenites allowed for collecting runoff water into special carved environment used by people living in the artificial caves. For these reasons during the centuries, the presence of rupestrian human settlements had several developments, with an almost continuous use of these caves up to the twentieth century (Grassi et al. 2004, 2007). The presence of people caused a continuous digging and reshaping of the flanks of ravines, together with a continuous maintenance (Cotecchia and Grassi 1975; Cherubini et al. 1996; Simeone et al. 2019). In fact, due to the instability of calcarenites related to the digging, there were occasional collapses, which were immediately repaired in order to limit the propagation of dams as well as to preserve the functionality of the dwellings. In addition, the flanks of ravines were used for agriculture, growing small areas on artificial terraces. These activities managed the presence of vegetation, thus limiting the proliferation of roots, which could potentially crack the calcarenites.

Starting from the second part of the twentieth century, the rupestrian settlements and the agricultural activities on the flanks of ravines were progressively abandoned, in order to gain more comfortable and modern houses (Cotecchia and Grassi 1975). In addition, people were fostered to leave rupestrian caves, since those were considered unhealthy and not adequate to the modern housing standard. The almost complete abandonment of ravines caused an interruption of the maintenance of their flanks. This implied that the weathering effects and the proliferation of vegetation roots progressively damaged the calcarenites, in particular those artificially hollowed. In addition, the drainage system was progressively degraded, thus the runoff inordinately flows, causing further problems to the calcarenites and grottoes. Therefore, after few decades, localized sparse collapses are occurring. These are destroying the ancient heritage of rupestrian settlements, as well as modifying the sides of ravines. Therefore, the effects are twofold: loss of cultural heritage and increase of geomorphological hazard.

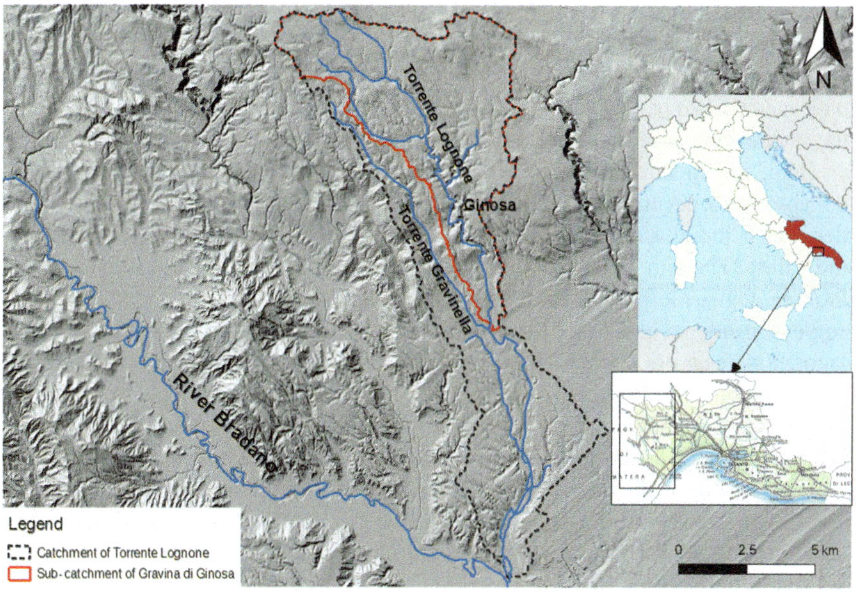

Fig. 6.1 The catchment of Gravina di Ginosa and its mainstream network

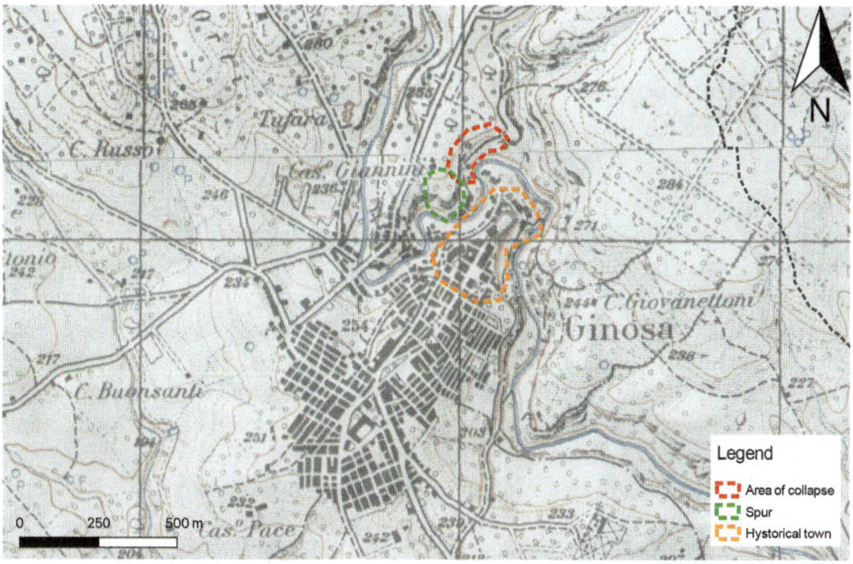

Fig. 6.2 Details of the investigated area

This work investigates a particular ravine crossing the town of Ginosa (Parenzan 1992), see Figs. 6.1 and 6.2, which is particularly rich with rupestrian settlements on its flanks. Moreover, the modern town, since the late medieval age, developed on the sides of the ravine, thus originating a twofold urbanization made by caves and buildings, often integrated, that is, the building was built starting from the caves. In this context, the progressive abandonment of the ancient rupestrian settlements as well as of the old part of the town caused serious damages, which altered the sides of the ravine and caused collapses of both buildings and caves.

Here, the most archaic side of the settlements of Ginosa is investigated. A collapse was occurred in December 2017, damaging the rupestrian caves and destroying the abandoned construction on the lower levels, toward the bottom of the valley. The aim of this work is to describe and analyze this particular collapse looking at the morphological features of the flanks, determined by advanced 3D surveys using an integrated approach, and at the rock features, determined by laboratory tests on samples. It is shown how the use of relatively cheap remote sensing approaches can help in identifying in details the landforms and features of the collapse. In particular, on one hand, these investigations return the accurate reconstruction of the morphology of the site, in order to identify the effects that the collapse caused on the flank. On the other hand, the in situ surveys and the laboratory tests on the rock samples collected from the flank aim at defining a preliminary characterization of the rock, in order to interpret the collapse. These analyses can be helpful in delineating the areas prone to the next collapses or clearly weakened by the previous collapse and then planning the early safety actions on the flank as well as the next monitoring strategies.

6.2 Methodological Approach for the Investigation of the Site

The collapsed flank was investigated using a 3D survey, based on photogrammetry, in order to survey the entire flank beside the collapse. In particular, a high-definition 3D model was constructed by a photogrammetry, using the images taken by an unmanned aerial vehicle (UAV), flying at low altitude, 50 m on the upper part of the slide. In particular, photos were collected by a DJI Phantom 3 standard. Images were taken according a uniform grid, with a 75% overlapping and with the camera aligned on the vertical direction, that is, nadiral photos. This allowed for reconstructing a grid model of the site with a resolution of 11 cm, starting from a point cloud extracted by a structure from motion algorithm (Westoby et al. 2012) implemented into the commercial software AgiSoft PhotoScan Professional version 1.3.4 (http://www.agisoft.com/downloads/installer/). In order to enhance the accuracy of the aerial survey, 20 Ground Control Points (GCPs) were positioned in the surveyed area. The placement of these points was done by a total station, TOPCON IS-3, according to a local reference system, which was successively referred to a topographic coordinated system, in particular the EPSG 32633, corresponding to the WGS84, UTM 33N. Note that the use of GCPs highly enhanced the accuracy of the survey, besides reducing deformations related to the image processing by the structure from motion algorithm (Westoby et al. 2012). Two hundred eighty-six images were processed, returning a point cloud made of 69,327 points, from which a dense point cloud was interpolated. The dense point cloud was constituted by 12,031,224 points; in order to avoid distortion, a mesh grid of triangular elements was constructed, starting from the dense cloud, and this was constituted by 2,406,238 triangular faces, which was used for extracting the geometry of the flank. Finally, a Digital Surface Model of the area was generated; it is a raster grid with a resolution of 11 cm/pix.

Successively, the collapsed zone was furtherly surveyed by terrestrial laser scanning, in order to have a better detailing of the collapse. A robotized total station, TOPCON IS-3, was used, which produced a point cloud, on a grid where each point was measured every 20 cm, both in vertical and horizontal directions. This returned a cloud of 204,129 points, detailing those zones of the collapse, which were shadowed or covered during the flight of the UAV.

Finally, some rock samples were collected from the flank, whereas the collapse occurred, and then their strength was measured in laboratory by unconfined compression load tests.

6.3 General Description of the Investigated Site

The investigated collapse is on a side of a valley named Gravina di Ginosa. It is located in Apulia, southeast Italy, on the southwest boundary of Murgia platform (Fig. 6.1). This is a large outcrop of the carbonatic Apulian foreland domain, which is characterized on its SSW boundary by deep narrow valleys, that is, ravines. Ginosa is on this boundary, and the site is characterized by direct contact between the Cretaceous limestone platform and Plio-Pleistocenic calcarenites. Calcarenites are white to yellowish, sometimes coarse grained. Moreover, they are not uniform, showing some sandy layers. In addition, there are several cracks, visible on the flanks of the ravine, which may be possibly related to tension release or to stress generated during the genesis of the ravine, as well as to decompression phenomena, which created toppling effects toward the void of the ravine.

Collapses involve part of the ancestral site of Ginosa as well as part of the historical town (Fig. 6.2). The investigated collapse exclusively involves the upper calcarenitic layers, which are naturally poorly cemented and deteriorated by weathering and anthropic activities. In particular, several artificial caves and hollows exist on the side, which were presumably dug and reshaped on a timespan ranging between the prehistoric age and the nineteenth century. The site is not new to collapses, since there are clear evidences of past collapses. These are constituted by medium-large blocks, which randomly lay on the side of the valley, showing states of deterioration due to a long exposition to weather actions. A further

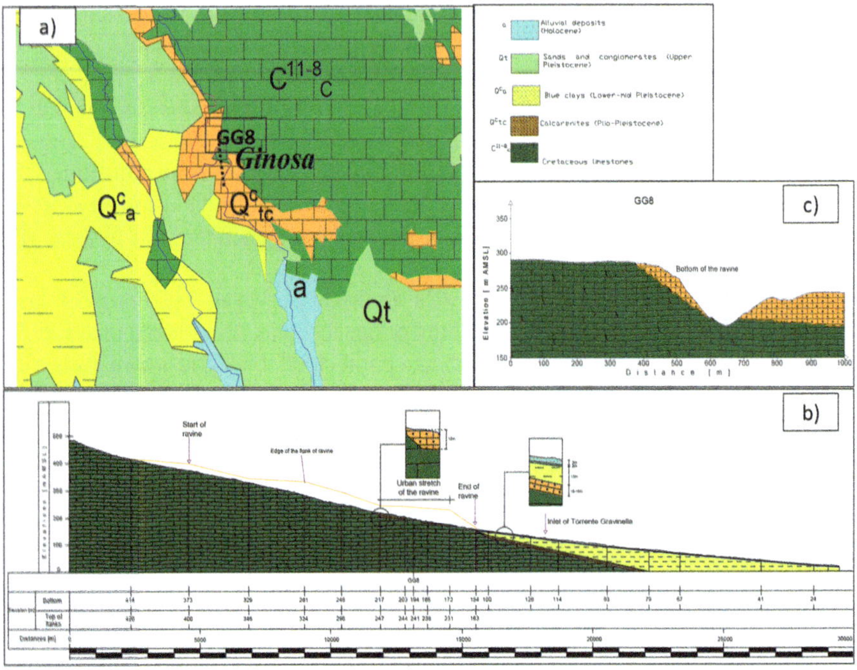

Fig. 6.3 (**a**) Geological map of the area; (**b**) longitudinal profile of Torrente Lognone; (**c**) transect of the ravine in the urban area of Ginosa

interesting feature of the site is the characteristic shape similar to a meander of the valley. Indeed, even if the paths of ravines are mainly oriented according the NNW–SSE direction, they are quite tortuous, and occasionally, there are some anomalous pseudo meanders. This is the case of Ginosa, where, coming from north, the typical path immediately before of the town of Ginosa diverts to the NNE–SSW direction and after about 1 km diverts again according to the SW–NE direction and, again, after about 850 m, it diverts according to the N–S direction. These diversions generate a sort of spur, oriented according to the NNE–SSW direction. The collapse is located on the SE side of this spur (Fig. 6.2).

The ravine of Ginosa is a part of an ephemeral stream, named Torrente Lognone, which is 29.4 km long, being a tributary of river Bradano (Fig. 6.1). The total catchment of Torrente Lognone is about 80 km^2, including the subcatchment of a right-side tributary, named Torrente Gravinella. The catchment originates from Murgia platform, where Cretaceous limestones diffusively outcrop, then, arriving on the southern edge of Murgia, it passes through Plio-Pleistocenic calcarenites, where the morphology becomes a ravine. Downstream, the river passes through upper Pleistocenic deposits, up to the confluence with river Bradano. Figure 6.3a shows the geological map of the entire catchment. Figure 6.3b shows the longitudinal profile of the entire Torrente Lognone, with the main geological features. It is possible to observe that the ravine occupies the central part of the profile; looking at

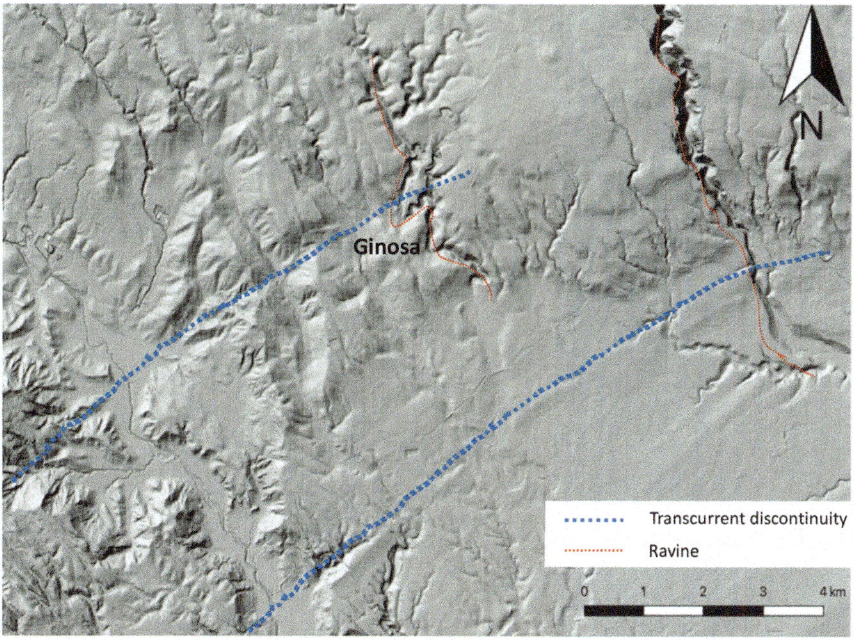

Fig. 6.4 Transcurrent discontinuities in the region of the south boundary of Apulian foreland

the detail focused on the town of Ginosa, it is possible to see the high thickness of Plio-Pleistocenic calcarenites. The transect GG8, in Fig. 6.3c, corresponds to the area of the investigated collapse. It shows the layer of calcarenites, which sometimes exceeds 50 m of thickness, as well as the shape of the ravine. It is not particularly deep and impervious; this fostered the rupestrian settlement, implying a diffuse artificial hollowing. However, this part of the ravine is also tectonically stressed: the meanders of the ravine with particularly tortuous path around the northern and east limit of the town and may be associated with the presence of distensive and transcurrent tectonic structures (Guerricchio and Melidoro 1986; Bentivenga et al. 2004), as shown in Fig. 6.4. These structures appear at large scale looking at the digital terrain model, at the edge of Murgia plateau, as well as in the alluvial deposits of the plain downstream Murgia toward Ionian sea (Fig. 6.4).

6.4 Analyses of the Flank Collapsed in December 2017

During the night between December 2nd and 3rd 2017, a collapse in the northern calcarenitic flank of the ravine occurred (Fig. 6.5). This damaged the rupestrian grottoes dug along this flank, as well as some old constructions at the toe of the flank. The collapse involved the upper part of the flank, from which some large calcarenitic blocks detached from a zone close to the edge. The total length of the involved front

Fig. 6.5 Image of the collapsed flank and of the evidences of past collapses

of the collapse is 80 m, while the presumable thickness of the collapsed calcarenite levels is 20 m. The flank underwent further collapses in the past, as it is possible to see from the presence of weathered calcarenitic blocks in the lower part of the flank (Fig. 6.5). Therefore, this failure is a recession of a past fall. The involved area is completely abandoned since some decades; therefore, there were no fatalities or injured people. The consequence of the collapse was an irregular retreat of the flank, with the partial destruction of some rupestrian caves. After the collapse, the flank showed an extremely irregular surface with several unstable blocks and jutting blocks, which are prone to further collapses. The artificial grottoes on the flanck are in a really dangerous situation, since a large part of their sustaining elements were damaged and reduced, thus making them prone to collapses.

After that the described event occurred, the site was surveyed, in order to reconstruct the morphology of the site after the collapse. In particular, according to the approach presented at paragraph 6.2, a high-definition 3D model was constructed by a photogrammetry, using the images taken by the UAV, flying at low altitude, 50 m on the upper part of the slide. The photogrammetry was integrated by a terrestrial laser scanner survey, made by a robotized total station, which produced a point cloud, on a grid where each point was measured every 20 cm, both in vertical and horizontal directions. Finally, calcarenitic samples were collected from the zone where the collapse started, in order to test the mechanical characteristics of the calcarenite, in terms of compressive strength to crushing loads. It is noteworthy that the calcarenites do not appear uniform on the flank, there are some coarse-grained layers and some finer layers, and this may suggest a nonuniform sedimentations and different ages of the layers. In particular, coarse-grained calcarenitic layers seem more similar to upper Pleistocenic deposits, while fine-grained layers are consistent with the features of the Plio-Pleistocenic deposits. Coarse-grained layers appear very weakly cemented and prone to disaggregation, which can be fostered by roots and adverse weather conditions, viz. rain and wind.

6.5 Discussion of the Surveys

6.5.1 3D Photogrammetric Model

3D photogrammetry allowed for constructing a high-resolution digital surface model (DSM) of the collapse area with an accuracy of 11 cm. In particular, this permitted a careful characterization of the site and of the geometrical features of the collapse. Moreover, it was possible to observe some interesting details of the site. Figure 6.6 shows the high-resolution DSM as well as a cross section of the collapsed flank. It is clear that the edge of the flank is lowered and irregular, while on the west side of the collapse, there is a spur extended toward the ravine, which appears lowered and with a clear detachment front. This spur has many caves, which are undergoing a sort of packing and then a lowering and decompression of the upper layers, which are evolving toward possible collapses. It is also evident that the investigated flank underwent several collapses during the century, which caused its retreat as well as a characteristic shape with a steep upper part of the flank and a moderate-to-low slope

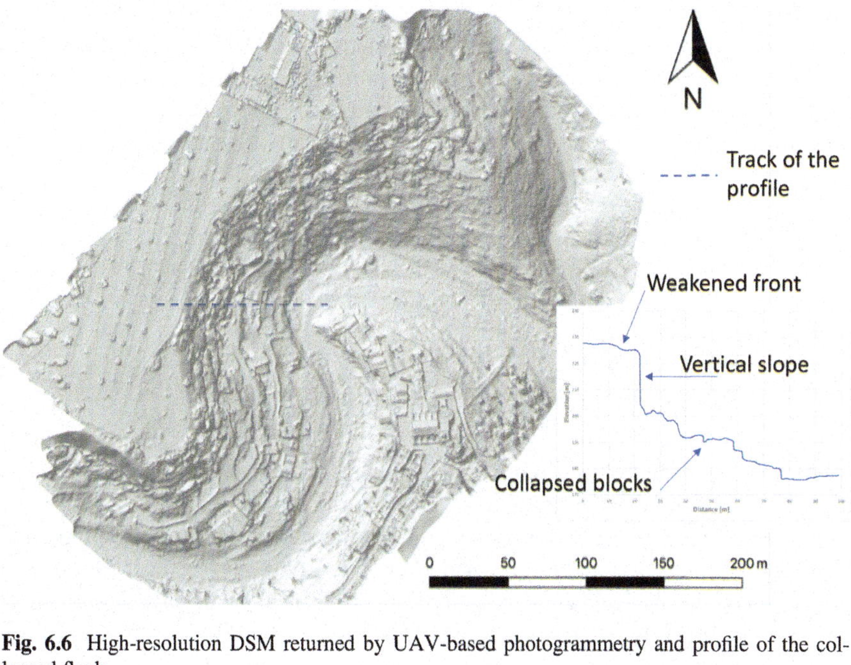

Fig. 6.6 High-resolution DSM returned by UAV-based photogrammetry and profile of the collapsed flank

zone, going toward the bottom of the ravine. It is also interesting to observe that the photogrammetry did not allowed for carefully reconstructing the caves on the vertical flank, in particular when these were hidden by jutting strata or blocks. Therefore, photogrammetry, as performed, was useful for a general high-resolution characterization of the collapse area, but it failed in reconstructing the details of the vertical flank.

6.5.2 *Terrestrial Laser Scanner of the Site*

The collapsed flank was surveyed by a terrestrial laser scanner from the bottom of the ravine. The scan of the flank was performed by a single position, and this implied some shadowed areas, which were not scanned, in particular at lower elevations. However, the upper part of the flank, whereas the blocks detached, was reasonably reconstructed. The accuracy of the scan was of one point every 20 cm at most. Even if the accuracy of the laser scansion seems lower than the photogrammetry, it is noteworthy that here points are actually measured from the instrument, while points of photogrammetry are reconstructed by a structure from motion algorithm. Therefore, the laser scansion can be generally considered more reliable than photogrammetry, unless artifacts related to obstacle between the scanned surface and the scanner exist. In particular, the scan of the collapse evidenced the irregularities of the surface after the collapse and the presence of caves, which were partly damaged by the collapse (Fig. 6.7). In addition, the scan permitted to identify and shape the jutting elements of the flank, which constitute calcarenitic blocks potentially prone to the collapse. Figure 6.8 shows some cross sections of the point cloud returned by the scan, where the jutting elements are clearly visible. It is also interesting to observe that the potentially prone-to-collapse area is the east part of the collapse flank, which is the side of the aforementioned spur, which is partly lowered and tilted toward NNW. This is likely related to the intensive presence of artificial caves, which hollowed the flank, thus causing the mentioned packing effects, which progressively compacted down the calcarenites.

6.5.3 *Laboratory Tests on the Calcarenitic Samples*

A number of samples were collected form the area where the collapsed block detached. Apparently, the calcarenite is supposed to be uniform, and then the collected samples were expected to show similar mechanical and physical properties. However, the laboratory tests returned some interesting results, in terms of both physical and mechanical characterization. In particular, the collected samples were characterized in terms of specific weight and water content. Moreover, unconfined

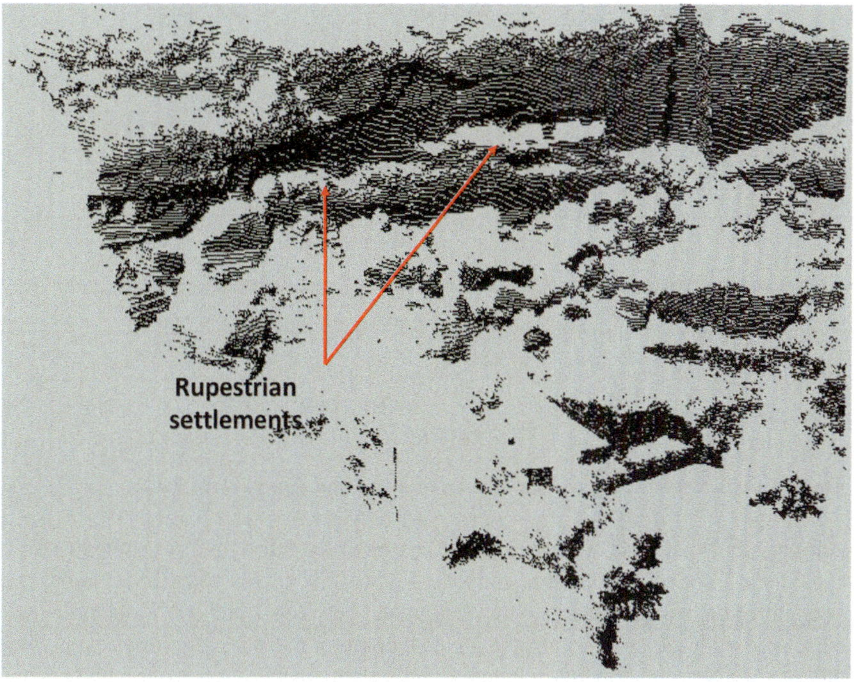

Fig. 6.7 Point cloud obtained by the terrestrial laser scanning of the collapsed flank

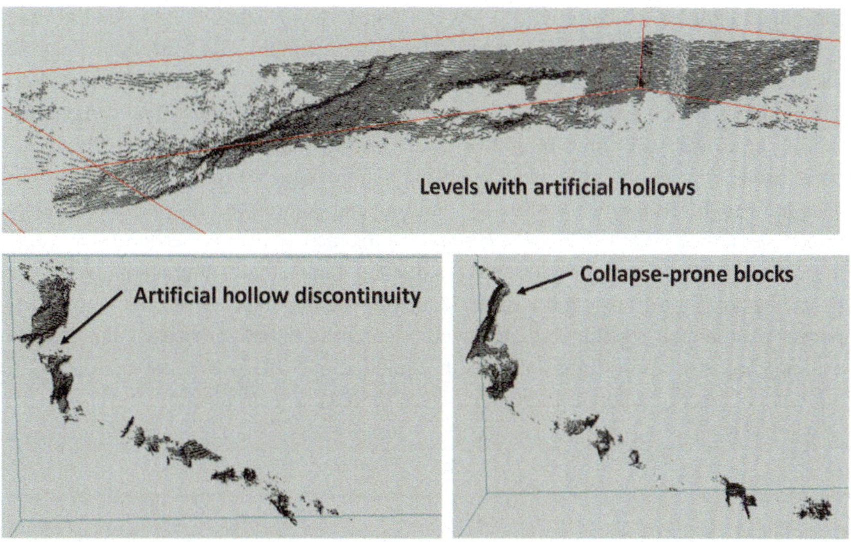

Fig. 6.8 Details of the flank returned by laser scanning, not visible by UAV-based photogrammetry

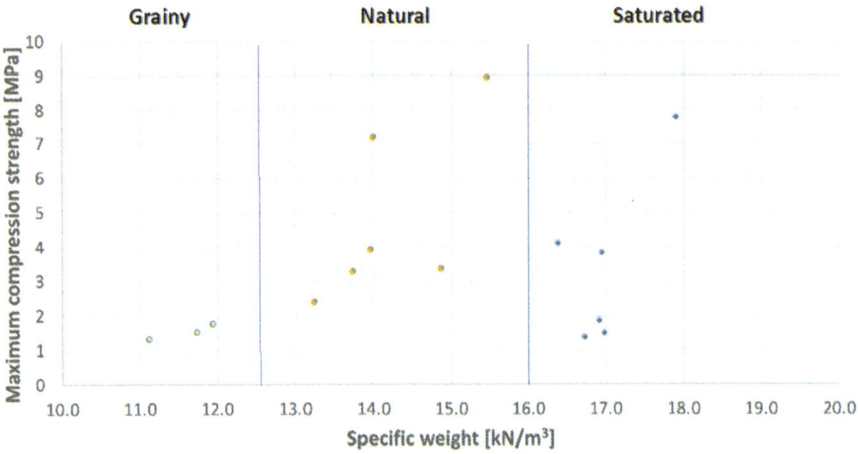

Fig. 6.9 Compression strength of the samples collected from the collapsed flank

compression load tests were performed, in order to determine the strength of the specimens extracted from the samples. Figure 6.9 shows that the compressing strength of the samples varies on a wide range, between 1 and 9 MPa being poorly correlated with the specific weight. Indeed, compressive strength varies even when the specific weight remains unaltered. This particularly happens for natural and saturated samples, where there is very high variability, even two times the minimum value, of the compressive strength. In order to explore the effects of the water content, some samples were saturated before the tests. The results show that the saturated samples show slightly lower values of resistance than natural ones (Fig. 6.9); but despite the increase of the unitary weight, their strength does not decrease. This particular behavior may be related to variable features of the calcarenites, which show different grain sizes and broadly variable cementation rates. The lower value of strength is in the grainy samples, where cementation is poor. Some very poorly cemented layers are visible from the images collected on the site, see Fig. 6.10, these resemble to sandy layers. Therefore, the presence of these poorly cemented layers affects the stability of the cliff of the ravine flack. Moreover, there may be layers which sedimented in different ages, that is, the upper layers may be of the mid-late Pleistocene, rather than Plio-Pleistocenic, thus characterized by poorer mechanical resistance than lower calcarenitic layers. However, this hypothesis needs further investigation, since no detailed age analyses were performed, as well as no specific paleontological analyses on fossils were possible.

Fig. 6.10 Details of the collapsed calcarenites, red arrows indicate discontinuities and grainy/sandy layers

6.6 Conclusions

The collapse of a flank in a ravine, namely Gravina di Ginosa, is here preliminary analyzed and presented. Two 3D geometrical models, the earlier based on UAV photogrammetry and the latter on terrestrial laser scanning, are analyzed in order to reconstruct the peculiar landforms and details, which characterize the flank of the ravine. Finally, the results of laboratory tests on samples collected on the site of the collapse are presented and discussed. The results of the specific investigation together with the large-scale morphological and geological analysis of the site where the ravine is located allow for conjecturing some interesting hypotheses about the geological and geomorphological evolution of the site. Ravines are complex morphological elements, which probably originated when the uplift of the Apulian foreland domain, due to the tectonic of the Apennine, caused some deep cracks on the borders of the platform. These deep fractures become a preferential flow path of water, and then they started working as ephemeral streams. The tectonics of the ravine possibly affected the stability and the resistance of the upper calcarenitic layers on the flank of the ravine, which may be weakened by the stresses and fracturing. In addition, the soft consistence of calcarenites made them prone to be carved and dug, in order to obtain dwellings. These activities continued for centuries, thus creating a network of artificial caves, which further weakened the flanks of the ravine, already tectonically stressed. For these reasons, those flanks where rupestrian settlements existed were traditionally prone to collapses, which systematically occurred, implying a continuous maintenance of the flanks done as digging new caves and repairing the damages. After the abandonment of the rupestrian settlements, their evolution continued without control, and this implied uncontrolled collapses and progressive loss of the artificial caves. Such phenomenon is now completely ungoverned, thus compromising not only the rupestrian caves but also the historical town of Ginosa, which developed on the flank of the ravine. For these reasons, continuous monitoring of the site is of paramount significance. For such monitoring approaches, like those herein presented, as well as an advanced modeling of the stability of the caves and of the flank, approaches based on the use of monitoring data will be very helpful. Finally, this work can be considered as an example of an early study integrating modern survey approaches for the study of the stability of ravines flanks. These are important geomorphological elements and anthropized natural environments, and then the study of their stability is of large interest for the preservation of both the natural environment and the cultural heritage.

References

Bentivenga M, Coltorti M, Prosser G, Tavarnelli E (2004) A new interpretation of terraces in the Taranto Gulf: the role of extensional faulting. Geomorphology 60(3–4):383–402

Cherubini C, Di Cuia N, Pagliarulo R, Ramunni FP (1996) Caratteri petrografici e meccanici delle calcareniti di Matera. Mem Soc Geol It 51:761–769

Cotecchia V, Grassi D (1975) Stato di conservazione dei "sassi" di Matera (Basilicata) in rapporto alle condizioni geomorfologiche e geomeccaniche del territorio e al-le azioni antropiche. Geol Appl ed Idrog X:55–105

Doglioni A, Simeone V (2014) Data-driven modeling of the dynamic response of a large deep karst aquifer. Procedia Eng 89:1254–1259

Doglioni C, Mongelli F, Pieri P (1994) The Puglia uplift (SE Italy): an anomaly in the foreland of the Apenninic subduction due to buckling of a thick continental lithosphere. Tectonics 13:1309–1321

Doglioni A, Simeone V, Giustolisi O (2012) The activation of ephemeral streams in karst catchments of semi-arid regions. Catena 99:54–65

Doglioni A, Galeandro A, Simeone V (2015) The generation of runoff through ephemeral streams. In: Lollino G, Arattano M, Rinaldi M, Giustolisi O, Marechal JC, Grant GE (eds) Engineering geology for society and territory, Vol. 3 River Basins, Reservoir Sedimentation and Water Resources

Grassi D, Grimaldi S, Simeone V (2004) On the causes of the instability affecting rupestrian urban centres of Apulia Region (Southern Italy) – 32° Intentional Geo-logical Congress Firenze – paper A32IGC9KF3 – Session T16.06 – Geoscience for Cultural Heritage – Natural hazard and cultural heritages

Grassi D, Grimaldi S, Simeone V (2007) Geological and geomorphological conditioning in localization of Apulian rupestrian settlements – Geophysical Research Abstract Vol. 9, 2007 EGU General Assembly

Guerricchio A, Melidoro G (1986) Problematiche di geologia applicata lungo la fascia costiera ionica del Golfo di Taranto. In: Atti del Convegno ENEA: Evoluzione dei litorali. Problematiche relative al Golfo di Taranto; Policoro (MT), 16–17 ottobre'86, pp 263–294

Lionetti G, Simeone V (2015) Condizionamenti geologici e geomorfologici nella localizzazione, nello sviluppo e nella salvaguardia degli insediamenti rupestri. GTA Geologia Territorio Ambiente – Rivista dell'Ordine dei Geologi di Basilicata 25:23–35

Parenzan P (1992) La "Gravina di Ginosa". Dalla preistoria ad oggi. Aspetti naturalistici, Congedo Editore, Galatina, (Le), 191 p

Simeone V, Doglioni A, Lacertosa RM, Sdao F (2019) Environmental and geological characters and stability problems in the historic centre of Matera (South Italy). In: Shakoor A, Cato K (eds) IAEG/AEG annual meeting proceedings, San Francisco, California, 2018 – Vol. 2. Springer, Cham, pp 161–168

Westoby M, Brasington J, Glasser N, Hambrey M, Reynolds J (2012) 'Structure-from-motion' photogrammetry: a low-cost, effective tool for geoscience applications. Geomorphology 179:300–314

Chapter 7
Relation Between On-Field and InSAR Data on Landslide-Induced Damage

Matteo Del Soldato, Silvia Bianchini, Pantaleone De Vita, Diego Di Martire, Roberto Tomás, Domenico Calcaterra, and Nicola Casagli

Abstract The consequences of landslide events, which are among the most widespread natural hazards worldwide, are relevant in both economic and civil issues. The Italian territory is one of the main representatives. For this reason, the study of landslide-induced damage and its monitoring by remote sensing data are effective tools to improve the knowledge of events and to avoid adverse consequences. The present study focused on the efficiency analysis of the relation between damage classes identified on edifices by field campaigns, and landslide kinematic parameters recorded by InSAR techniques were investigated in two Italian sites. Persistent scatterers (PS) derived from the processing of ENVISAT and COSMO-SkyMed radar images were considered to analyse the behaviour of buildings to be compared with their rate of displacement. The two study areas were the Colle Lapponi – Piano Ovetta basin, strongly affected by a landslide in Agnone (Molise region, southern Italy), and the southern sector of the Volterra town (Tuscany region, central Italy), extensively affected by soil erosion and mass movements. In order to validate the reliability of the relationship found between the classes of damage affecting the structures and the velocities re-projected along the steepest slope direction, a symmetric matrix was created. As result, a map of the efficiency of the relation between recorded damage and deformation velocities was tested in two different areas. The resulting maps can be useful to better localize the possible area prone to be damaged in case of recorded displacement and consequently for managing the territory and the urban expansion.

M. Del Soldato (✉) · S. Bianchini · N. Casagli
Department of Earth Sciences, University of Firenze, Florence, Italy
e-mail: matteo.delsoldato@unifi.it

P. De Vita · D. Di Martire · D. Calcaterra
Department of Earth, Environment and Resources Sciences, Federico II University of Naples, Complesso Universitario Monte Sant'Angelo, Naples, Italy

R. Tomás
Departamento de Ingeniería Civil, Escuela Politécnica Superior, Universidad de Alicante, Alicante, Spain

M. De Maio, A. K. Tiwari (eds.), *Applied Geology*,
https://doi.org/10.1007/978-3-030-43953-8_7

Keywords Landslide · Damage · InSAR · Agnone · Volterra · Building classification

7.1 Introduction

Landslide phenomena, both induced by natural or anthropic causes, are considered one of the main significant geohazards affecting the entire world. Landslides severely affect all the states of Europe, and the Italian territory is one of the most severely affected (Herrera et al. 2018), with a relevant social and economic impact, as well as consequences on constructions (Schuster and Highland 2001; Kjekstad and Highland 2009). Investigating and monitoring the landslide effects is fundamental to improve the knowledge and to avoid further recurrences (Confuorto et al. 2017, 2019). Relevant damage on structures is mainly caused by differential displacements affecting them (Bianchini et al. 2015a; Infante et al. 2016), but the availability of diffuse radar benchmarks on each construction depends on the features of the satellite (e.g. wavelength and ground resolution), the orientation of the slopes and the applied A-DInSAR (advanced differential interferometry synthetic aperture radar) processing techniques. The differential settlement is difficult to investigate since it could be a relevant time-consuming process to isolate two or more PS (persistent scatterers) data for each building. First, it could be conducted only with the use of high resolution, i.e. X-band radar images, and then the PS for each building have to be coupling analysed to assess the differential settlement. A more rapid process can be conducted by investigating the displacement of a single structure, even if it is an approximation, to carry out analyses at municipality scale.

The use of radar data for the back-analysis of landslide-prone areas is largely known in the scientific community (Bayer et al. 2018; Del Soldato et al. 2018b; Intrieri et al. 2018; Zhang et al. 2018), as well as for civil protection purposes (Casagli et al. 2018; Dias et al. 2018; Solari et al. 2018). Recently, the free availability of Sentinel-1 products allowed a continuous monitoring over wide areas with short revisiting time, e.g. 6 days (Raspini et al. 2018).

The investigation of the landslide-induced damage on the buildings could play a key role in the analysis of the effects on structures and infrastructures (Tomás et al. 2012; Infante et al. 2017). Several authors worked on this topic to find the best way to recognize and categorize the severity of the damage on buildings affected by a mass movement. In the 1950s, Skempton and MacDonald (1956) divided the recognizable damage on structure according to its severity and the involved elements in: (i) architectural, (ii) functional and (iii) structural ones. Then, in the 1970s (Burland and Wroth 1974), other authors began to take into consideration landslide effects on structures and infrastructures. Burland (1977), based on previous experiences of Jennings and Kerrich (1962) or MacLeod and Littlejohn (1974), developed a simple classification scheme, distinguishing the opening and the frequency of visible fractures, for consequences of settlements affecting foundations in different soils, then improved by Boscardin and Cording (1989). Successively, a huge landslide occurred in Ancona (Marche region, central Italy) in 1983 (Alexander 1983) suggested a new classification of landslide-induced damage recorded on buildings (Alexander 1986). Damaged buildings located in landslide areas were categorized into six levels, based on the mutual comparison of several landslide-affected buildings by the description of internal and external evidences, as well as the discrimination based on tilting and fractures expressed in degrees and centimetres, respectively. In addition, a checklist of relevant and further information to collect for better characterizing the occurred phenomena was suggested. After several applications (e.g. Crescenzi et al. 1994; Iovine and Parise 2002), Chiocchio et al. (1997) realized a new, more complete classification scheme, involving geologists, geomorphologists, and civil engineers, to provide quantitative reference values and to improve the categorization and suggestions for the rehabilitation measures. Recently, Cooper (2008) developed an improved classification for landslide-induced and subsidence-induced damage, also considering the possible cracks and fractures on ground, on the basis of several existing methods, e.g. Handbook (1975).

In the present study, the results of a PS radar analysis on buildings, considering an average velocity of displacement (mm/year), were compared with the damage recorded in situ on structures. The aim of the work was to quickly seek a relation between the classes of damage surveyed on buildings and the kinematic parameters recorded by remote sensing, e.g. velocity and maximum displacement, considering constructions as rigid elements. The results could be useful for local administrators, civil protection intents, architectural and engineering designs of buildings, as well as scientific purposes (Alexander 1986), even if several approximations were done. The objective of the work is to rapidly assess areas where more landslide-induced damage after an event occurred during the PS data spanning time, in order to design possible landslide effects knowing the velocity of displacement.

Two sites with landslide and landslide-induced consequences were investigated:

- Colle Lapponi – Piano Ovetta (CL-PO hereafter) basin in the Agnone municipality, Molise region (southern Italy), with several structures and infrastructures affected by landslide-induced damage (Calcaterra et al. 2008; Del Soldato et al. 2018a);
- Southwestern sector of the Volterra town where slope instability is diffused, and numerous buildings are damaged (Bianchini et al. 2016b, 2017).

The analysis of the kinematic parameters of landslides was conducted using SAR (synthetic-aperture radar) data with ENVISAT, C-band, and COSMO-SkyMed, X-band, constellations in both sites from 2002 to 2010 and from 2012 to 2015, respectively. Landslide-induced damages on structures and infrastructures were assessed by Cooper (2008) and Del Soldato et al. (2017) approaches.

First, the correlation was developed for the CL-PO landslide area in Agnone, where the more suitable kinematic parameters between the velocity of displacement and cumulative displacement were defined, then validated by testing in the southern sector of Volterra town.

7.2 Study Area

The first area of interest is located in the northwest part of Molise Region (southern Italy), which is a territory strongly affected by landslides and erosional processes. The morphology is gentle and covered by shrubs in the valley portion with sparse areas of brushwood. The CL-PO landslide is located in the western portion of the Agnone Municipality (Fig. 7.1a). It is characterized by a total length of about 1500 m

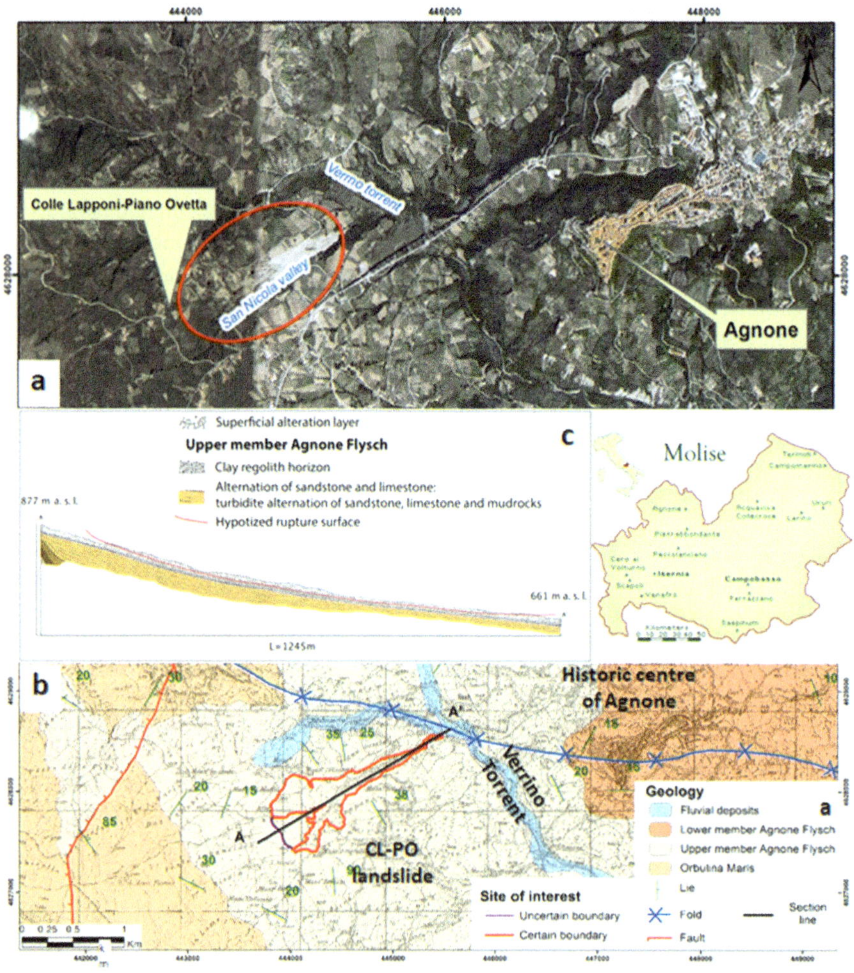

Fig. 7.1 Colle Lapponi – Piano Ovetta localization (**a**), geological map (**b**) and cross section (**c**)

and an estimated area reaching approximatively 280,000 m^2 (Calcaterra et al. 2008) continuously in evolution, causing reshaping of the earth surface and modifying the morphology of the area of interest.

The CL-PO landslide is a complex mass movement affecting a vast area characterized by structurally complex formations (Esu 1977). Geologically, the area is featured by two members of the Agnone Flysch, Sannitico-Molisane Formation, dated Upper Miocene (Vezzani et al. 2004; Filocamo et al. 2015): the lower member is mainly characterized by alternation of marls, calcarenites, and turbidite silicoclastic deposits; the upper member, directly involved in the complex sliding, is constituted of an alternation of altered marly clayey and sandy levels, intercalations of sandstone and limestones (Fig. 7.1b). The formations are generally mantled by an altered regolith horizon composed of clays. The geological relationship between the members and the shallow cover involved in the landslide are visible in the section A–A' of Fig. 7.1.

The investigated landslide has about one century of documented history as testified by old literature information (Almagià 1910) and by the application of the Structure from Motion (SfM) (Pollefeys et al. 1999; Westoby et al. 2012; Fonstad et al. 2013; Lucieer et al. 2013) techniques applied on sets of historical aerial images (1945, 1954, 1981, 1986, 1991, 2003) for reconstructing old 3D models and the more recent evolution investigated by the persistent scatterers (PS – from 1992 to 2019 considering the ERS, ENVISAT and COSMO-SkyMed data) analysis (Del Soldato et al. 2018b). The major reactivation occurred in January 2003 in correspondence of heavy rainfalls combined with snow melting. Besides several hydraulic interventions were fulfilled, during the following years, 2004, 2006 and 2007, further movements of lower intensity were registered and continue to cause important damage on constructions. The severity of the damage on some buildings located on the crown and the possible recurrences forced the local administrators to promulgate restriction rules for several structures and some roads involved by the landslide.

The second test site is in Tuscany Region (central Italy), at 400–500 m a.s.l. in a territory delimitated by two rivers: Era and Cecina. The urban area of Volterra is located in a tectonic depression generated during the post-orogenic extensional stage (Neogene), and the surrounding area is mainly hilly, with gentle slope ranges between 5° and 20°. Geologically, the southern sector of the Volterra town is characterized by 'blue clays' overlapped by the cemented 'Villamagna sands' and by the 'Volterra limestone', on which the historic Volterra town is built (Bazzoffi et al. 1997; Pascucci et al. 1999). These formations are mantled by shallow colluvial deposits mainly deriving from the gradual weathering of the upper units and contributing to the diffuse ground instability and shallow landsliding (Pratesi et al. 2015; Bianchini et al. 2015b, 2016b). Moreover, shrinkage–swelling phenomena of clayey lithologies dominate the hydrogeological behaviour of soils and contribute to mass movements. The combination of hydrogeological features and setting of these formations caused the undermining of the basal clays of the Volterra hills with consequent retrogressive slope failure (Bianchini et al. 2016b). The southwestern

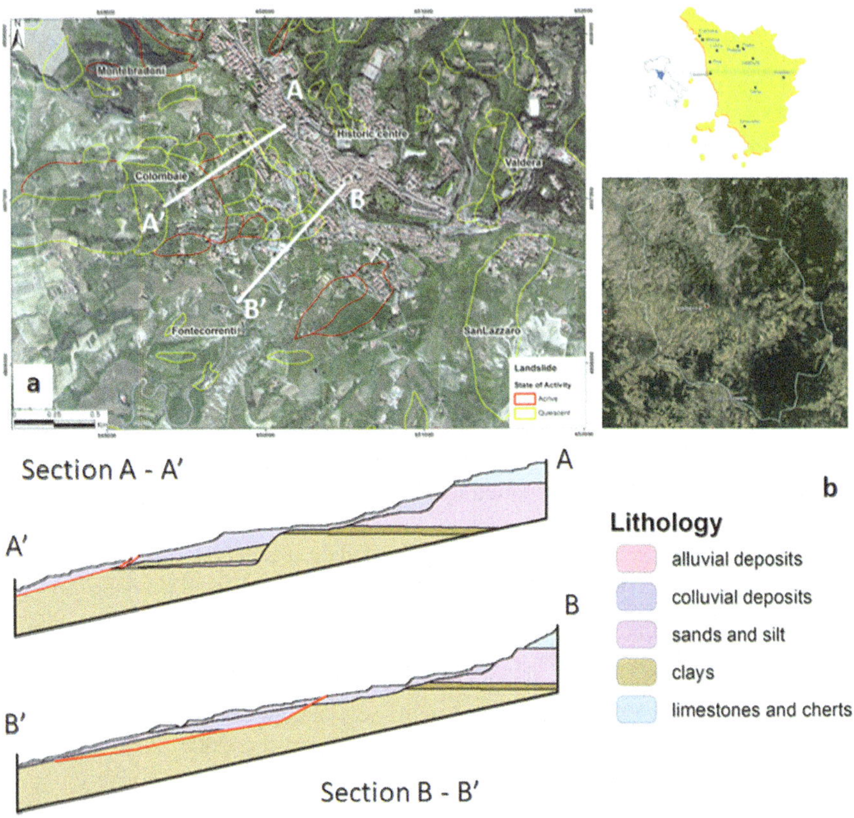

Fig. 7.2 Volterra localization, definition of sectors, landslide inventory (**a**) and two geological section of the Le Colombaie and the Fontecorrenti sectors (**b**)

territory of the Volterra town is the one mostly affected by slope instabilities (Fig. 7.2) and badlands phenomena (Bianchini et al. 2016a). This is mainly because in this part the colluvial debris coverage, made up of chaotic detritus, reaches the higher thickness of up to 20 m.

The pre-existing information on landslides in the study area relies on two landslide inventories, one provided by Tuscany region at regional scale and one by GEOPROGETTI company that refers to a detailed survey performed in 2010 on the southwestern portion of the Volterra area (GEOPROGETTI 2010).

Differently to the CL-PO landslide, for the Volterra southern sectors, no relevant information on historical slope instabilities were found in literature, as well as by the SfM analysis of 3D reconstruction by historical aerial images.

7.3 Methods and Available Data

7.3.1 SAR Data

Ancillary data, e.g. 10-m cell DEM (digital elevation model) (Tarquini et al. 2012) and derived maps as *Aspect*, exposure map, and *Slope*, gradient map, were used in addition to geological map. For investigating the kinematic parameters, two different data sets were used, covering a long period. The oldest one is the ENVISAT data set, acquired in C-band with 5.6 cm of wavelength and 24 days of revisiting time, and the youngest is the COSMO-SkyMed data set, recording information in X-band with 3.1 cm of wavelength and generally 16 days of revisiting time. SAR products derived by the processing of ENVISAT and COSMO-SkyMed data cover the areas of interest from November 2002 to May 2015 and from February 2003 to January 2015 for Agnone and Volterra sites, respectively.

ENVISAT images were processed, for both Agnone and Volterra, by the PSP-DIFSAR (persistent scatterers pairs differential InSAR) (Costantini et al. 2008) and the PSInSAR (persistent scatterers InSAR) techniques (Ferretti et al. 2001). PS products are available from Portale Cartografico Nazionale (PCN) of the Italian Environmental Minister (http://www.pcn.minambiente.it/), thanks to a specific agreement with the Italian Ministry for the Environment, Territory and Sea (Costantini et al. 2017). COSMO-SkyMed products were processed with two different approaches: (i) the Coherent Pixel Technique (CPT – Mora et al. 2003; Blanco-Sanchez et al. 2008) algorithm for the data set covering the CL-PO basin of Agnone; (ii) the SqueeSAR algorithm (Ferretti et al. 2011) for the COSMO-SkyMed images covering the Volterra area (Table 7.1).

Table 7.1 Main features of the PSI data set in the CL-PO area and Volterra site. CSK indicates the COSMO-SkyMed sensor

	Features	**ENVISAT**		**CSK**	
	Wavelength	C (~5.6 cm)		X (~3.1 cm)	
	Incident angle	θ =~23°		θ = 26.6°	
	Ground spatial resolution	4 × 20m		3 × 3m	
	Revisit time	35 days		16 days	
	Orbit	*Ascending*	*Descending*	*Ascending*	*Descending*
CL-PO area	Period (mm/yy)	11/02-07/10	11/02-06/10	10/12-05/15	02/12-01/14
	Processing method	PSInSAR/PSP	PSInSAR/PSP	CPT	CPT
	n° images	50	45	41	26
Volterra	Period (mm/yy)	08/03-07/10	02/03-06/10	02/10-11/13	01/11-01/15
	Processing method	PSInSAR/PSP	PSInSAR/PSP	SqueeSAR	SqueeSAR
	n° images	41	35	25	41

PS products derived by the processing of ENVISAT and COSMO-SkyMed data sets in both orbits were further elaborated by re-projecting the maximum displacement (Dlos) and the velocity of displacement (Vlos) measured along the LoS (Line of Sight) to the steepest slope.

Notti et al. (2014) re-projected the velocity recorded along the LoS into velocity along the steepest slope to investigate the real velocity of movement according to the Eq. (7.1). Following the same approach, the maximum displacement values were re-projected along the slope direction, by modifying the Eq. (7.1) in Eq. (7.2).

$$V_{\text{slope}} = \frac{V_{\text{los}}}{C} \tag{7.1}$$

$$D_{\text{slope}} = \frac{D_{\text{max}}}{C} \tag{7.2}$$

where D_{max} and V_{los} are parameters measured by the satellite along the LoS and C represents a coefficient that is based on the directional cosines and DEM-derived parameters and represents the percentage of movement that the SAR sensor can register. In this way, the real movement suffered by the investigated element identified. The parameters projected along the slope were plotted using the absolute values to look for a relation, assuming that the detected movements cannot be upwards with respect to the slope direction.

7.3.2 Damage Classification

To classify the building damage, two different approaches were applied: Cooper (2008) and Del Soldato et al. (2017).

The first approach derives from an analysis of several previous methods made by American and European researchers and allows categorizing the landslide-induced damage, the subsidence-induced damage and, for the first time, the ground cracks and fractures. The severity of landslide-induced damage can be categorized in eight different classes from *grade 0*, hairline cracks not visible from outside, to *grade 7*, representing the total collapse. Ground damage, instead, is classified only from *grade 3*, tight cracks only in hard surfaces, to *grade 8*, extensive fractures with major scarps (Cooper 2008). Ground damage can be classified only for high grades to be comparable with landslide-induced damage on structures.

The more recent approach combines several previous methods, including the Cooper (2008) categorization and Baggio et al. (2009) approach developed by the Italian Department of Civil Protection for the analysis of the effects of seismic events, where for the first time the extension of the damage were considered. To consider the extent of the damage, the building under investigation has to be ideally divided into three portions, and for each one the severity of the fractures has to be assessed. In this way, the classification approach is divided into two phases (Del Soldato et al. 2017): (i) classification of the cracks recognized on buildings; (ii) a posteriori categorization based on the severity of the cracks and their extension according to a symmetric matrix. The first phase allows classifying the fractures and cracks on buildings in six classes, dividing the description and the referring crack-width for different load-bearing structure, from *no damage (G0)* to *Very severe (G5)*. This classification, as focuses on fractures and cracks, does not include classes regarding collapse. The same classification, from *Weak* (G2) to *Very severe* (G5), is applied to ground fractures as in Cooper (2008). The second phase, combining the severity of the categorized cracks and their extension, is dedicated only to structures and infrastructures, categorizing them in eight classes from *no damage* to *Unusable*, for collapsed buildings.

7.3.3 SAR Parameters – Building Damage Combination

Values of D_{slope} and V_{slope} for ENVISAT and COSMO-SkyMed constellations were combined with the classes of landslide-induced damage recorded during two field surveys and classified according to the Cooper (2008) and Del Soldato et al. (2017) approaches. All the combinations between V_{los}, D_{los}, V_{slope}, D_{slope} and the two classifications were plotted and a possible correlation was looked for. The investigation was first conducted on the Agnone site and then the found relationship was validated testing the same approach in the two sectors of Volterra.

To assess the reliability of the correlation, a symmetric matrix in which the velocity and the damage classes were grouped both in four classes was built, in addition to the no data (ND). The velocity parameter was grouped as follows:

- Stability range is defined up to 2 mm/year, defined by the standard deviation of the processing chain
- Low range is from 2 mm/year to 5 mm/year
- Medium range is from 5 mm/year to 10 mm/year
- High range is major than 10 mm/year

Levels of damage of Del Soldato et al. (2017) and Cooper (2008) were gathered as follows:

- Low class – No damage and negligible, Class0 and Class1
- Medium class – Weak and moderate, Class2 and Class3
- Severe class – Severe and very severe, Class4 and Class5
- High class – Potential collapse and no-habitable, Class6 and Class7

7.4 Results and Discussion

The analysis of the PS of ENVISAT and COSMO-SkyMed satellites shows that both sites were affected by ground displacements in the period 2002–2014.

The ENVISAT displacement map of the CL-PO basin shows for the period 2002–2010 high velocities (approximatively of −6.5 mm/year) in the landslide crown. In addition, relevant velocities were recorded also in the territory above the crown, probably influenced by the relevant landslide occurred in 2003 and the further reactivations, and in the left flank. In the other areas, no relevant displacements were recorded. Analysing the data acquired by the COSMO-SkyMed satellites from 2012 to 2015, the high velocities on the landslide crown and the above territory were confirmed, even if no sudden displacement or reactivation were recorded. The buildings in the left flank, instead, show an inversion of the velocities (approximatively of +2.5 mm/year) in the 2012–2015 period. This inversion could have been caused by a retrogression of the landslide insisting in the closest basin and involving this area. Further like velocities were not recorded even if the surrounding territory is known to be severally afflicted by ground deformation and erosional phenomena (Table 7.2).

The southwestern sectors of Volterra during the period between 2002 and 2010 covered by ENVISAT images have a density of PS data near 200 points per km^2, while in the subsequent period 2012–2014, approximately, 1200 points per km^2 are presented, acquired by the COSMO-SkyMed satellites. The two sectors show a similar ground displacement in both deformation maps with few areas exhibiting

Table 7.2 Reliability matrix of the relationship between V_{slope} and classes of damage

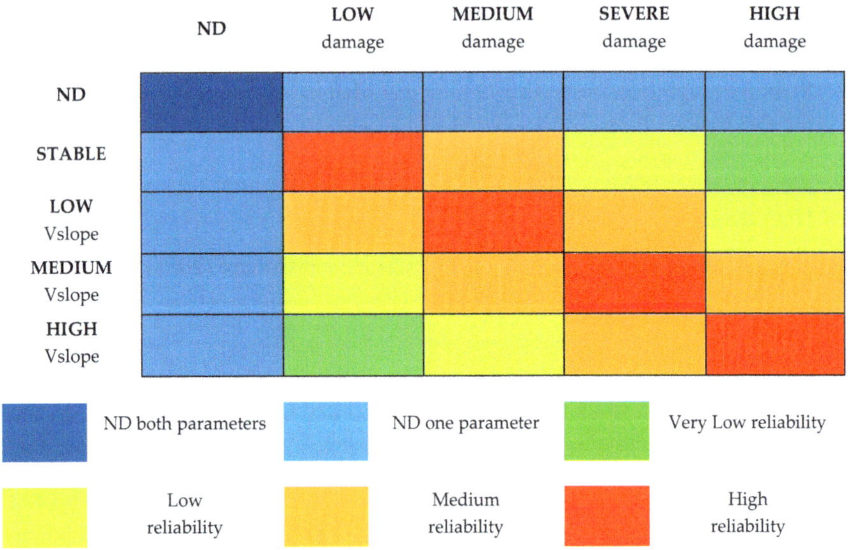

changes. The main relevant differences can be identified in the northeastern part of *Le Colombaie* sector that shows stability during the ENVISAT spanning time, while relevant ground velocities, up to −7.5 mm/year, are recorded in the more recent period. In the central portion of the *Fontecorrenti* sector, positive values approximatively 5 mm/year are measured in the oldest period, while in the most recent period covered by COSMO-SkyMed data, stable values were recorded. In both ground deformation maps, the stability in the historic centre of Volterra can be easily recognized (Figs. 7.3 and 7.4).

The damage to the buildings located in the CL-PO landslide in Agnone and *Le Colombaie* and *Fontecorrenti* sectors of Volterra was classified using two approaches, Cooper (2008) and Del Soldato et al. (2017).

In CL-PO basin, 56 buildings in November 2015 and July 2016 were classified according to the Cooper (2008) and Del Soldato et al. (2017) approaches showing a good similarity for numerous constructions, with some exceptions. The two side-walks with Cooper (2008) are categorized in the higher class (Fig. 7.5 – sectors 1b and 3b), while according to Del Soldato et al. (2017), they were classified as 'potential collapse' and 'very severe', which mainly reflect the real situation of the structures (Fig. 7.5 – sectors 1c and 3c). While the buildings in the left flank were classified into the same classes in both approaches (Fig. 7.5 – sector 2), some buildings in the crown were differently assessed (Fig. 7.5 – sectors 1 and 3). For example, a better categorization, with respect to the real conditions of the buildings, can be recognized for structures severely damaged by the landslide occurred and, sometimes, due to their abandonment.

In the two sectors of Volterra, approximatively 320 structures between the early 2015 and September 2016 were categorized according to Cooper (2008) and Del Soldato et al. (2017) methodologies (Fig. 7.6). By both approaches, few damaged buildings were identified mainly localized in the upper portion of both sectors (Fig. 7.6 – sectors 1 and 3) and in the middle of the *Le Colombaie* slope (Fig. 7.6 – sector 2). It is worth noticing as sector 2 is a new complex of buildings that already exhibits few recognizable cracks. The similarities are evident, but the methodology developed by Del Soldato et al. (2017) allows better discriminating the few differences of severity of damage.

In Table 7.3 shows the summary and the classification of Agnone and Volterra sites by Cooper (2008) and Del Soldato et al. (2017). The comparison shows that Del Soldato et al. (2017) approach distributes more edifices in the middle classes due to a better discrimination of the structures in the ranks.

The possible combinations between the classes of damage assessed by the approach developed by Del Soldato et al. (2017) and the satellite parameters extracted by the ENVISAT and COSMO-SkyMed sensors (red line in Fig. 7.3) were plotted in a graph. This was first made for the CL-PO basin in Agnone and then applied, as validation, in the Volterra southwestern sectors. In this section, only the results showing a regression line are shown. Furthermore, the reliability of the found relationship by means of the symmetric matrix was assessed.

In the CL-PO landslide, the damage classes were combined with the ENVISAT and COSMO-SkyMed parameters, according to a symmetric matrix to have as

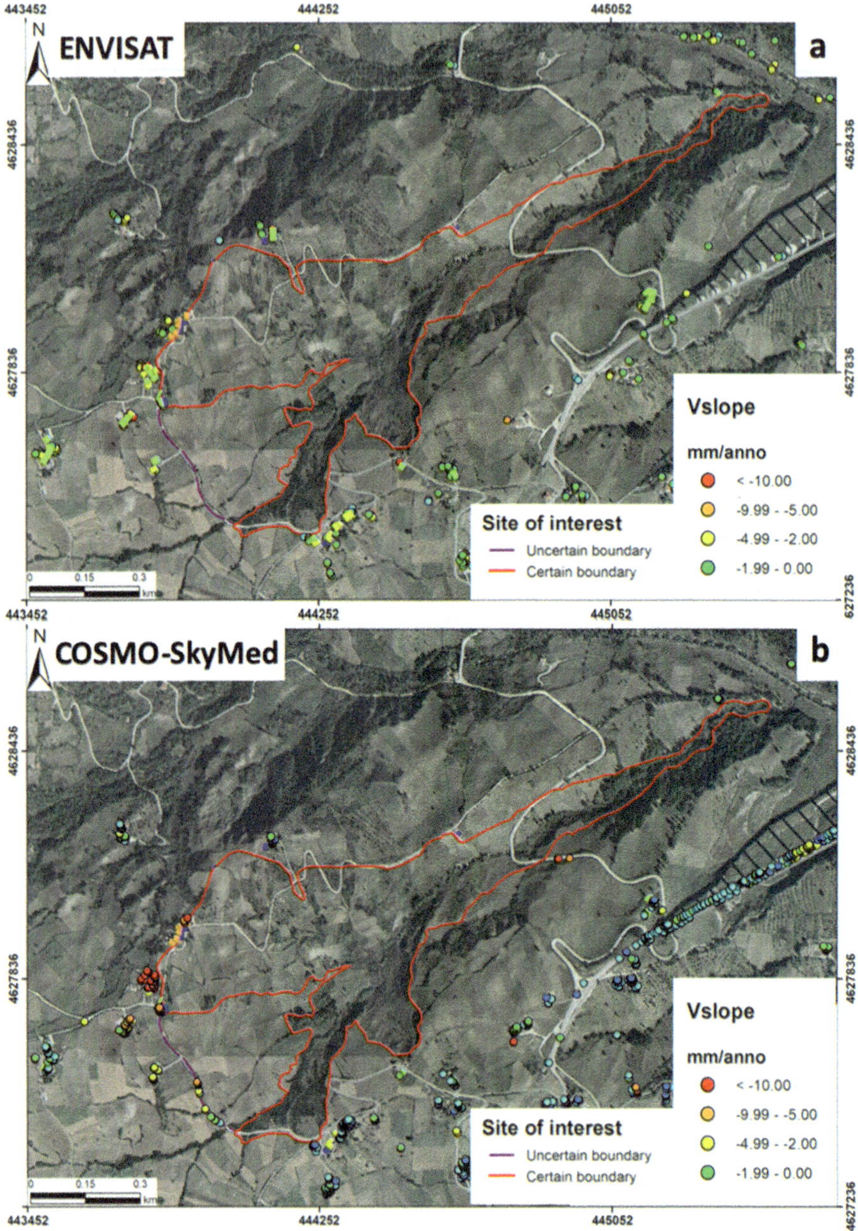

Fig. 7.3 V_{slope} of the CL-PO basin in Agnone calculated for the ENVISAT (a) and COSMO-SkyMed (b) data

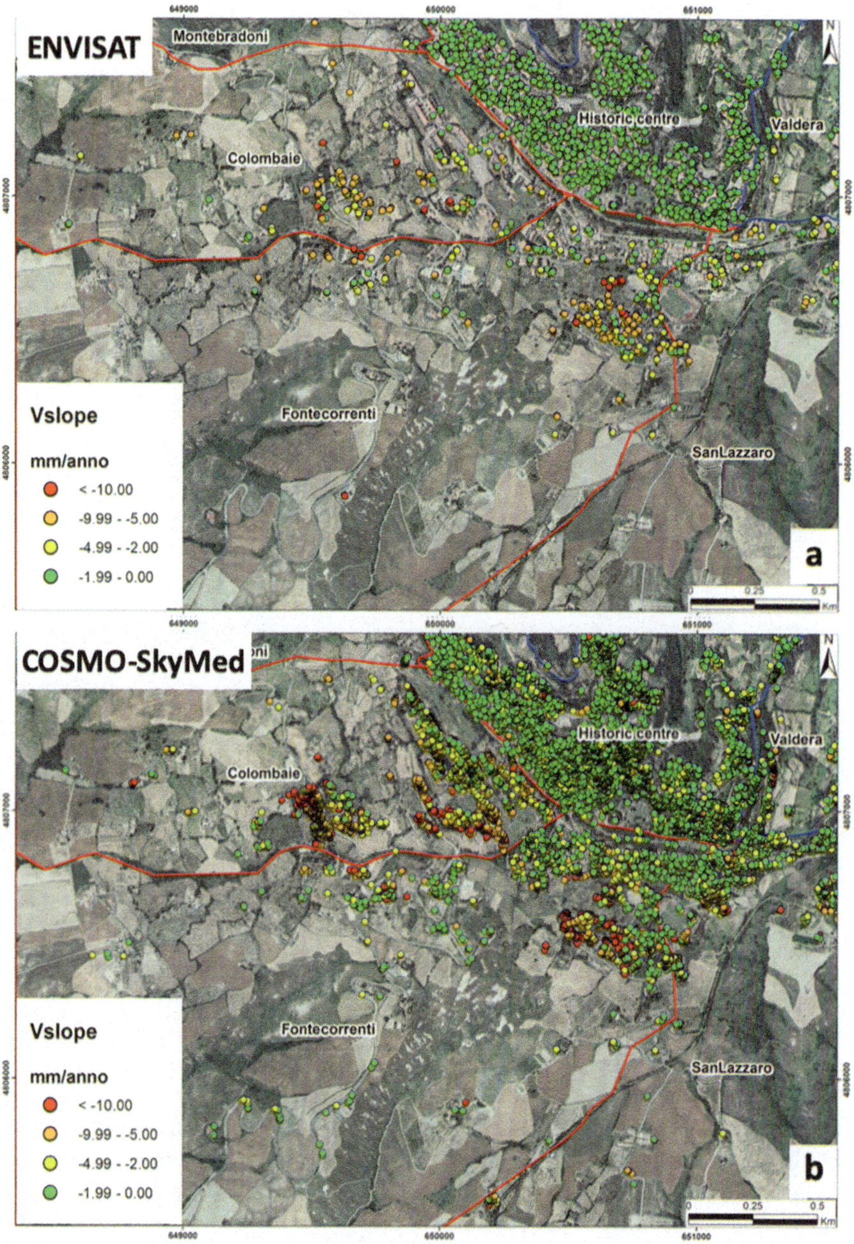

Fig. 7.4 V_{slope} of the southwestern of Volterra calculated for the ENVISAT (**a**) and COSMO-SkyMed (**b**) data

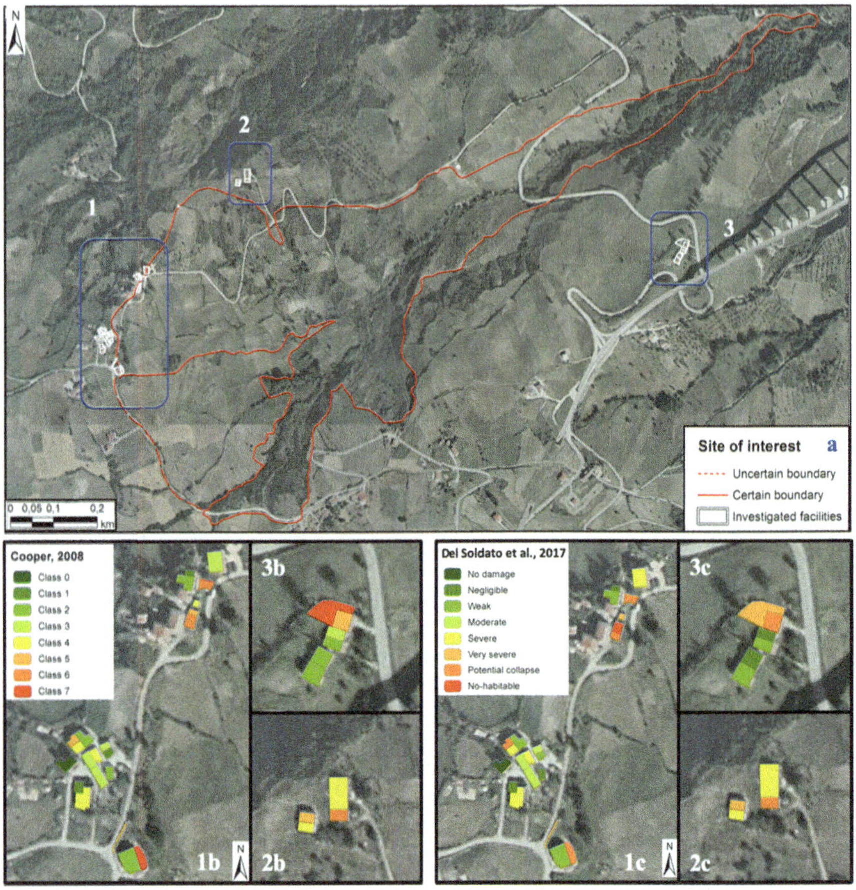

Fig. 7.5 Map of the investigated buildings in the CL-PO basin in Agnone (**a**) and the classification by the Cooper (2008) (**b**) and Del Soldato et al. (2017) approaches (**c**)

results a map of reliability, divided in classes, between the velocity of displacement suffered to an entire edifice and the recorded damage. Some outliers recognizable for the 'Very low reliability' can be identified and justified considering that buildings affected by damage, not only landslide-induced, were taken into consideration. Furthermore, new and restored constructions were considered. It is worth noticing as the better correlation was found for the COSMO-SkyMed images because the damage on the structures were recorded in the similar period.

For the first site of CL-PO landslide area, despite the limited number of buildings, mainly in the crown area, more of the buildings show high rates of reliability (Fig. 7.8a and c). Some exceptions were analysed singularly, e.g. in the crown area, and some structures exhibit low reliability due to no relation between the landslide phenomena and the recorded damage, i.e. rural buildings abandoned, or featured by recent restoration.

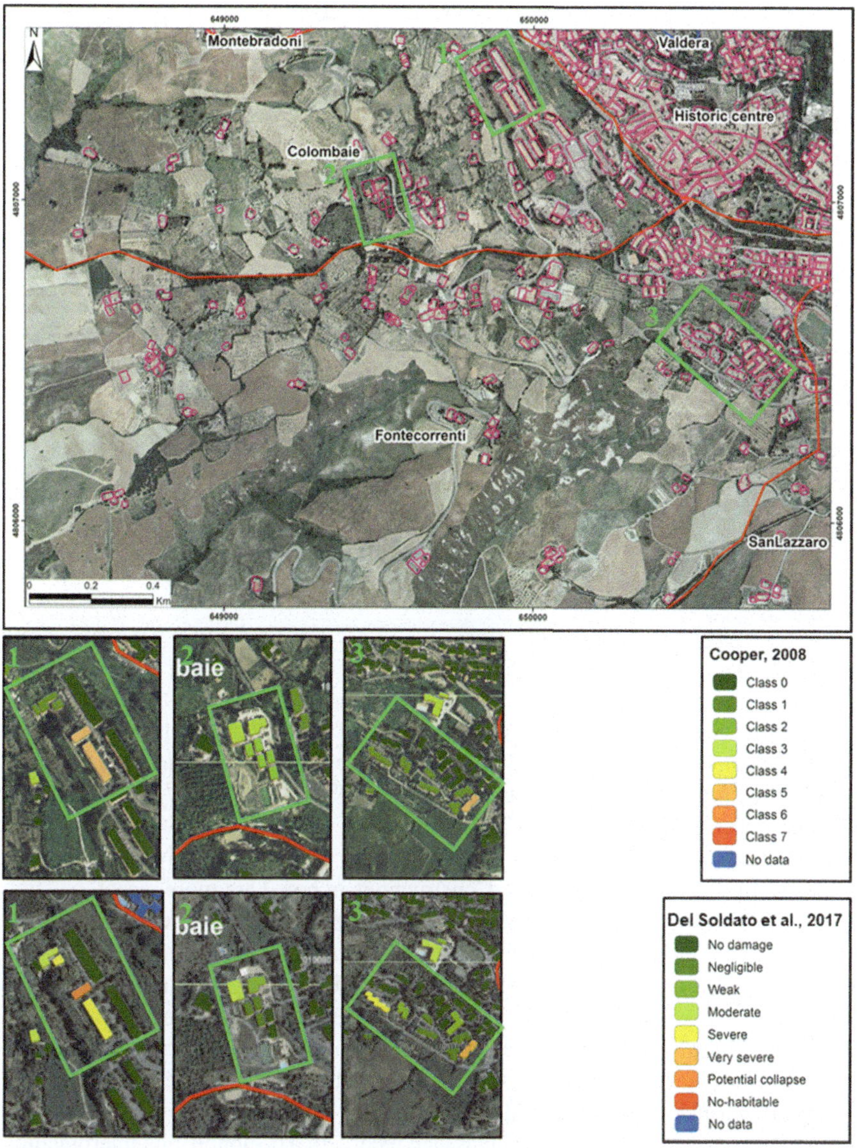

Fig. 7.6 Classification of the buildings of the two southwestern sectors of Volterra by the Cooper (2008) and Del Soldato et al. (2017) approaches

The same correlations were checked and tested for the site of Volterra, confirming the better fitting with some outliers singularly analysed. The ground resolution of the SAR sensors, 3 × 3 m of COSMO-SkyMed with respect to the 28 × 28 m of ENVISAT, their wavelength and the differences between Del Soldato et al. (2017) and Cooper (2008) classification damage adopted have to be considered. High

Table 7.3 Summary and comparison of Cooper (2008) and Del Soldato et al. (2017) approaches of Agnone and Volterra buildings

	Classes	Agnone (number of buildings)	Volterra (number of buildings)
Cooper 2008	0	21	280
	1	3	17
	2	7	11
	3	5	9
	4	8	3
	5	4	3
	6	4	0
	7	4	0
Del Sodlato et al. 2017	No damage	16	288
	Negligible	9	10
	Weak	6	11
	Moderate	4	7
	Severe	7	3
	Very severe	5	4
	Potential collapse	6	0
	Total collapse	3	0

reliability is evident for a large part of the constructions (no investigation of damage was made into the historic centre of Volterra since it was not affected by landslides) testing the same parameters. Considering the COSMO-SkyMed V_{slope} and the Del Soldato et al. (2017) approach, some structures featured by low reliability were highlighted (boxes 2 and 3 in Fig. 7.7). These two areas were deeply investigated by field surveys and ENVISAT data. Constructions in box 2 are more recent, certainly after 2003, and no visible damage is recognizable, except fine cracks and some bumps in the grounds, even if placed in an unstable sector characterized by high velocities. A similar circumstance occurred for the buildings placed in box 3 of Fig. 7.7. Low reliability between COSMO-SkyMed V_{slope} and damage ranks can be justified by analysing the ENVISAT V_{slope}. In this way, good correlation and high reliability were shown allowing to suppose that the main damage were caused by landslide displacement occurred in the period 2003–2010, when high rates of velocity of displacement on this area were recorded. To support this interpretation, no important increase of the damage severity was recorded during the various field campaigns.

Considering the maximum velocities for each class, we can define an envelope from which we can extract a maximum value of velocity related to the classes of damage because no differential settlements in this combination were considered, but

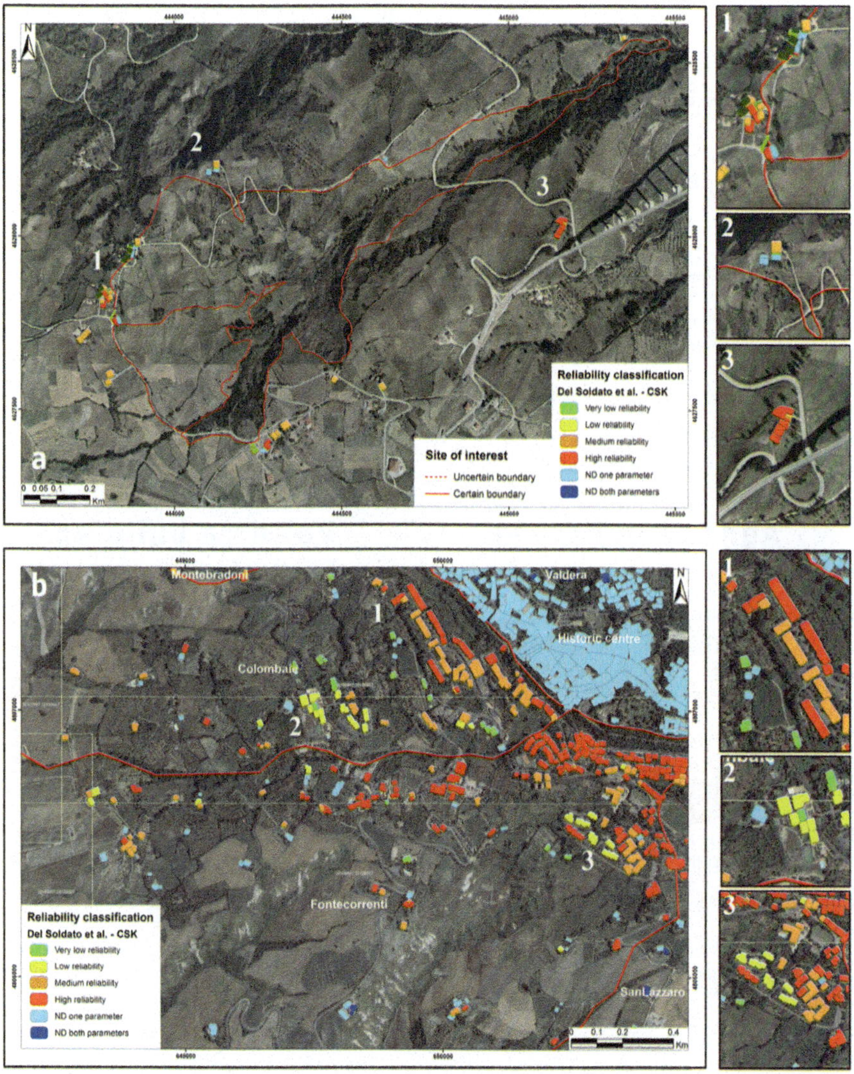

Fig. 7.7 Plot of the velocity re-projected along the slope (V_{slope}) gathered by COSMO-SkyMed data related with the building damage classes recorded by Del Soldato et al. (2017) in the CL-PO area (**a**) in the two southwestern sectors of Volterra in Volterra (**b**)

only velocity of movement affecting the entire structure. The implementation of the differential settlement in the relation between classes of damage and V_{slope} can be a further development of the presented work.

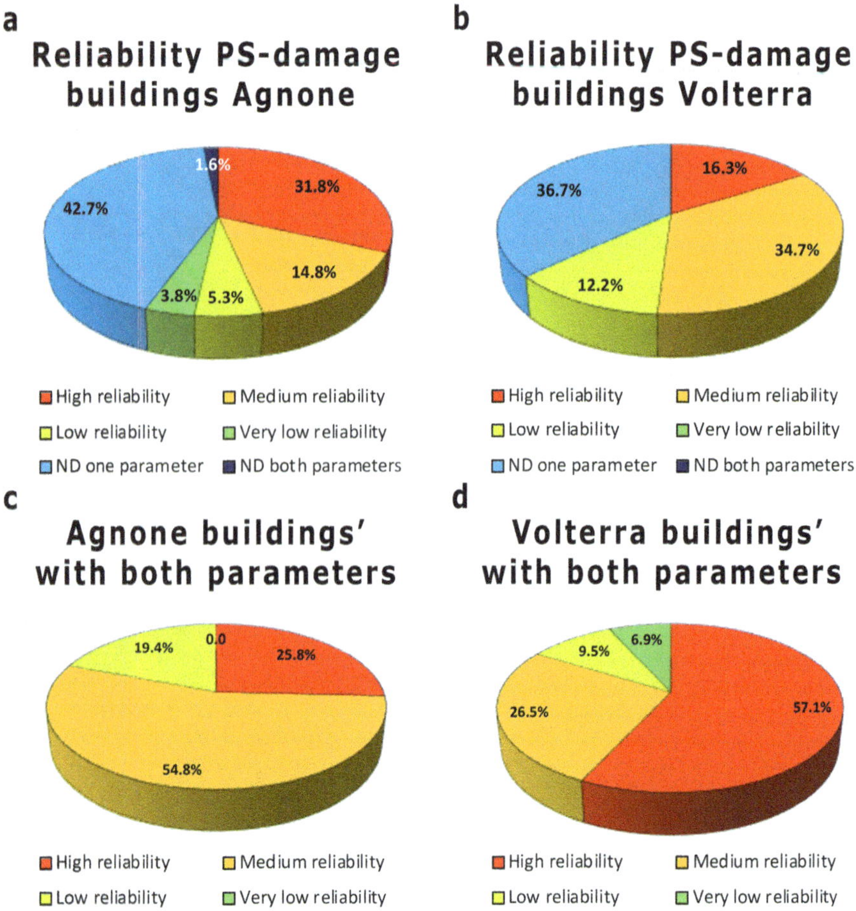

Fig. 7.8 Percentage of the reliability of the PS-damage combination divided per class for both Agnone (**a**) and Volterra (**b**) and considering only the edifices with both parameters (**c** and **d**, respectively)

The percentage of the classes of the reliability PS-damage was plotted in the graph in order to have an idea of the efficiency of the adopted approach (Fig. 7.8). It was applied separately for Agnone (Fig. 7.8a) and Volterra (Fig. 7.8b), and it is possible to extract relevant information about the obtained results:

- Approximatively half of the total buildings are classified with high classes of reliability: 'High reliability' and 'Medium reliability'.
- The 'Low reliability' and 'Very low reliability' are approximatively 10% of the total of the investigated buildings. This percentage is mostly referring to new or restored constructions in areas with relevant velocity of displacement, or to abandoned edifices in stable areas.

- For several structures, one of the two parameters is missing. The PS data can be missing because of geometrical problems (layover, shadowing or foreshortening) or for too low coherence due to several factors, e.g. the presence of diffuse vegetations. Damage information are missing mainly because of difficult access in several areas.

In order to have a better idea of the performance of the presented approach, the same analysis was made considering only the buildings with both parameters, PS velocity and damage classes (Fig. 7.8c and d). The results reveal how in both sites, Agnone, beside the low number of edifices, and the southwestern sectors of Volterra, approximatively the 75% of the structures show a good reliability, 'High reliability' and 'Medium reliability'. It can be assessed as a good result considering that in both areas of interest abandoned, new constructed and restored structures are present.

The continued development on the satellite issue bodes well that in a closed future more structures could be covered with PS velocity data and the number of ND could be decreased.

Furthermore, it can help in further analyses, e.g. in a better assessment of the fragility curves and the quick design of damage probability maps (Del Soldato et al. 2019), with a view to improve the urban and environmental management and for helping the local authorities in the plan of mitigation strategies.

7.5 Conclusions

The challenge of this chapter was to find a rapid method to study a possible relation between ground velocities by remote sensing data and landslide-induced damage on structures and infrastructures. Statistically, the results demonstrate as more than half of structures can be involved in this type of analysis with the available data and as more than 75% of the structures with both parameters show a good reliability of the applied approach.

The investigation of the displacement evolution and induced effect on the structures were conducted by the analysis of ENVISAT and COSMO-SkyMed PS data and by field surveys on the CL-PO landslide of Agnone (Molise region, southern Italy) and on two urban sectors of Volterra (Toscana region, central Italy). The SAR data to which the cinematic parameters (velocity and cumulated displacement of the landslide) were extracted, are referred to November 2002 – May 2015 and February 2003 – January 2015 for Agnone and Volterra, respectively. The landslide-induced damage were identified and classified according to the Cooper (2008) and Del Soldato et al. (2017) approaches by means of several field surveys in 2015 and 2016. Then the ground rates recorded by PS ENVISAT and COSMO-SkyMed re-projected along the local steepest slope direction (V_{slope}) in order to have a more real rate of displacement. These two data, V_{slope} and classes of damage, were combined by means of a symmetrical matrix in order to look for a possible direct relation. This approach was applied first in the Agnone basin and then in Volterra for validating the results obtained in the first test site. To assess the efficiency of the applied approach, the percentage of structures showing good reliability were estimate considering the structures with the lack of at least one parameter, then taking into consideration only the edifices where both parameters, PS velocity and damage classes, were collected. The better correlation, confirmed also by the reliability matrix, was found with the V_{slope} of the COSMO-SkyMed data and the Del Soldato et al. (2017) classification for both sites. The obtained maps show a good potentiality considering that in both case studies, edifices with low reliability were singularly analysed, and they were abandoned or recently restored constructions.

The relationship found needs to be improved with future analyses including, e.g. lithological and geomorphological parameters. In any case, this preliminary result can be useful for local administrators to focus attention on the spatial extension of affected areas and on the possible effects and serious consequences on structures and infrastructures, when wide areas are involved in slope instabilities. In this way, the local administrators can spend economical and technical resources and efforts in order to improve the prevention for further events or for restoring structures more consciously. This combination could be extrapolated in order to extract information for further damaging in case of future recurrences.

Acknowledgements The COSMO-SkyMed images were collected by an ad-hoc 465 project entitled 'Ground deformation monitoring of slow-moving landslides in Agnone (Molise region, Italy) for building damage assessment' carried out using CSK® Products, © ASI (Italian Space Agency), delivered under an ASI license to use.

References

Alexander D (1983) The landslide of 13 December 1982 at Ancona. Central Italy: Report to the Internatinal Disaster Institute, London

Alexander D (1986) Landslide damage to buildings. Environ Geol Water Sci 8:147–151

Almagià R (1910) Studi geografici sulle frane in Italia, vol II. L'Apennino centrale e meridionale. Conclusioni generali. Società Geografica Italiana (IS)

Baggio C, Bernardini A, Colozza R, Corazza L (2009) Manuale per la compilazione della scheda di 1 livello di rilevamento danno, pronto intervento e agibilità per edifici ordinari nell'emergenza post-sismica. Editrice Italiani nel Mondo srl-Roma

Bayer B, Simoni A, Mulas M, Corsini A, Schmidt D (2018) Deformation responses of slow moving landslides to seasonal rainfall in the Northern Apennines, measured by InSAR. Geomorphology 308:293–306

Bazzoffi P, Boscagli A, Brandi G, Busoni E, Calzolari C, Chiarucci A, Chiaverini I, Chisci G, Colica A, De Dominicis V (1997) Badland processes and significance in changing environments. Supplementi di Geografia Fisica e Dinamica del Quaternario 3:151–166

Bianchini S, Ciampalini A, Raspini F, Bardi F, Di Traglia F, Moretti S, Casagli N (2015a) Multitemporal evaluation of landslide movements and impacts on buildings in San Fratello (Italy) by means of C-band and X-band PSI data. Pure Appl Geophys 172:3043–3065

Bianchini S, Pratesi F, Nolesini T, Casagli N (2015b) Building deformation assessment by means of persistent scatterer interferometry analysis on a landslide-affected area: the Volterra (Italy) case study. Remote Sens 7:4678–4701

Bianchini S, Del Soldato M, Solari L, Nolesini T, Pratesi F, Moretti S (2016a) Badland susceptibility assessment in Volterra municipality (Tuscany, Italy) by means of GIS and statistical analysis. Environ Earth Sci 75:1–14

Bianchini S, Pratesi F, Nolesini T, Del Soldato M, Casagli N (2016b) A PSI-based analysis of landslides in the historic town of Volterra (Italy), landslides and engineered slopes. In: Experience, theory and practice: proceedings of the 12th international symposium on landslides (Napoli, Italy, 12–19 June 2016). CRC Press, p 411

Bianchini S, Nolesini T, Del Soldato M, Casagli N (2017) Evaluation of building damages induced by landslides in Volterra area (Italy) through remote sensing techniques, workshop on world landslide forum. Springer, pp 111–120

Blanco-Sanchez P, Mallorquí JJ, Duque S, Monells D (2008) The coherent pixels technique (CPT): an advanced DInSAR technique for nonlinear deformation monitoring. Pure Appl Geophys 165:1167–1193

Boscardin MD, Cording EJ (1989) Building response to excavation-induced settlement. J Geotech Eng 115:1–21

Burland JB (1977) Behavior of foundations and structures on soft ground. In: Proceedings of 9th ICSMFE, pp 495–546

Burland JB, Wroth CP (1974) Settlement of buildings and associated damage. State of the art view. In Conf. on settlement of structures. Cambridge, Pentech Press, London

Calcaterra D, Di Martire D, Ramondini M, Calò F, Parise M (2008) Geotechnical analysis of a complex slope movement in sedimentary successions of the southern Apennines (Molise, Italy). Landslides and Engineered Slopes, pp 299–305

Casagli N, Morelli S, Frodella W, Intrieri E, Tofani V (2018) TXT-tool 2.039-3.2 ground-based remote sensing techniques for landslides mapping, monitoring and early warning, landslide dynamics: ISDR-ICL landslide interactive teaching tools. Springer, pp 255–274

Chiocchio C, Iovine G, Parise M (1997) A proposal for surveying and classifying landslide damage to buildings in urban areas. Eng Geol Environ:553–558

Confuorto P, Di Martire D, Centolanza G, Iglesias R, Mallorqui JJ, Novellino A, Plank S, Ramondini M, Thuro K, Calcaterra D (2017) Post-failure evolution analysis of a rainfall-triggered landslide by multi-temporal interferometry SAR approaches integrated with geotechnical analysis. Remote Sens Environ 188:51–72

Confuorto P, Di Martire D, Infante D, Novellino A, Papa R, Calcaterra D, Ramondini M (2019) Monitoring of remedial works performance on landslide-affected areas through ground- and satellite-based techniques, Catena, vol 178, July 2019, pp 77–89

Cooper AH (2008) The classification, recording, databasing and use of information about building damage caused by subsidence and landslides. Q J Eng Geol Hydrogeol 41:409–424

Costantini M, Falco S, Malvarosa F, Minati F (2008) A new method for identification and analysis of persistent scatterers in series of SAR images. In: Geoscience and remote sensing symposium, 2008. IGARSS 2008. IEEE International. IEEE, pp II-449–II-452

Costantini M, Ferretti A, Minati F, Falco S, Trillo F, Colombo D, Novali F, Malvarosa F, Mammone C, Vecchioli F (2017) Analysis of surface deformations over the whole Italian territory by interferometric processing of ERS, Envisat and COSMO-SkyMed radar data. Remote Sensing of Environment

Crescenzi E, Iovine G, Parise M (1994) Analysis of landslide damage in a village in southern Italy: a preliminary report. Incontro Internazionale dei Giovani Ricercatori in Geologia Applicata, Lausanne Suisse, pp 68–72

Del Soldato M, Bianchini S, Calcaterra D, De Vita P, Martire DD, Tomás R, Casagli N (2017) A new approach for landslide-induced damage assessment. Geomat Nat Haz Risk 8(2):1–14

Del Soldato M, Di Martire D, Bianchini S, Tomás R, De Vita P, Ramondini M, Casagli N, Calcaterra D (2018a) Assessment of landslide-induced damage to structures: the Agnone landslide case study (southern Italy). Bull Eng Geol Environ:1–22

Del Soldato M, Riquelme A, Bianchini S, Tomàs R, Di Martire D, De Vita P, Moretti S, Calcaterra D (2018b) Multisource data integration to investigate one century of evolution for the Agnone landslide (Molise, southern Italy). Landslides 15(11):1–16
Del Soldato M, Solari L, Poggi F, Raspini F, Tomás R, Fanti R, Casagli N (2019) Landslide-induced damage probability estimation coupling InSAR and field survey data by fragility curves. Remote Sens 11(12):1486
Dias P, Catalao J, Marques FO (2018) Sentinel-1 InSAR data applied to surface deformation in Macaronesia (Canaries and Cape Verde). Proc Comput Sci 138:382–387
Esu F (1977) Behaviour of slopes in structurally complex formations. In: Proceedings of the international symposium on the geotechnics of structurally complex formations, pp 292–304
Ferretti A, Prati C, Rocca F (2001) Permanent scatterers in SAR interferometry. IEEE Trans Geosci Remote Sens 39:8–20
Ferretti A, Fumagalli A, Novali F, Prati C, Rocca F, Rucci A (2011) A new algorithm for processing interferometric data-stacks: SqueeSAR. IEEE Trans Geosci Remote Sens 49:3460–3470
Filocamo F, Rosskopf CM, Amato V, Cesarano M, Di Paola G (2015) The integrated exploitation of the geological heritage: a proposal of geotourist itineraries in the Alto Molise area (Italy). Rend Online Soc Geol Ital 33:44–47
Fonstad MA, Dietrich JT, Courville BC, Jensen JL, Carbonneau PE (2013) Topographic structure from motion: a new development in photogrammetric measurement. Earth Surf Process Landf 38:421–430
GEOPROGETTI studio associato company (2010) Indagini geognostiche e sismiche per l'analisi dell'assetto geologico e geomorfologico del versante Sud di Volterra. Report for the Volterra Municipality, http://www.comune.volterra.pi.it
Handbook SE (1975) National coal board. Production Department, London
Herrera G, Mateos RM, García-Davalillo JC, Grandjean G, Poyiadji E, Maftei R, Filipciuc T-C, Auflič MJ, Jež J, Podolszki L (2018) Landslide databases in the geological surveys of Europe. Landslides:1–21
Infante D, Confuorto P, Di Martire D, Ramondini M, Calcaterra D (2016) Use of DInSAR data for multi-level vulnerability assessment of urban settings affected by slow-moving and intermittent landslides. Proc Eng 158:470–475
Infante D, Di Martire D, Confuorto P, Ramondini M, Calcaterra D, Tomàs R, Duro J, Centolanza G (2017) Multi-temporal assessment of building damage on a landslide-affected area by interferometric data. In: Research and Technologies for Society and Industry (RTSI), 2017 IEEE 3rd international forum on (pp 1–6). IEEE
Intrieri E, Raspini F, Fumagalli A, Lu P, Del Conte S, Farina P, Allievi J, Ferretti A, Casagli N (2018) The Maoxian landslide as seen from space: detecting precursors of failure with Sentinel-1 data. Landslides 15:123–133
Iovine G, Parise M (2002) Schema illustrato per la classificazione ed il rilievo dei danni da frana in aree urbane. Mem Soc Geol Ital 57:595–603
Jennings J, Kerrich J (1962) The heaving of buildings and the associated economic consequences with particular reference to the Orange Free State goldfields. Civil Eng S Afr Trans South Afr Inst Civil Eng 4:221–248
Kjekstad O, Highland L (2009) Economic and social impacts of landslides
Lucieer A, de Jong S, Turner D (2013) Mapping landslide displacements using structure from motion (SfM) and image correlation of multi-temporal UAV photography. Prog Phys Geogr 38 (Issue 1):0309133313515293
MacLeod I, Littlejohn G (1974) Discussion on Session 5, In: Conference on settlement of structures, Cambridge. Pentech Press, London, pp 792–795
Mora O, Mallorqui JJ, Broquetas A (2003) Linear and nonlinear terrain deformation maps from a reduced set of interferometric Sar images. IEEE Trans Geosci Remote Sens 41(10):2243–2253. https://doi.org/10.1109/TGRS.2003.814657
Notti D, Herrera G, Bianchini S, Meisina C, García-Davalillo JC, Zucca F (2014) A methodology for improving landslide PSI data analysis. Int J Remote Sens 35:2186–2214

Pascucci V, Merlini S, Martini IP (1999) Seismic stratigraphy of the Miocene-Pleistocene sedimentary basins of the Northern Tyrrhenian Sea and western Tuscany (Italy). Basin Res 11:337–356

Pollefeys M, Koch R, Van Gool L (1999) Self-calibration and metric reconstruction inspite of varying and unknown intrinsic camera parameters. Int J Comput Vis 32:7–25

Pratesi F, Nolesini T, Bianchini S, Leva D, Lombardi L, Fanti R, Casagli N (2015) Early warning GBInSAR-based method for monitoring Volterra (Tuscany, Italy) city walls. IEEE J Sel Top Appl Earth Observ Remote Sens 8:1753–1762

Raspini F, Bianchini S, Ciampalini A, Del Soldato M, Solari L, Novali F, Del Conte S, Rucci A, Ferretti A, Casagli N (2018) Continuous, semi-automatic monitoring of ground deformation using Sentinel-1 satellites. Sci Rep 8:7253

Schuster RL, Highland L (2001) Socioeconomic and environmental impacts of landslides in the western hemisphere. US Department of the Interior, US Geological Survey, Menlo Park

Skempton AW, MacDonald DH (1956) The allowable settlements of buildings. Proc Inst Civ Eng 5:727–768

Solari L, Raspini F, Del Soldato M, Bianchini S, Ciampalini A, Ferrigno F, Tucci S, Casagli N (2018) Satellite radar data for back-analyzing a landslide event: the Ponzano (Central Italy) case study. Landslides:1–10

Tarquini S, Vinci S, Favalli M, Doumaz F, Fornaciai A, Nannipieri L (2012) Release of a 10-m-resolution DEM for the Italian territory: comparison with global-coverage DEMs and anaglyph-mode exploration via the web. Comput Geosci 38:168–170

Tomás R, García-Barba J, Cano M, Sanabria MP, Ivorra S, Duro J, Herrera G (2012) Subsidence damage assessment of a gothic church using differential interferometry and field data. Struct Health Monit 11(6):751–762

Vezzani L, Ghisetti F, Festa A, Follador U (2004) Carta geologica del Molise. SELCA, Firenze

Westoby M, Brasington J, Glasser N, Hambrey M, Reynolds J (2012) 'Structure-from-Motion'photogrammetry: a low-cost, effective tool for geoscience applications. Geomorphology 179:300–314

Zhang Y, Meng X, Jordan C, Novellino A, Dijkstra T, Chen G (2018) Investigating slow-moving landslides in the Zhouqu region of China using InSAR time series. Landslides:1–17

Chapter 8
A Hierarchical Model for the Rocca di Sciara Northeastern Slope Instabilities (Sicily, Italy)

Mario Valiante, Francesca Bozzano, Marta Della Seta, and Domenico Guida

Abstract In this chapter, an object-oriented hierarchical approach for the study of landslides is proposed. The model has been applied on the instabilities that affected the northeastern slope of the Rocca di Sciara relief on the left slope of the Imera river valley in northern Sicily. The last reactivation of the lower portion of this slope dates to April 2015; the event caused severe damages to the road network, also involving the Palermo – Catania highway leading to the failure of the Imera viaduct. The hierarchical model of landslides has been built considering the spatiotemporal relations of all the inventoried landslides along the slope in order to reconstruct the event series; it includes four levels of detail: the focal level composed by the landslides themselves, one level of decomposition storing their components and two levels of generalization. The last generalization defines the Rocca di Sciara landslide system, which groups all the landslides spatially connected and their corresponding evolution phases. This methodological approach for studying nested and superimposed landslide events can be useful for the reconstruction of the landsliding history of a slope and its evolution both in spatial and in temporal terms.

Keywords Landslide · Landslide system · Landslide hierarchy · Object-oriented modelling · Event-based · Multi-temporal inventory · Rocca di Sciara · Scillato

8.1 Introduction

Often landslide processes produce complex arrangements of features along a hill-slope; such complexity, made up by nested and overlapping landslides, results from a sequence of events (Guida et al. 1987, 1988, 2006; Corbi et al. 1996; Lee et al.

M. Valiante (✉) · F. Bozzano · M. Della Seta
Department of Earth Sciences, Sapienza University of Rome, Rome, Italy
e-mail: mario.valiante@uniroma1.it

D. Guida
Department of Civil Engineering, University of Salerno, Fisciano, Italy

M. De Maio, A. K. Tiwari (eds.), *Applied Geology*,
https://doi.org/10.1007/978-3-030-43953-8_8

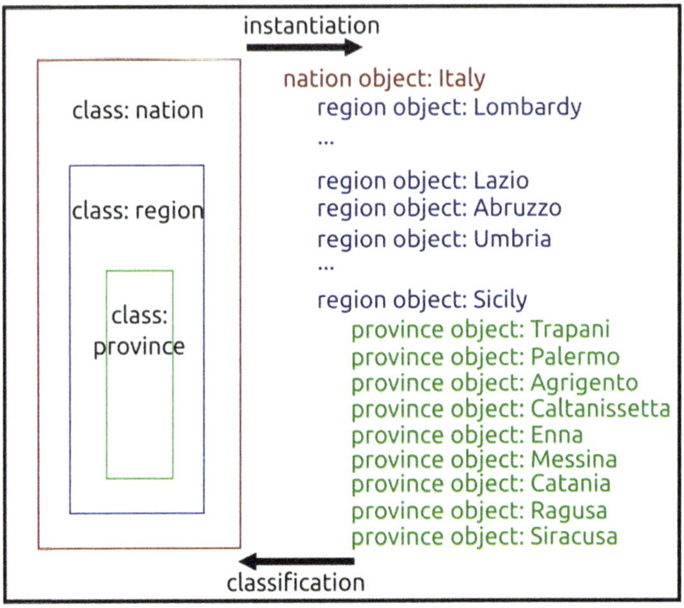

Fig. 8.1 Class – object relation. (Modified after Egenhofer and Frank 1992)

2001; Parise 2003; Valiante et al. 2016; Samia et al. 2017a, b). On the other hand, even dealing with a single landslide implies to face its inner complexity, i.e. describe its components (Parise 1994, 2003; Fiorillo et al. 2001; Niethammer et al. 2012; Tarolli et al. 2012; Dufresne et al. 2016; Morelli et al. 2018). In this framework, an object-oriented modelling could come to hand in the design of landslide inventories.

The object-oriented data model is based on the assumption that any entity can be represented by exactly one object regardless of its complexity or inner structure (Egenhofer and Frank 1992). Complexity is then handled through the classification process: any real-world feature and its behaviour can be described with a class definition, containing its attributes and the needed functions (called *methods*) to manipulate them, then any operation of simplification or generalization can be performed defining a set of sub-classes and super-classes. As described in Fig. 8.1 for different application, any feature described by a class definition is an object (an instance of that specific class); simplifying, a class is the description of a feature and its behaviour, while an object is the feature itself (Chaudhri 1993; Atkinson et al. 1995; Kösters et al. 1997).

The described classification process results in a set of classes linked by parent–child relations (generalized and specialized classes) and sibling relations (classes sharing a common super-class) in a hierarchical structure. Hierarchies are usually exploited to model and therefore decrease complexity of natural systems; in this perspective, a hierarchy is defined as a multi-level or layered system where each level can be decomposed in a number of interrelated subsystems until a non-decomposable elementary subsystem is defined (Simon 1962, 1995; Odum

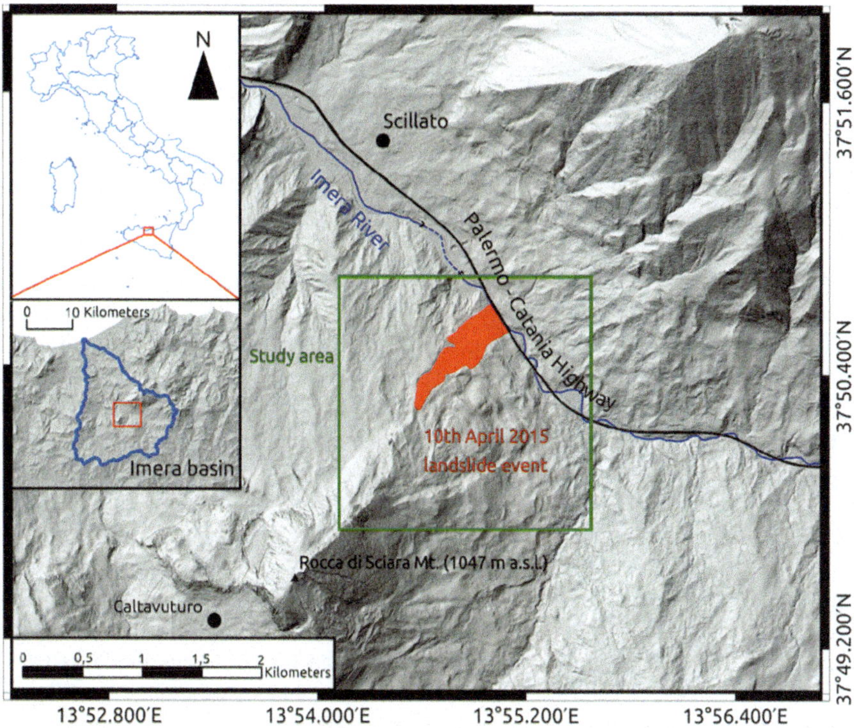

Fig. 8.2 Location of the study area. (Reference system WGS84, EPSG 4326)

and Barrett 2005; Wu 2013). Depending on the objectives of a particular study or analysis, the hierarchical level closer to the study object is called *focal level* or *level 0*, which sets the starting point for decompositions (*levels -x*) or generalizations (*levels + x*) (Wu 1999).

In this work, an object-oriented approach for the study of landslides will be presented. The model has been applied on a selected case study in Sicily (Italy) located on the northeastern slope of the Rocca di Sciara relief (Fig. 8.2). The attention to this site was driven by the landslide event occurred on 10th April, 2015, which involved the provincial road linking the Scillato and Caltavuturo villages and, mainly, the A19 Palermo-Catania Highway, producing the collapse of the Imera viaduct and the consequent highway interruption.

The proposed object-oriented hierarchical approach aims: (1) to the improvement of the existing inventories, defining a tool capable of storing both spatial and temporal relations between landslides in a single data set, avoiding data fragmentation; (2) to define a model for the description of complex arrangements of landslides and their evolution.

8.2 Methodology

The landslide inventory has been built collecting the available data such as previous inventories, both local (PAI – Northern Imera River Basin Plan) and national (IFFI Project), through the interpretation of a multi-temporal series of aerial photographs and by means of field survey (May 2017). After these operations, the collected features have been stored in the database following an object-oriented approach.

In the proposed model, the main (focal) classes represent single landslides as a whole, differentiated by the type of movement (Hungr et al. 2014), so that every landslide object within the inventory would be an instance of its corresponding class. Starting from this main level, both generalization (or aggregation) and decomposition operations can be performed.

Increasing the detail, each focal class can be decomposed in a set of sub-classes describing landslide components such as detachment area, transit zone, deposit zone, etc. The other way around, the aggregation of such component objects results in a landslide object. Otherwise, a first level of generalization defines an order of super-classes, here defined as *landslide complexes*. This set of super-classes is made up from the aggregation of all the landslides of the same type on a topological basis: given that at least two landslides share the movement type, the minimum requirement to form a landslide complex is that these two landslide objects have to be spatially connected; referring to the Region Connection Calculus model (RCC8) (Randell et al. 1992; Cohn and Bennett 1997), any relation is valid other than *Equals* (two landslides cannot be spatially coincident), *Externally Connected* (this could be considered a transitional phase from disconnected landslides to a landslide complex) and, of course, *Disconnected* (Fig. 8.3).

If a landslide complex overlaps with another landslide complex of different type or with a landslide with type of movement different from the ones making up the landslide complex itself, a further generalization can be performed, defining the *landslide system* class. A landslide system is thus defined as a group of heterogeneous landslides spatially connected (Fig. 8.4).

The object-oriented approach is not focused on the data themselves but on the model describing those data. In this perspective, classification means not only to describe real-world features but also to categorize their mutual relations.

In a static condition, just considering spatial arrangements, the hierarchical model proposed contains both vertical and horizontal relations. The vertical structure is defined by the asymmetrical *whole – part* relations among the different levels of the hierarchical system: a higher-level object cannot exist without its lower-level components. Horizontal relations describe how objects interact on the same level; they can be both symmetrical and asymmetrical. For instance, considering a debris flow object components, the source area *is connected* to the deposit area by a channel, hence the channel *connects* source area and deposit area, while the channel *touches* the source area and vice versa (*externally connected* relation referring to the RCC8 model). As described before, at the focal level horizontal relations are crucial for the

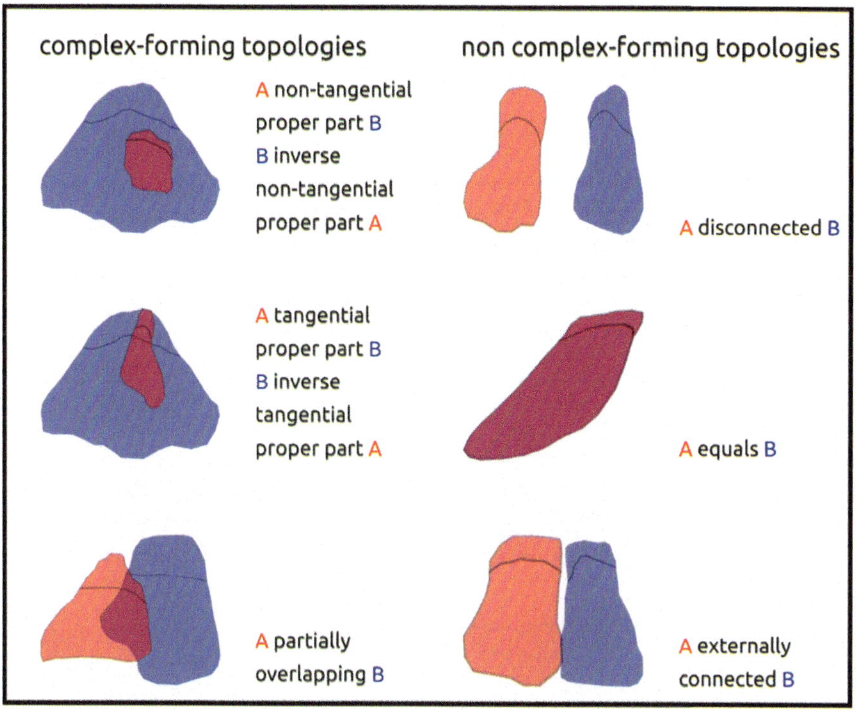

Fig. 8.3 RCC8 model for landslides

existence of the higher levels, which are defined after the analysis of the topological relations among the objects of such level.

8.3 The Rocca di Sciara Case Study

The Rocca di Sciara relief is located in southern Italy (Northern Sicily), along the left slope of the Imera river valley, close to the Caltavuturo village (PA). Several landslide processes have affected this area, and their features are still evident in the area of interest and on the slopes around the Imera basin. The last event, occurred on the northeastern slope on April 10th, 2015, had a great impact for the transportation network, as it severely damaged an important connection road and a highway.

In the investigated area, the Imerese sequence partially crops out (Fig. 8.5). Starting from the top of the Rocca di Sciara relief towards the Imera River valley, it is possible to identify (according to Catalano et al. 2011): (i) Crisanti Fm. (upper Jurassic – upper Cretaceous), limestone megabreccias and thick-bedded calcarenites, outcropping in the upper section of the slope; (ii) Caltavuturo Fm. (Eocene – lower Oligocene), reddish to greyish thin-bedded or laminated calcilutites, marly limestone ad clayey marls, outcropping in the middle section of the slope mainly with folded

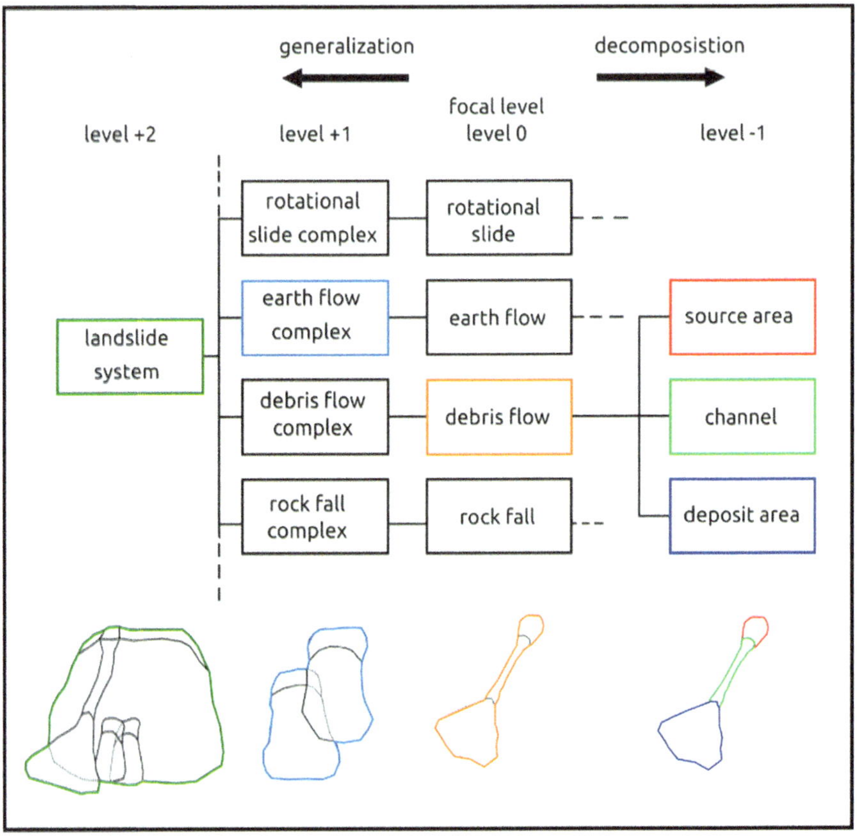

Fig. 8.4 Hierarchical model for landslides. (Modified after Wu 1999)

structures, often covered by boulders from the Crisanti Fm.; (iii) Numidian Flysch – Portella Colla member (upper Oligocene – lower Miocene), laminated brownish claystones with quartz sandstones intercalations. In the lower section of the slope, this formation is often involved in landslides, and its presence is inferred from boreholes log-stratigraphies (Fig. 8.5).

The present morphology of the northeastern slope results from the sum of several landslide events. At the bottom of the carbonate slopes, in the highest part of the slope, extensive limestone boulders deposits can be found; these deposits are interpreted as the result of impulsive events (rock avalanches), as in some areas they are found covering old soils. The middle section of the slope shows alternation

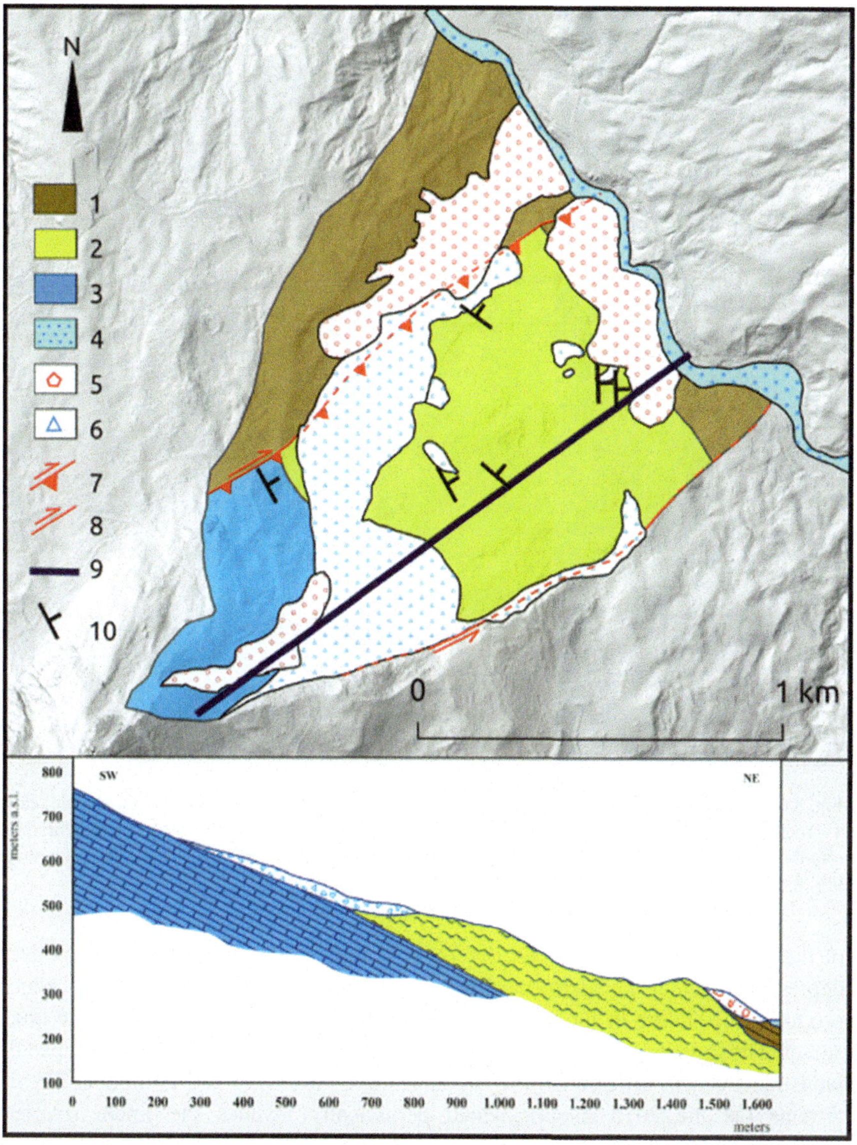

Fig. 8.5 Geological sketch and cross section of the study area: 1 – Numidian Flysch; 2 – Caltavuturo Fm.; 3 – Crisanti Fm.; 4 – alluvial deposits; 5 – landslide debris; 6 – boulder deposits; 7 – transpressional fault, inferred if dashed; 8 – strike-slip fault, inferred if dashed; 9 – geological cross section; 10 – bedding

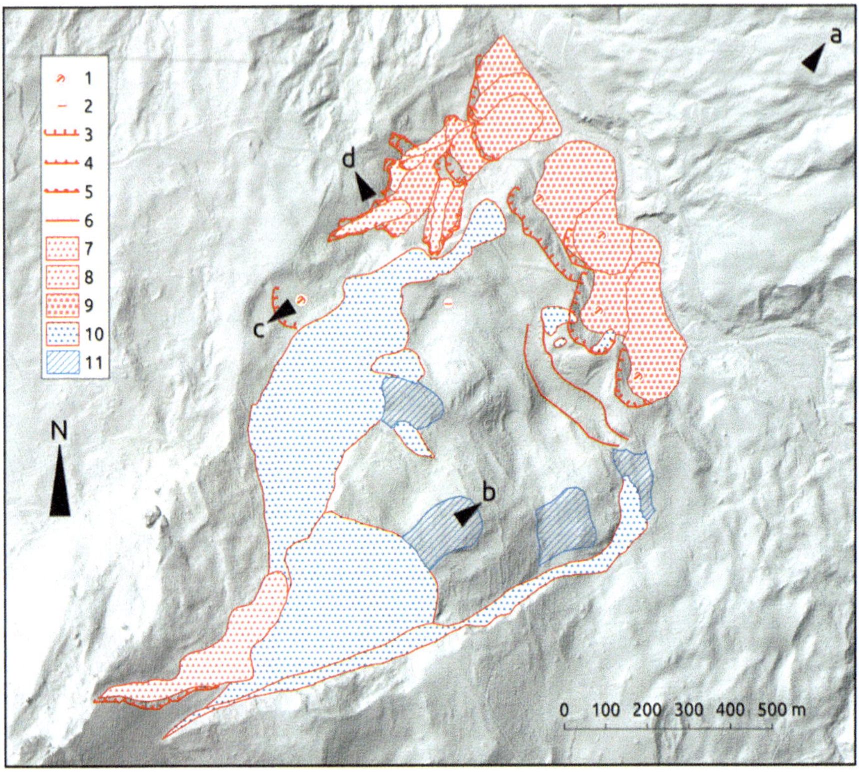

Fig. 8.6 Geomorphological evidences map. 1 – counterslope; 2 – hollow; 3 – rotational slide scarp; 4 – rockfall scarp; 5 – earthflow scarp; 6 – trench scarp; 7 – rockfall debris; 8 – earthflow debris; 9 – rotational slide body; 10 – boulder deposits; 11 – flat slope; a-d - points of view for pictures in Fig. 8.7

of flat surfaces or counterslopes, divergent-straight slopes and several meters deep depressions; all these landforms suggest an ancient gravitational activity which produced the sliding of the upper terrigenous formations over the underlying limestones. The lowest section of the slope, along the Imera River, is presently shaped by landslide events, with shallow rotational slides involving the Numidian Flysch terrains (as the 2015 event), which periodically invades the Imera riverbed (Figs. 8.6 and 8.7).

8.4 Hierarchical Model of the Rocca di Sciara Landslide System

All the landslide-related features collected have allowed the implementation of the object-oriented model, therefore the reconstruction of hierarchies.

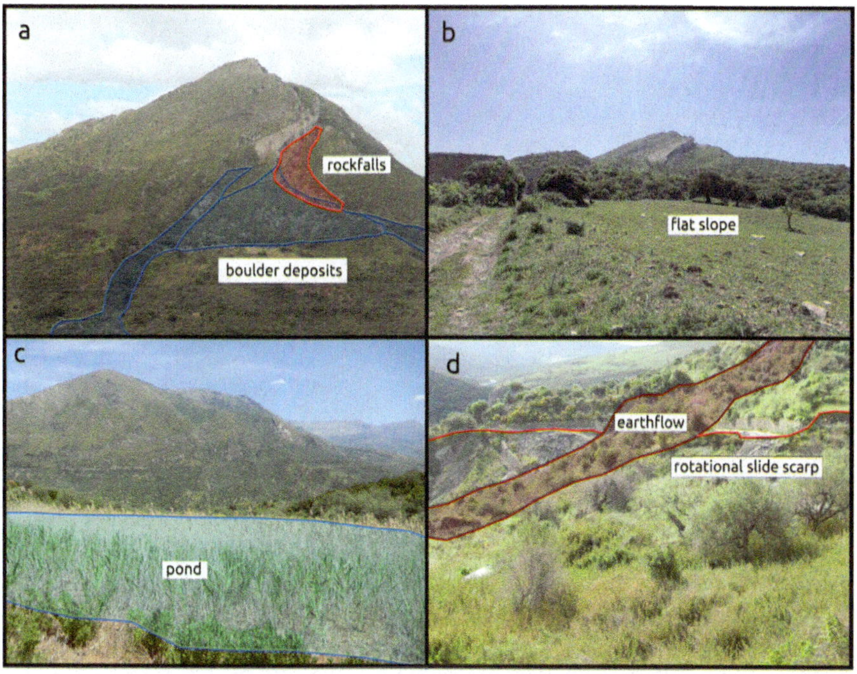

Fig. 8.7 Details of geomorphological features mapped in Fig. 8.6, letters from a to d refer to points of view also represented in Fig. 8.6

The focal class (level 0) contains all the inventoried landslides after the aggregation of the stored components (level −1) derived from the evidences collected in the field and by previous data analysis. In the lower portion of the slope, there are two clusters of rotational slides, which can be aggregated in two rotational slide complexes (level +1), the northern one contains the 2005 and 2015 rotational slides and interacts with some peripheral earth flows also related to the 2015 event. The southern rotational slide complex is developed above a spread-like movement. The three rock avalanche deposits in the upper portion of the slope can be also aggregated in a rock avalanche complex and on the scarps left by these phenomena currently occur rock falls. All these objects lay above the rock slope deformation. From the aggregation of all the mentioned objects results the Rocca di Sciara landslide system (level + 2) (Fig. 8.8).

8.5 Discussions

Usually spatial data sets are not allowed to store overlapping features, treating such an occurrence as a topological error (ESRI 2019); despite this is extremely helpful for various applications, dealing with landslides it could be a limitation. Representing nested or overlapping landslides as a coverage of adjacent tiles

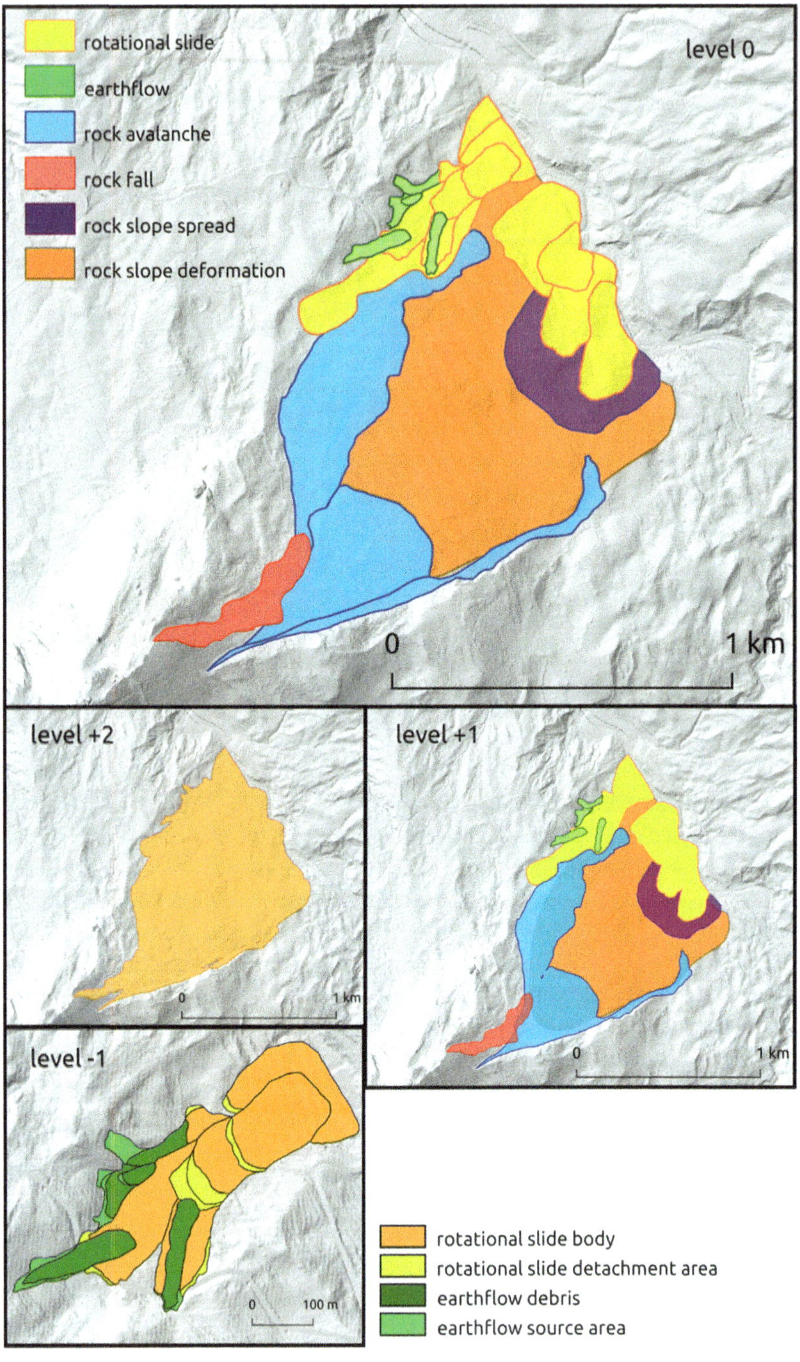

Fig. 8.8 Hierarchical model of the Rocca di Sciara landslide system represented at the level +2. Level −1 is focused on the 2015 event only

produce an inconsistency between the topology of the data and the topology of the real-world features causing a loss of information. In this object-oriented approach, the landslide inventory is built preserving spatial relations so that every inventoried landslide is stored entirely, and its topological connections can be easily retrieved through queries. Overlapping and nested landslides are then handled with the hierarchical classification, which lead to the simplification of a complex and heterogeneous gravitational history of a hillslope into one landslide system object.

Associating a timing to every landslide object, absolute or relative, allows also the definition of temporal relations within the landslide system, which leads to reconstruction of the events that affected the area of interest. Referring to the Rocca di Sciara landslide system, the oldest event recognized is the rock slope deformation, which presumably caused the sliding of the Caltavuturo Fm. deposits over the limestones of the Crisanti Fm. The upper portion of this movement has been then covered by the occurrence of the three rock avalanches. After these major events, the landslide system evolved with shallower phenomena such as rock falls in the upper section and rotational slides – earth flows in the lower portion favoured by the outcropping of the Numidian Flysch with a dip-slope bedding (Fig. 8.9).

Combining spatial and temporal storing capabilities, a landslide system can efficiently contain countless reactivation of the same landslide object, useful to perform frequency analysis or to investigate the development of that movement and to evaluate its evolution. Studying the history of every landslide within a landslide system allows also to reconstruct their interactions and to infer how one movement can influence the others.

8.6 Conclusions

A hierarchical model based on an object-oriented approach has been proposed for the study of landslides. The focal level of the hierarchical structure is built collecting all the landslides affecting or that affected the area of interest. A level of decomposition derives from the inventory of landslide components, while two levels of generalization are extracted based on the spatial relations between landslides: landslide complexes result from the aggregation of landslides with the same type of movement partially overlapping, while landslide systems are defined grouping all the connected landslide complexes and the landslides not forming a complex interacting with them.

The model has been applied on the northeastern slope of the Rocca di Sciara relief where a recent landslide event severely damaged the national road network. Geological and geomorphological settings of the slope allow the development of a complex and heterogeneous landsliding sequence which has been encapsulated in the definition of the Rocca di Sciara landslide system.

A landslide system built with an accurate spatio-temporal inventorying of landslides should be a tool for the fast retrieval of useful information such as how many events affected a slope and how they developed, their magnitude and frequency and

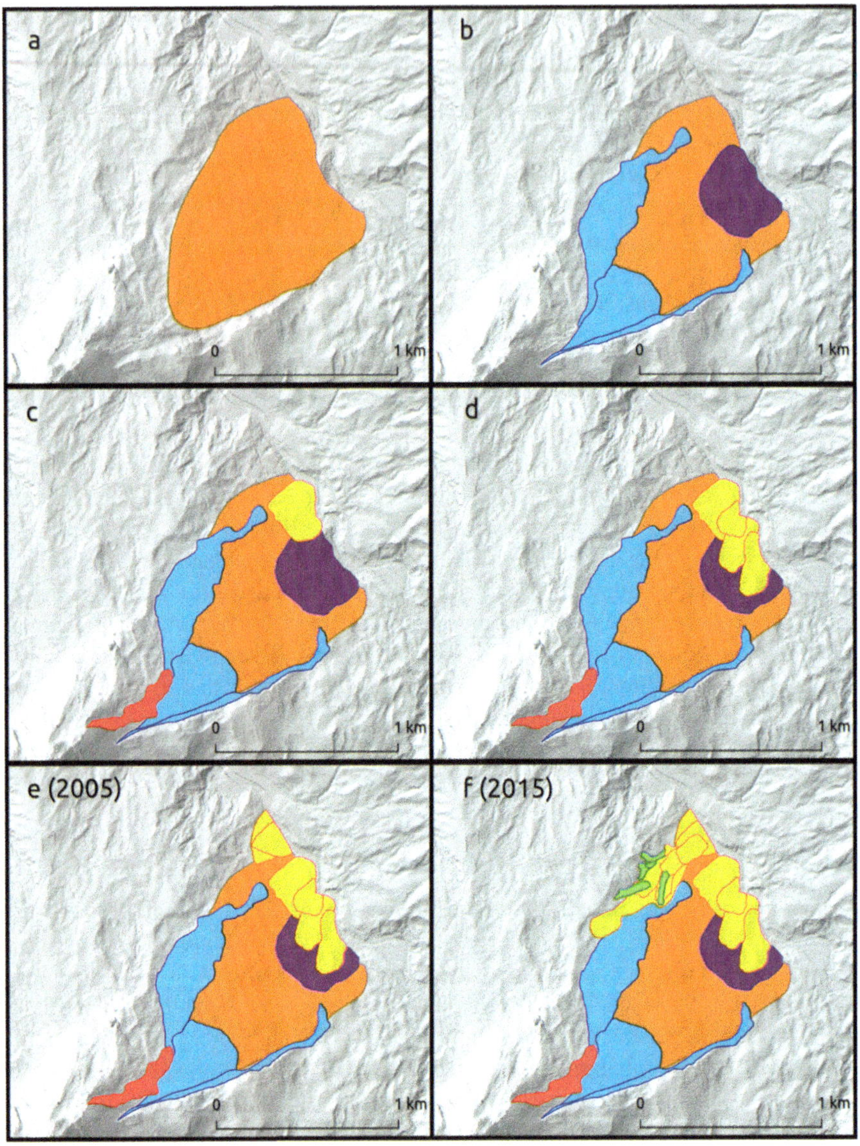

Fig. 8.9 Event series for the Rocca di Sciara landslide system. Please note that steps **b**, **c** and **d** are cumulated time steps and do not refer to just three events

how they interacted. All these data regarding the past and present activity of a slope are the assumption for understanding its most likely evolution, thus, to contribute to the formulation of reactivation scenarios.

References

Atkinson M, Bancilhon F, DeWitt D et al (1995) The object-oriented database system manifesto. In: Web page. http://www.cs.cmu.edu/afs/cs.cmu.edu/user/clamen/OODBMS/Manifesto/htManifesto/Manifesto.html

Catalano R, Avellone G, Basilone L et al (2011) foglio 609–596 TERMINI IMERESE – CAPO PLAIA. NOTE Illus. della Cart. Geol. D'ITALIA alla scala 150.000 224

Chaudhri AB (1993) Object database management systems: an overview. BCS OOPS Newsl 18:1–28

Cohn AG, Bennett B (1997) Qualitative spatial representation and reasoning with the region connection Calculus. GeoInformatica 1:1–44

Corbi I, Guida M, Tetamo G, Vallario A (1996) Considerazioni sul rischio a franare di aree campione nel bacino del Fiume Biferno (Molise). Mem Soc Geol Ital 51:1087–1100

Dufresne A, Bösmeier A, Prager C (2016) Sedimentology of rock avalanche deposits – case study and review. Earth Sci Rev 163:234–259. https://doi.org/10.1016/j.earscirev.2016.10.002

Egenhofer MJ, Frank AU (1992) Object-oriented modeling for GIS. URISA J 4:3–19

ESRI (2019) Topology in ArcGIS. http://desktop.arcgis.com/en/arcmap/latest/manage-data/topologies/topology-in-arcgis.htm. Accessed 20 Feb 2019

Fiorillo F, Guadagno FM, Aquino S, De Blasio A (2001) The December 1999 Cervinara landslides: further debris flows in the pyroclastic deposits of Campania (southern Italy). Bull Eng Geol Environ 60:171–184. https://doi.org/10.1007/s100640000093

Guida D, Guida M, Perriello Zampelli S et al (1987) Deformazioni gravitative al margine di morfostrutture carbonatiche: un esempio nel Monte Bulgheria (Campania). Mem Soc Geol Ital 37:363–373

Guida D, Guida M, Iaccarino G et al (1988) Tipologia e diffusione delle deformazioni gravitative profonde di versante nel settore meridionale dell'Appennino campano-lucano in relazione all'assetto stratigrafico-strutturale. Mem Soc Geol Ital 41:889–896

Guida D, Nocera N, Siervo V (2006) Analisi morfoevolutiva sulla riattivazione di sistemi franosi a cinematismo intermittente in Appennino campano-lucano (Italia meridionale). G Di Geol Appl 3:114–122. https://doi.org/10.1474/GGA.2006-03.0-15.0108

Hungr O, Leroueil S, Picarelli L (2014) The Varnes classification of landslide types, an update. Landslides 11:167–194. https://doi.org/10.1007/s10346-013-0436-y

Kösters G, Pagel BU, Six H-W (1997) GIS-application development with GEOOOA. Int J Geogr Inf Sci 11:307–335. https://doi.org/10.1080/136588197242293

Lee EM, Hall JM, Meadowcroft IC (2001) Coastal cliff recession: the use of probabilistic prediction methods. Geomorphology 40:239–269

Morelli S, Pazzi V, Frodella W, Fanti R (2018) Kinematic reconstruction of a deep-seated gravitational slope deformation by geomorphic analyses. Geosciences 8:26. https://doi.org/10.3390/geosciences8010026

Niethammer U, James MR, Rothmund S et al (2012) UAV-based remote sensing of the Super-Sauze landslide: evaluation and results. Eng Geol 128:2–11

Odum EP, Barrett GW (2005) Fundamentals of Ecology. Fifth edition. Thomson - Brooks/Cole. 624 pp

Parise M (1994) Il Rilevamento Geomorfologico a Grande Scala Nell'Interpretazione Della Cinematica Dei Movimenti Gravitativi. Geol Romana 30:273–282

Parise M (2003) Observation of surface features on an active landslide, and implications for understanding its history of movement. Nat Hazards Earth Syst Sci 3:569–580

Randell DA, Cui Z, Cohn AG (1992) A spatial logic based on regions and connection. In: 3rd International conference on knowledge representation and reasoning. Morgan Kaufmann, p 12

Samia J, Temme A, Bregt A et al (2017a) Characterization and quantification of path dependency in landslide susceptibility. Geomorphology 292:16–24. https://doi.org/10.1016/j.geomorph.2017.04.039

Samia J, Temme A, Bregt A et al (2017b) Do landslides follow landslides ? Insights in path dependency from a multi-temporal landslide inventory. Landslides 14:547–558. https://doi.org/10.1007/s10346-016-0739-x

Simon HA (1962) The architecture of complexity. Proc Am Philos Soc 106:467–482. https://doi.org/10.1080/02841850903061437

Simon HA (1995) Near decomposability and complexity: how a mind resides in a brain. In: Morowitz HJ, Singer JL (eds) The mind, the brain, and complex adaptative systems. Addison-Wesley, Reading, pp 25–44

Tarolli P, Sofia G, Dalla Fontana G (2012) Geomorphic features extraction from high-resolution topography: landslide crowns and bank erosio. Nat Hazards 61:65–83. https://doi.org/10.1007/s11069-010-9695-2

Valiante M, Bozzano F, Guida D (2016) The Sant'Andrea-Molinello landslide system (Mt. Pruno, Roscigno, Italy). Rend Online Soc Geol Ital 41:214–217. https://doi.org/10.3301/ROL.2016.132

Wu J (1999) Hierarchy and scaling: extrapolating information along a scaling ladder. Can J Remote Sens 5:367–380

Wu J (2013) Hierarchy theory: an overview. In: Rozzi R, Pickett S, Palmer C et al (eds) Linking ecology and ethics for a changing world: values, Philosophy, and actions, ecology and ethics. Springer, Dordrecht

Part III
Geomechanics

Chapter 9
Comparing Direct and Indirect Methods to Estimate Uniaxial Compressive Strength of Rocks Belonging to the Dolomites Sequence (NE Italian Alps)

Elia Longo, Ennio Chiesurin, and Mario Floris

Abstract The main rock mass classifications consider uniaxial compressive strength (UCS) as a fundamental input parameter. The estimation of UCS in geomechanical field surveys is often carried out by Schmidt Hammer tests, so that it is very important to evaluate the accuracy and precision of the results derived from the application of this method. To this end, some of the most representative formations belonging to the Dolomites sequence (NE Italian Alps) were considered to carry out a comparison between in situ and laboratory tests (point load and uniaxial tests). The great lithological variability of the sequence allowed to include in the analysis different lithotypes: chalks, rhyolites, basalts, limestones, marly limestones, dolomites and sandstones. UCS values derived from the different methods were compared to analyse the correspondence of the results for each lithotype. A particular attention was paid to UCS values calculated using R index of Schmidt Hammer to estimate the degree of reliability of the method. The findings of this study provide a large database of compressive strength values for the lithotypes belonging to the Dolomites sequence and can help scientists and professionals to better understand the limits in using Schmidt Hammer method for geomechanical rock mass classification.

Keywords UCS · Schmidt hammer · Laboratory tests · Rock mass classification · Dolomites sequence · Italian Alps

---

E. Longo · M. Floris (✉)
Department of Geosciences, University of Padua, Padua, Italy
e-mail: mario.floris@unipd.it

E. Chiesurin
Geo-engineering studio, Feltre, Italy

M. De Maio, A. K. Tiwari (eds.), *Applied Geology*,
https://doi.org/10.1007/978-3-030-43953-8_9

9.1 Introduction

Uniaxial compressive strength (UCS) is a fundamental mechanical property to assess the quality and strength of rock masses, so it is of primary importance in rock mechanics and rock engineering design (Dehghan et al. 2010; Liu et al. 2015). The main rock mass classifications, such as rock mass rating (Bieniawski 1989), rock mass index (Palmström 1995) and slope mass rating (Romana 1985), and the evaluation of jointed rock masses strength through Hoek-Brown failure criterion (Hoek and Brown 1980) require the estimation of UCS.

As well known, UCS can be estimated through laboratory tests, such as point load test (PLT) and unconfined compression test (UCT), or in situ through Schmidt Hammer (SH) test. UCT directly measures UCS, while PLT and SH are indirect methods because the resulting Is and R indexes from these tests provide UCS estimation by empirical correlations (ISRM 1985; Thuro and Plinninger 2001; ASTM 2002; Aydin and Basu 2005). UCT is time-consuming and expensive, it is not easy to collect and prepare samples for the testing (Lee et al. 2016), so that few data can be obtained, which often are not sufficiently representative of rock strength properties. For these reasons, UCT is rarely performed especially in the professional field, while PLT and SH are commonly used to estimate UCS.

The main aim of this work is to compare the results between direct and indirect methods to estimate UCS and to verify the reliability of PLT and especially SH test, the most used technique in research and professional fields. Samples from the most representative formations belonging to the Dolomites sequence in the northeastern Italian Alps were collected and tested. Lithotypes difficult to sample due to high fracturation and anisotropy, such as phyllites of the crystalline basement, and soft rocks with UCS < 20 MPa were excluded. On the outcrops of the selected lithotypes, R index was measured through SH tests, and samples for PLT and UCT testing were collected. UCS values have been derived by applying several empirical correlations proposed in the scientific literature and finally compared to those measured by UCT calculating two statistics: variance accounted for (VAF) and root mean square error (RMSE).

9.2 Study Area

Dolomites area is located at the border between Belluno (Veneto Region) and Trento (Trentino-Alto Adige region) provinces in the northeastern Italian Alps (Fig. 9.1).

From a geological point of view, Dolomites belong to the eastern sector of South Alpine, between Periadriatic lineament and *Valsugana* fault. They form a large syncline linked to a Neogene pop-up delimited by Valsugana fault and its back-thrusts to the south and by *Funes* and *Passo delle Erbe* lineaments to the north (Doglioni 1987). The stratigraphic sequence of the Dolomites mainly consists of Triassic and Permian rocks, locally the Paleozoic crystalline basement outcrops. The

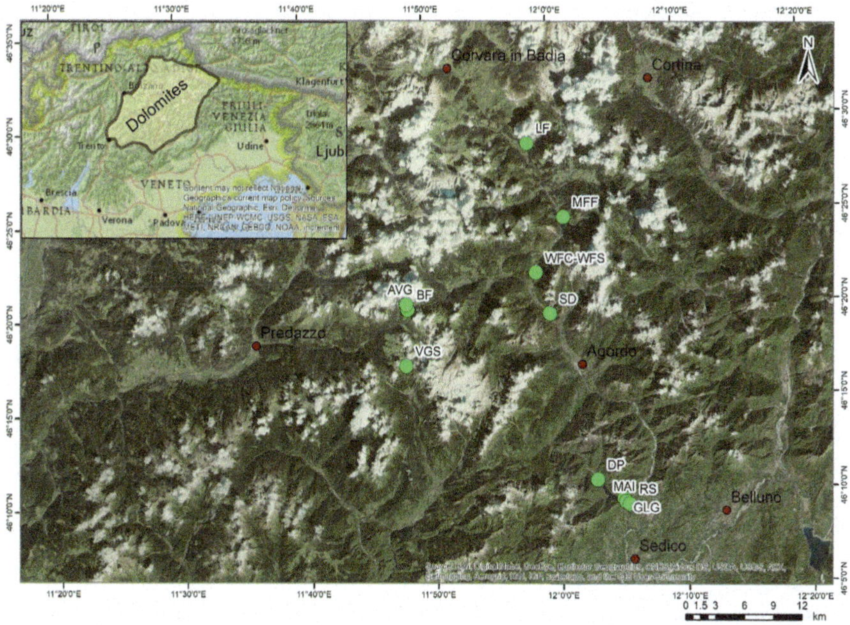

Fig. 9.1 Location of the Dolomites area (inset) and of sampling sites. (See the text for the explanation of labels)

most recent part of the sequence consists of Jurassic and Cretaceous formations outcropping in the southernmost sector of the study area (*Vallone Bellunese*) (Bosellini et al. 2003).

In order to test the reliability of indirect methods in UCS estimation of different lithotypes, not all the formations of the Dolomites sequence but only the most representatives, in terms of extension of outcrops and thickness, were considered. Lithotypes characterized by high fracturation and anisotropy were excluded. The formations included in the analysis are briefly described below (Avanzini et al. 2010; Barbieri and Grandesso 2007; Neri et al. 2007)

- *Atesino* volcanic group (AVG) (Permian): thick succession of lava, tuffs and ignimbrites interbedded with thinner volcanic and sedimentary layers. It mainly consists of dacitic and rhyolitic ignimbrites.
- *Val Gardena* Sandstone Formation (VGS) (Upper Permian): succession of continental medium-to-coarse-grained sandstone, interbedded with conglomerates and mudstones in the upper and lower part, respectively. Sandstones consist of fragments of Permian volcanites, quartz, feldspars, biotite and fragments of basement.
- *Bellerophon* Formation (BF) (Upper Permian): cyclic sequence of dolomites, marls, shales and chalks (lithotype considered in the investigation) in the lower part, dolomites, limestones and marls in the upper part.

- *Werfen* Formation (WF) (Lower Triassic: Induan-Olenekian): sequence of calcareous and terrigenous deposits subdivided in ten members, nine of which are present in the Dolomites. The two members with greater thicknesses were selected for the analysis: Campil Member (WFC), consisting of sandstones interbedded with mudstones and siltstones, and Siusi Member (WFS), consisting of grey micritic limestones interbedded with calcareous siltstones and sandstones, in the lower part, marls and marly limestones, in the upper part.
- *Sciliar* Dolomite (SD) (Middle Triassic: Anisian-Ladinic): platform carbonate rocks almost totally dolomitized.
- *Livinallongo* Formation (LF) (Middle Triassic: Upper Anisian-Lower Ladinic): micritic limestones, bituminous limestone, intercalations of turbidites and volcanoclastic sandstones (*Pietra Verde*, PV), the latter were considered in the analysis.
- *Monte Fernazza* Formation (MFF) (Middle Triassic: Upper Ladinic): decimetric to metric layers of volcanoclastic sandstones consisting of lava clasts and fragments of glass and crystals.
- *Dolomia Principale* Formation (DP) (Upper Triassic: Upper Carnic-Noric): dolomitized peritidal platform carbonates with fossiliferous and stromatolitic levels.
- *Calcari Grigi* Group (GLG) (Lower Jurassic: Hettangian-Lower Aalenian): sequence of wackestones and packstone, micritic limestones, and oolitic-pisolitic grainstones.
- *Maiolica* Formation (MAI) (Upper Jurassic-Lower Cretaceous: Upper Tithonian-Upper Barremian): decametric layers of micritic limestones, with conchoidal fracture, often flint nodules are present.
- *Scaglia Rossa* Formation (RS) (Upper Cretaceous-Paleocene): thin-layered calcilutites and marly limestones interbedded with thin pelitic layers and flint nodules.

The investigated lithotypes consist of dolostones (SD, DP), limestones (WFS, GLG, MAI), marly limestones (RS), sandstones (VGS, WFC), chalks (BF), porphyries (AVG) and volcanoclastic rocks (PV, MFF).

9.3 Methods

UCS can be estimated through direct and indirect methods. Directs methods, such as UCT, allow to determine UCS by laboratory tests, while indirect methods are based on the evaluation of indexes, R (SH) and I_{50} (PLT), which can be correlated to UCS through empirical relationships.

SH is a rapid and non-destructive technique to estimate UCS both in laboratory and in situ directly on rock outcrops. In this work, we used SH type '*L*', which is designed to test rocks with UCS greater than 20 MPa. R index is obtained from the elastic rebound on clean and polished surfaces with emery stone. Each test consists of 20 determinations of R taking into account the orientation of SH (Fig. 9.2). Then,

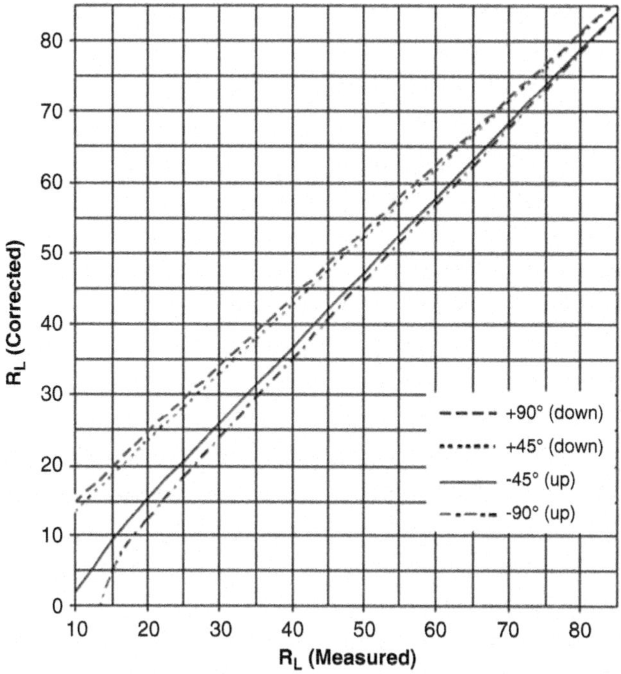

Fig. 9.2 Correction curves for SH orientation. (From Basu and Aydin 2004)

Table 9.1 Empirical equations to estimate UCS through SH test, using the corrected values of *R* and the unit weight of the selected lithotypes (ρ)

Equation		Lithotypes
$UCS\ [MPa] = 9.97e^{0,\ 02 \times R \times \rho}$	Deere and Miller (1966)	All
$UCS\ [MPa] = 0.000004R^{4.29}$	Yasar and Erdogan (2004)	Carbonate rocks (MAI, GLG, WFS, RS, RS, DP, SD); basalt
$UCS\ [MPa] = 0.33(R \times \rho)^{1.35}$	Aufmuth (1973)	All
$UCS\ [MPa] = 12.74e^{0.02 \times R \times \rho}$	Beverly et al. (1979)[a]	All
$UCS\ [MPa] = 3.32e^{0.04 \times R \times \rho}$	Cargill and Shakoor (1990)	Sandstones (WFC, VGS, PV, MFF)
$UCS\ [MPa] = 18.17e^{0.02 \times R \times \rho}$	Cargill and Shakoor (1990)	Carbonate rocks (MAI, GLG, WFS, RS, RS, DP, SD);
$UCS\ [MPa] = 4.45R - 76.18$	O'Rourke (1989)	Sandstones (WFC, VGS, PV, MFF), chalks (BF)

[a]In Xu et al. (1990)

UCS is calculated through empirical formulations reported in Aydin and Basu (2005) (Table 9.1), using the mean value of all the corrected measures of *R* and of the ten higher values.

PLT allows to estimate UCS by determining the point load strength index (I_{50}), which is derived by the point load strength I_S. The test can be carried out by loading diametrically or longitudinally cylindrical rock specimens, but it can also be performed with irregular specimens. In the present work, PLT were carried out on 23 irregular specimens, obtaining a statistically reliable result. To estimate UCS, two methods were applied: 'Size Correction Factor' and 'Logar' method, using the sample thickness (D) and the equivalent diameter (D_e) (ISRM 1985; Thuro and Plinninger 2001; ASTM 2002).

I_S is calculated using irregular specimens as the ratio between breaking load (F) and the area of the cross section of the sample (A):

$$I_S = \frac{F}{A} = \frac{F}{WD} \tag{9.1}$$

where W and D are the width and thickness of the specimen, respectively.

Considering D_e, Eq. 9.1 become:

$$I_S = \frac{F}{D_e^2} = \frac{\pi F}{4WD} \tag{9.2}$$

being $A = WD = \frac{\pi D_e^2}{4}$.

Because I_S strictly depends on the thickness of specimen, it must be corrected to the standard equivalent diameter of 50 mm to determine I_{50}. To this end, 'Size Correction Factor (F)' and 'Logar' methods were used.

In the case of Size Correction Factor:

$$I_{50} = f\frac{F}{WD} = \left(\frac{D_e}{50}\right)^{0.45}\frac{F}{WD} = \left(\frac{D_e^2}{2500}\right)^{0.225}\frac{F}{WD} \tag{9.3}$$

or, using equivalent diameter:

$$I_{50} = f\frac{F}{D_e^2} = f\frac{\pi F}{4WD} = \left(\frac{D_e}{50}\right)^{0.45}\frac{\pi F}{4WD} = \left(\frac{D_e^2}{2500}\right)^{0.225}\frac{\pi F}{4WD} \tag{9.4}$$

In the Logar method, the pairs of values I_S-D^2 or I_S-D_e^2 from each test are plotted in a semi-logarithmic diagram and linearly interpolated. From the regression line, it is possible to obtain I_{50}, which is I_S corresponding to $D^2 = 2500$ mm ($D = 50$ mm).

In both methods, PLT has to be carried out using specimens with thickness between 15 and 85 mm (ISRM 1985) or 30 and 85 (ASTM 2002), and D/W between 0.3 and 1 (best condition).

Using Logar method, just one I_{50} value is obtained. In the case of Size Correction Factor, I_S from each test is converted into I_{50}; UCS is calculated using the mean value of the entire data set subtracting the two lower and two higher values (ISRM 1985; ASTM 2002). Then, UCS is obtained from:

$$\text{UCS} = 23 \times I_{50} \tag{9.5}$$

UCT allows the direct measurement of UCS. The laboratory test is carried out placing a cylindrical rock specimen in a pressure chamber where it is compressed at a constant speed until the failure. UCS is calculated as the ratio between breaking load and cross-section area of specimen:

$$\text{UCS}[MPa] = \sigma_a = \frac{F}{A} \tag{9.6}$$

From rock samples collected during in situ survey, three cylindrical specimens were obtained, four in the case of BF due to the low quality of the rocks. Volume and unit weight of specimens were calculated; the latter is required also to estimate UCS through SH. Finally, UCS was calculated as the mean value of all the tests performed for each lithotype.

UCS values obtained from indirect methods were compared to those from UCT, which is considered the more accurate method. The comparison was carried out plotting on predicted vs. observed scatter charts the pairs of UCS values PLT-UCT and SH-UCT for each lithotype, considering PLT and SH values as predicted and UCT as observed values. In addition, two statistics, Variance Accounted For (VAF) and Root Mean Square Error (RMSE), were calculated:

$$\text{VAF} = \left[1 - \frac{\text{var}(y - y')}{\text{var}(y)}\right] \times 100 \tag{9.7}$$

$$\text{RMSE} = \sqrt{\frac{1}{n}\sum_{i=1}^{N}(y - y')} \tag{9.8}$$

VAF and RMSE allow to evaluate the reliability of forecasting models. In Eqs. 9.7 and 9.8, *y* are the measured values, *y'* the predicted ones, *n* the number of samples and var the variance. The higher the VAF and the lower the RMSE, the lower the prediction error and the better the model.

9.4 Results

Table 9.2 reports UCS values estimated through SH tests. The empirical equations of Table 9.1 were applied using the mean value of all the corrected R determinations and of the ten higher. As expected, the calculated UCS is higher in the second case. Large differences can be noted between the results from the different equations proposed in the scientific literature. Such differences are very large in the case of limestones (MAI, GLG, WFS) and sandstones (WFC, VGS). In particular, the application of the empirical formulation proposed by Cargill and Shakoor (1990)

Table 9.2 UCS [MPa] estimated through SH tests using the mean value of all the corrected measures of R (1) and of the ten higher values (2)

	Deere		Aufmuth		Beverly		O′Rourke		Cargill		Yasar	
	UCS [MPa]		*UCS [MPa]*		*UCS [MPa]*		*UCS [MPa]*		*UCS [MPa]*		*UCS [MPa]*	
Litho.	(1)	(2)	(1)	(2)	(1)	(2)	(1)	(2)	(1)	(2)	(1)	(2)
SD	120	153	222	252	154	195	143	164	219	279	50	74
DP	223	285	300	332	284	364	191	212	406	520	117	163
WFS	117	130	219	231	149	166	145	154	213	236	52	62
GLG	77	88	171	186	99	113	117	129	141	161	29	38
MAI	88	102	186	203	113	131	126	140	161	186	36	47
RS	101	112	202	214	129	143	135	145	184	204	43	52
VGS	67	83	155	178	86	105	103	123	150	227	21	33
WFC	117	136	219	237	149	173	148	161	456	613	55	71
BF	26	30	62	74	34	38	29	43	–	–	–	–
AVG	110	131	212	233	141	168	–	–	–	–	–	–
PV	154	169	253	265	197	216	179	188	793	958	97	113
MFF	135	152	237	251	173	194	163	173	611	767	73	88

Table 9.3 UCS [MPa] estimated through PLT. Size correction factor and Logar method were applied using the thickness of the specimen (D) and the equivalent diameter (De)

	Size correction factor		Logar method	
	UCS [MPa]		*UCS [MPa]*	
Litho.	D	De	D	De
SD	225	186	224	203
DP	262	217	270	235
WFS	168	139	162	160
GLG	172	143	170	149
MAI	151	125	157	144
RS	197	163	175	162
VGS	59	49	58	55
WFC	187	155	173	133
BF	56	47	50	46
AVG	113	94	111	100
PV	427	354	387	356
MFF	203	168	216	196

resulted in very high UCS for some of the tested lithotypes (WFC, PV, MFF and DP), while using the equation proposed by Yasar and Erdogan (2004), the lower UCS values were obtained.

UCS values estimated through PLT are reported in Table 9.3. Size Correction Factor and Logar method were applied considering both the thickness of the tested specimen (D) and its equivalent diameter (D_e). Obtained results do not show significant differences in the UCS values calculated using the two methods. For all the considered lithotypes, UCS values estimated using the thickness of the specimens are higher than those estimated using the equivalent diameter.

Table 9.4 UCS values measured through UCT

Litho.	SD	DP	WFS	GLG	MAI	RS	VGS	WFC	BF	AVG	PV	MFF
UCS [MPa]	225	336	–	64	247	102	97	139	24	128	240	207

Table 9.5 RMSE and VAF statistics calculated considering UCS values estimated through PLT and SH as predicted and those measured by UCT as observed

Point load test	**RMSE**	**VAF (%)**
Size Correction Factor using D	83	21
Size Correction Factor using D_e	73	34
Logar method using D	70	42
Logar method using D_e	66	45
Schmidt Hammer test	**RMSE**	**VAF (%)**
Deere and Miller (1966) using the ten higher R determinations	59	71
Deere and Miller (1966) using all R determinations	77	64
Yasar and Erdogan (2004) using the ten higher R determinations	123	46
Yasar and Erdogan (2004) using all R determinations	141	43
Aufmuth (1973) using the ten higher R determinations	75	68
Aufmuth (1973) using all R determinations	64	65
Beverly et al. (1979) using the ten higher R determinations	46	74
Beverly et al. (1979) using all R determinations	53	71

UCS values measured through UCT are reported in Table 9.4. SD and DP (Dolostones), MAI (limestone), PV and MFF (volcanoclastic rocks) have the higher UCS; BF (chalks) and GLG (limestone) the lower, possibly due to the texture and composition and to the high microfracturing. In the case of WFS, it was no possible to prepare the specimens for the test due to the high fracturation degree.

The large differences in the estimation of UCS through indirect methods, especially in the case of SH, require to compare the calculated values to those measured by UCT, to verify the reliability of the results and identify the best empirical formulation to estimate UCS. To this end, the statistics VAF and RMSE were calculated, considering UCS values estimated through indirect methods as predicted and those through UCT as observed (Table 9.5).

In the case of PLT, the best correlation was obtained through the Logar method using the equivalent diameter. In the case of SH, the best correspondence between predicted and observed values was obtained using Deere and Miller (1966) and Beverly et al. (1979) empirical formulations using the ten higher R determinations. These correspondences are clearly evidenced in the predicted vs. observed scatter plots of Figs. 9.3 and 9.4. In the plots, the black lines indicate a perfect correspondence between predicted and observed values: points above the line indicate an overestimation, while those under it an underestimation of UCS.

Figure 9.3 shows a good correlation between values predicted through PLT Logar method using equivalent diameter and those measured through UCT up to UCS equal to 200 MPa; for higher values, over/underestimation can occur. In the case of

limits of SH estimations are overcome carrying out numerous determinations of R, which generally allow to estimate UCS with good approximation. As shown by the results obtained in this work, the most common empirical correlations proposed in the literature lead to a high variability of UCS estimation, a more reliability can be reached by the combined use of different methods (i.e. SH and PLT).

In the case of PLT, its accuracy in predicting the UCS depends on the ratio between the UCS and the tensile strength, and tensile strength in turn is highly conditioned by the fracturing degree of the rock matrix or by the presence of anisotropy. As the SH, these limiting factors can be overcome by increasing the number of tests. On the basis of our statistical analysis, the Logar method is recommended to estimate UCS.

9.6 Conclusions

The main aim of this work was to compare uniaxial compressive strength (UCS) estimated through indirect methods to that measured by direct ones. The two most common used indirect methods, point load test (PLT) and Schmidt Hammer (SH) test, were performed, and the main empirical formulations proposed in the literature were used to estimate UCS. The obtained UCS values were then compared with those measured through uniaxial compressive test (UCT) carried out at the laboratory of rock mechanics of the Department of Geosciences (University of Padua, Italy). To the purposes of this work, the most representative, in terms of extension of outcrops and thickness, lithotypes belonging to the Dolomites sequence (northeastern Italian Alps) were considered.

Obtained results allow to point out the main pros and cons of the cited methods. UCT is the most rigorous method but requires a difficult procedure for preparation of specimens and often does not allow a statistical data processing. Indirect methods, even if less rigorous, allow to carry out numerous tests, so that the results can be statistically representative of the strength properties at site conditions.

The most used indirect method in the professional and research fields is SH. Obtained results show that it is a reliable method depending on the empirical formulation adopted to estimate UCS. In the case of this work, equations proposed by Beverly et al. (1979) and Deere and Miller (1966) showed the best performance when UCS values estimated through these equations are compared to UCT values. Undoubtedly, further tests in different geological environments are required to improve the reliability of this method and for extending it to the determination of strength properties of both weak rocks (UCS < 20 MPa) and highly anisotropic rocks, such as the metamorphic ones.

References

ASTM (2002) D5731–02, standard test method for determination of the point load strength index of rock. ASTM International, West Conshohocken

Aufmuth RE (1973) A systematic determination of engineering criteria for rocks. Bull Assoc Eng Geol 11:235–245

Avanzini M, Bargossi MG, Borsato A, Selli L (2010) Note Illustrative della Carta Geologica d'Italia alla scala 1:50,000 – Foglio 60 Trento (Explanatory notes for the Geological Map of Italy at 1:50,000 scale – Sheet 60 Trento). Provincia Autonoma di Trento and APAT, Firenze, 2010, 244 p

Aydin A, Basu A (2005) The Schmidt hammer in rock material characterization. Eng Geol 81:1–14

Barbieri G, Grandesso P (2007) Note Illustrative della Carta Geologica d'Italia alla scala 1:50,000 – Foglio 82 Asiago (Explanatory notes for the Geological Map of Italy at 1:50,000 scale – Sheet 82 Asiago). Regione Veneto and APAT, Firenze, 2007, 135 p

Basu A, Aydin A (2004) A method for normalization of Schmidt hammer rebound values. Int J Rock Mech Min Sci 41:1211–1214

Beverly BE, Schoenwolf DA, Brierly GS (1979) Correlations of rocks index values with engineering properties and the classification of intact rocks. Brierley Associates, Washington, DC

Bieniawski ZT (1989) Engineering rock mass classifications. Wiley, New York, 251 p

Bosellini A, Gianolla P, Stefani M (2003) Geology of dolomites. Episodes 26:181–185

Cargill JS, Shakoor A (1990) Evaluation of empirical methods for measuring the uniaxial compressive strength. Int J Rock Mech Min Sci 27:495–503

Deere DU, Miller RP (1966) Engineering classification and index properties for intact rocks. Technical report. Air force weapons laboratory, New Mexico, No. AFNL-TR. Kirtland, pp 65–116

Dehghan S, Sattari GH, Chehreh Chelgani S, Aliabadi M (2010) Prediction of uniaxial compressive strength and modulus of elasticity for Travertine samples using regression and artificial neural networks. Min Sci Technol 20:41–46

Doglioni C (1987) Tectonics of the Dolomites (Southern Alps, Northern Italy). J Struct Geol 9:181–193

Heuze FE (1980) Scale effects in the determination of rock mass strength and deformability. Rock Mech Rock Eng 12:167–192

ISRM (1985) Suggested method for determining point load strength. ISRM commission in testing methods, working group on revision of the point load test method. Int J Rock Mech Min Sci Geomech Abstr 22:51–60

Hoek E, Brown ET (1980) Empirical strength criterion for rock masses. J Geotech Eng Div 106:1013–1035

Lee JS, Kieschnick J, Geyer C, Brumley J, De Spain L (2016) Comparison of Different Methods to Estimate Uniaxial Compressive Strength in a Barnett Shale. In: Proceedings of the 50th US Rock Mechanics/Geomechanics Symposium, ARMA, 26-29 June 2016, Houston, Texas, USA, 2279–2287 pp

Liu Z, Shao J, Xu W, Wu Q (2015) Indirect estimation of unconfined compressive strength of carbonate rocks using extreme learning machine. Acta Geotech 10:651–663

Neri C, Gianolla P, Furlanis S, Caputo R, Bosellini A (2007) Note Illustrative della Carta Geologica d'Italia alla scala 1:50,000 – Foglio 29 Cortina D'Ampezzo (Explanatory notes for the Geological Map of Italy at 1:50,000 scale – Sheet 29 Cortina D'Ampezzo). Regione Veneto and APAT, Firenze, 2007, 208 p

O'Rourke JE (1989) Rock index properties for geoengineering in underground development. Min Eng:106–110

Palmström A (1995) A rock mass characterization system for rock engineering purposes. J Rock Mech Tunnelling Technol 1:1–40

Romana M (1985) New adjustment ratings for application of Bieniawski classification to slopes. In: Proceedings of International Symposium on the role of rock mechanics, ISRM, September 1985, Zacatecas, Mexico, 49–53 pp
Thuro K, Plinninger R (2001) Scale effects in rock strength properties. Part 2: Point load test and point load strength index. In: Proceedings of the ISRM Regional Symposium EUROCK 2001, Espoo, Finland, June 3–7, 2001, 175–180 pp
Thuro K, Plinninger R, Zäh S, Schütz S (2001) Scale effects in rock strength properties. Part 1: Unconfined compressive test and Brazilian test. In: Proceedings of the ISRM Regional Symposium EUROCK 2001, Espoo, Finland, June 3–7, 2001, 169–174 pp
Xu S, Grasso P, Mahtab A (1990) Use of Schmidt hammer for estimating mechanical properties of weak rock. In: Proceedings of the 6th international IAEG congress, vol 1. Balkema, Rotterdam, pp 511–519
Yasar E, Erdogan Y (2004) Estimation of rock physiomechanical properties using hardness methods. Eng Geol 71:281–288

Chapter 10
CO_2 Sequestration and Enhanced Coalbed Methane Recovery: Worldwide Status and Indian Scenario

Bankim Mahanta and Vikram Vishal

Abstract In the current world, deep-seated coal deposits are considered the high-capacity sink site for the storage of anthropogenic carbon dioxide (CO_2). For CO_2 sequestration in coalbeds, a lot of things need to be considered. At present CO_2-ECBM recovery techniques are regarded quite a promising technology to mitigate the atmospheric CO_2, which are the primary causes of global warming. For the last two decades, a lot of researches have been carried out worldwide. These studies include investigation in the micro/nanoscale to large field scale. Considering the vastness of CO_2-ECBM technology, it is quite difficult to summarize its whole process. Here, this chapter provides a detailed study and the status of the CO_2-ECBM recovery worldwide and in the Indian scenario. The first part of the chapter summarizes the importance of CO_2-sequestration on a worldwide scale, followed by the mechanism of CO_2 trapping in the deep-seated coalbeds. Then, this chapter summarizes the global scenario and the Indian scenario of CO_2 sequestration in coal beds.

Keywords CO_2 sequestration · Mechanism of CCS · Coalbed methane · CO_2 – ECBM · CO_2 emission

B. Mahanta
Department of Earth Sciences, Indian Institute of Technology Bombay, Mumbai, India

Department of Civil Engineering, Monash University, Melbourne, VIC, Australia

IITB-Monash Research Academy, Indian Institute of Technology Bombay, Mumbai, India

V. Vishal (✉)
Department of Earth Sciences, Indian Institute of Technology Bombay, Mumbai, India

Department of Civil Engineering, Monash University, Melbourne, VIC, Australia
e-mail: v.vishal@iitb.ac.in

M. De Maio, A. K. Tiwari (eds.), *Applied Geology*,
https://doi.org/10.1007/978-3-030-43953-8_10

10.1 Introduction

Coal is considered as the primary energy provider to industries like power, cement, steel, and alumina refineries (Saini et al. 2016). More than 40% of the total production of electricity in the world comes from the thermal power plants (The World Bank 2017). According to the International Energy Agency (IEA), the global electricity production from coal is expected to rise about 24% by 2040 (IEA 2015).

Coal combustion releases various greenhouse gases (GHG) into the atmosphere, primarily CO_2 that is considered as one of the significant causes of global warming. Retrieved data of the atmospheric CO_2 concentration from open sources like National Oceanic and Atmospheric Administration-Earth System Research Laboratory (NOAA-ESRL) and European Environmental Agency (EEA) has been complied in Figure 10.1 that represents the CO_2 concentration from the pre-industrial time to the present time. Due to the increased consumption of fossil fuel, currently, the atmospheric CO_2 concentration has jumped up to 410 ppm, as compared to the preindustrial level of 280 ppm (EEA 2013; NOAA-ESRL 2017).

To avoid the toxic gas emissions into the atmosphere as a result of the burning of fossil fuels, various studies are being carried out to establish clean coal technologies, and those include underground coal gasification (UCG), oxyfuel combustion, SOx and NOx reduction in boilers, and chemical looping combustion (Prabu and Mallick 2015). Not only coal combustion, but coal mining also introduces these GHGs into the atmosphere. The methane gas entrapped in the structure of coal is called coalbed methane (CBM) or coal seam gas. The capture of this entrapped gas from coal has

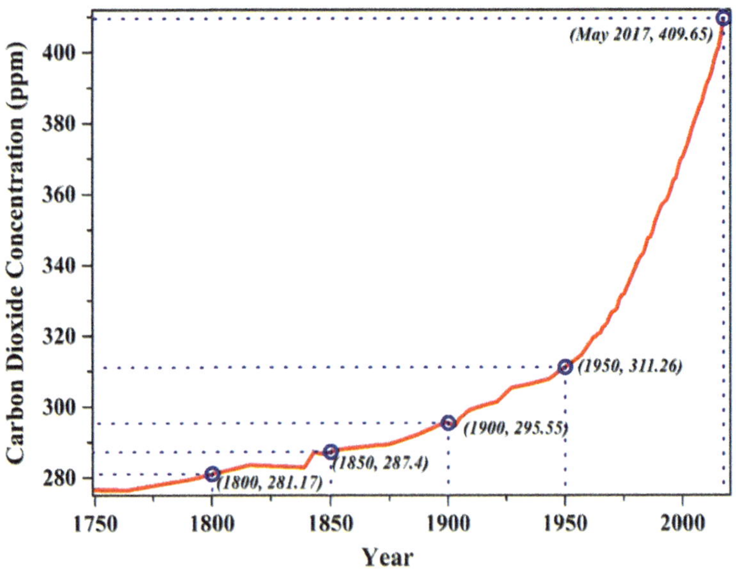

Fig. 10.1 Year-wise atmospheric CO_2 concentration (EEA 2013; NOAA-ESRL 2017)

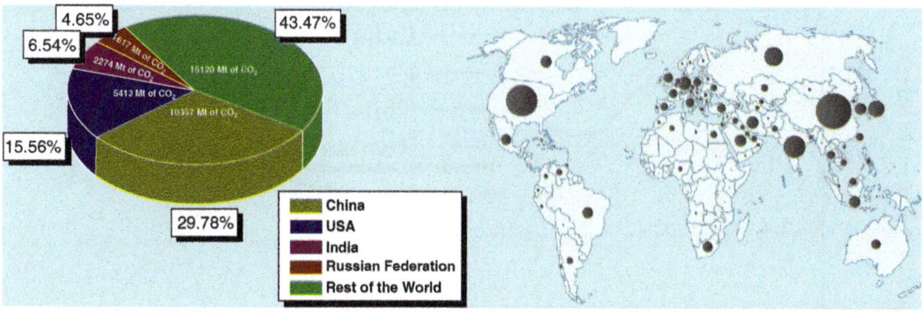

Fig. 10.2 Percentage of CO_2 emission out of the total production in the world in the year 2015 (Boden et al. 2016; Global Carbon Atlas 2016)

been proved to be a source of energy production, thereby meeting the energy demand. CBM can be used for domestic use and for the production of electricity in power plants (Prabu and Mallick 2015). Worldwide coal seam gas reserve is estimated to be 256 trillion m^3 and hence is considered a major source of energy (Hamawand et al. 2013).

To reduce the emission of CO_2 into the atmosphere, CO_2 can be stored in geological formations, which included depleted oil and gas reservoirs, deep saline aquifers, basalt formations, and unmineable coal formation (Vishal et al. 2013). Of these mentioned sink sites for the CO_2 sequestration, coalbeds are preferred over other sites as injected CO_2 gets preferably adsorbed onto the coal micro/mesopores, thereby enhancing the desorption of CH_4. To meet the growing energy demands, unconventional sources of energy need to be investigated with the enhanced coalbed methane (ECBM) being a better alternative that also stores anthropogenic CO_2.

After China and the USA, India ranks third in coal production (Fig. 10.2). For power generation, India solely relies on coal resources. As a developing country, India encounters a massive amount of energy demand to achieve considerable growth in every sector. India's more than 70% of total produced electricity comes from coal-fired plants (Ministry of Coal 2016; The World Bank 2017). Also, to continue the successful growth, nearly a dozen ultra-mega coal-fired power plants are planned to be installed over the next decade. These coal-fired plants result in the emission of a large amount of CO_2, which is considered a major greenhouse gas responsible for the climatic changes (Akash et al. 2016).

Figure 10.3 represents the percentage of the global CO_2 emission of the world with the four countries having made a maximum contribution over 54 years. In the year 1961, the contribution of India to the global CO_2 emission is around 1.32% of the total emission, which reached up to 6.54% by 2015 (Boden et al. 2016; Global Carbon Atlas 2016).

In a time duration of 15 years from 1990 to 2005, India's CO_2 emission almost doubled, which is primarily due to the fossil fuel combustion from domestic sources, as well as various industries such as steel, cement, power, petroleum refineries, and fertilizer (Garg et al. 2006; Vishal et al. 2013). Through modeling, it has been estimated that the total emission of CO_2 in India will increase from 1229 million

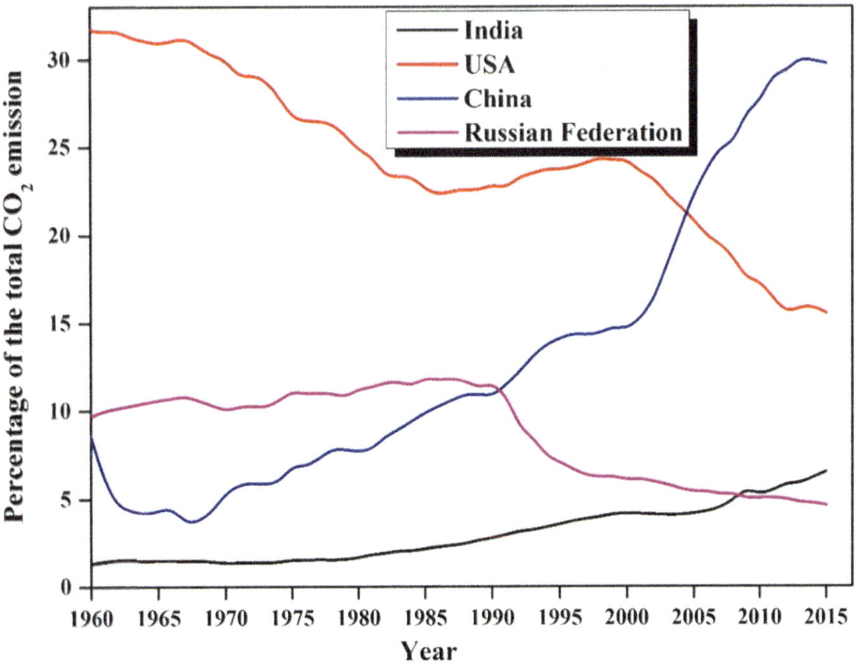

Fig. 10.3 CO_2 emission over the period 1961–2015 (Boden et al. 2016; Global Carbon Atlas 2016)

tons to approximately 3000 million tons in time duration from 2005 to 2030 (Nair et al. 2003). Therefore, to avoid the rise of the CO_2 concentration in the atmosphere, detailed research should be carried out regarding the CO_2 sink capacity in the Indian scenario. Once the sink sites are identified, a further feasibility study must be carried out.

As CO_2 concentration is increasing from time to time, to control the concentration of CO_2 in the atmosphere, various mitigation techniques need to be implemented. Carbon capture, utilization, and storage (CCUS) are deemed an effective method to achieve the goal of reducing the CO_2 concentration in the atmosphere. Geological carbon sequestration is a technique used for the mitigation of global climate change where the CO_2 emitted by the burning of fossil fuels in electricity generation and various industries is captured, followed by their transportation to a storage site where the CO_2 can be stored for a considerable period. CCUS may lead to meeting the objective of reducing greenhouse gas (GHG) emissions by 50–85% by 2050, and its successful implementation can preclude the increment of global average temperature by a scale of 2 °C compared to the preindustrial time by 2100 (IPCC 2007; IEA 2012; Viebahn et al. 2014). To address the clean source of energy, attention should be provided to the ECBM technology that will facilitate enhanced coalbed methane production as well as mitigation of large-scale anthropogenic CO_2 (Vishal et al. 2013, 2015b).

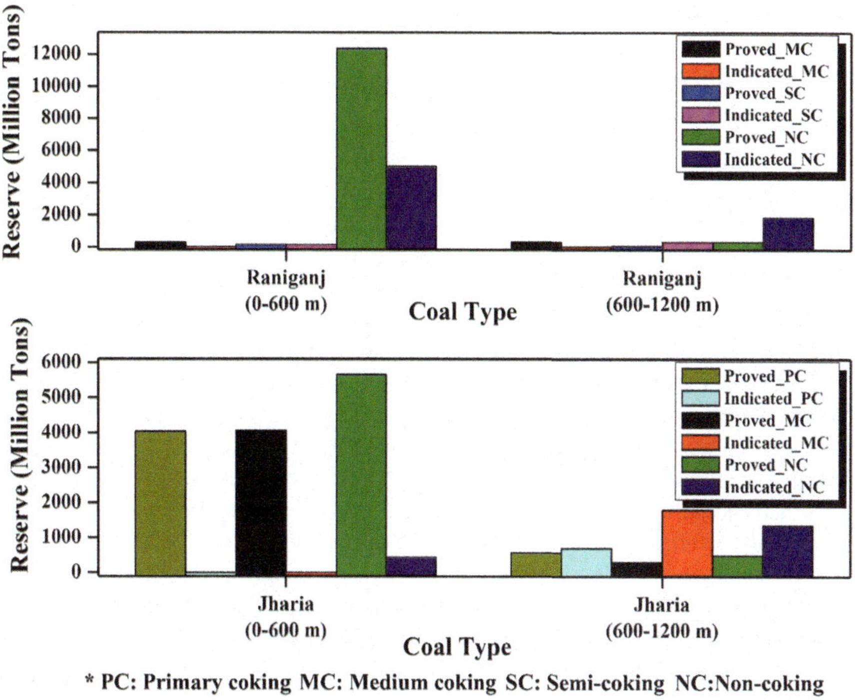

Fig. 10.4 Depth-wise estimated prime coking coal reserves in Raniganj and Jharia coalfield (Ministry of Coal 2016)

As per the annual report published in April 2017 by the Ministry of Coal, Government of India has estimated a cumulative 308.802 billion tons of in situ geological resources of coal in India up to a depth of 1200 m. Nearly 27% of the total estimated coal reserves is present in the state of Jharkhand (Ministry of Coal 2016). Coal exploration from Raniganj coalfield and Jharia coalfield in West Bengal and Jharkhand, respectively, provides a major contribution towards the coal exploration in India. Figure 10.4 illustrates an estimation of proved and indicated coal reserves present in both the coalfields. With the understanding on the viability of available mining technologies and limitation in mining beyong 300 m depth, the unminable coal deposits may be considered for carbon sequestration.

In this chapter, a summary of India's dependence on coal for the energy production, importance of CCUS worldwide and in India, and the status of ECBM from an Indian point of view have been presented.

10.2 Background Knowledge

10.2.1 Mechanism of CO_2 Sequestration and ECBM Recovery

Primarily, CBM extraction from coalbeds can be achieved in two different ways; the dewatering process and the process involved with the injection of CO_2 into the coalbeds. The dewatering process reduces the hydrostatic pressure of the coalbed, allowing the entrapped gases to move from the coal surface. In contrast, the second method is involved with the CO_2 injection into coalbeds, thereby enhancing the desorption of methane gas from the porous coal surface.

In case of CBM and ECBM recovery, parameters such as temperature, pressure along with the rank of coal, moisture content, and maceral composition of coal become significant (Azmi et al. 2006; Kedzior 2009). Some coals may have a higher amount of sorption capacity despite being rich in inorganic content (mineral matters). This is due to the presence of certain maceral types and clay minerals that also have a larger surface area compared to other minerals that facilitate adsorption. The adsorption capacity of coal is predominantly dependent on the coal rank or its maturity, along with its type and maceral composition (Hildenbrand et al. 2006; Butland and Moore 2008). The process of CO_2 adsorption and CH_4 desorption are the foremost mechanism of CO_2 sequestration and ECBM recovery (Wang et al. 2016b). During the process of interaction of CH_4 and CO_2 with the coal surface molecules, changes in adsorption energy, equilibrium distance, and molecular bond occur. Chemical reactions leading to the generation of new substance does not occur, which indicates the adsorption between these gases and coal surfaces to be physical (Baisheng et al. 2013). Primarily, coal is composed of carbon atoms that are of organic origin. Inside the coal structure, an equilibrium force is maintained between these carbon atoms. However, on the surface of the coal, without the attractive effect from one side, a force imbalance results. As a result of which the surficial coal atoms lack to maintain the balanced attractive force as compared to the internal carbon atoms. Hence, with strong free energy, these surficial coal atoms act as potential adsorption sites for gases like CO_2 and CH_4, depending upon their adsorption affinity. (Stevens et al. 1998; Wang et al. 2014, 2016a; Vishal et al. 2018). The injection of CO_2 gas into the cleats of the coal results in a reduction in the CH_4 partial pressure in the free gas phase without affecting the reservoir pressure and thereby promoting the desorption of CH_4 from the coal matrix (Mazumder et al. 2006).

On these sites, adsorption affinity varies from gas to gas. In a comparison of CO_2 and CH_4, the earlier has an advantage of preferential adsorption on the coal surfaces. Being nonpolar, the weak van der Waals force or the London dispersion forces is the only intermolecular force that acts in the molecules of CH_4 and CO_2.

Table 10.1 Physical/chemical characteristics of CH_4 and CO_2 in context to their adsorption ability (modified after Wang et al. 2016a)

Physical/chemical parameters	CH_4	CO_2
Boiling point (° C)	−161.49	−78.48
Critical temperature (° C)	−82.01	31.04
Critical pressure (MPa)	4.6407	7.386
Critical density (kgm^{-3})	426	466
Ionization potential (eV)	13.79	15.6
Effective diameter (nm)	0.414	0.456
van der Waals constant (a) (bar L^2/mol^2)	2.30	3.658

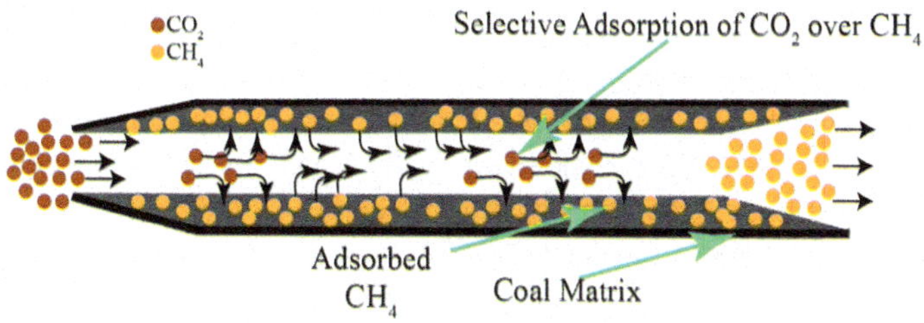

Fig. 10.5 Schematic diagram representing selective adsorption of CO_2 over CH_4 on coal surface (Vishal et al. 2018)

$$\left(P + \frac{a}{V^2}\right)(V - b) = RT \tag{10.1}$$

In van der Waals equation, the van der Waals constant 'a' signifies a measure of the intermolecular force of attraction (Eq. 10.1). With a higher 'a' value, gases can be easily liquefiable and adsorbed onto the solid-surface. Table 10.1 illustrates a comparison of various characteristics of CH_4 and CO_2 for their adsorption ability. In contrast to CH_4, CO_2 has a higher 'a' value that suggests its higher adsorption affinity towards coal during ECBM recovery. Higher intermolecular forces result in a higher critical temperature in case of CO_2, which facilitates its easier adsorption compared to CH_4. Hence, upon injection into the coal matrix, preferentially, CO_2 gets adsorbed on the coal surfaces/matrix, replacing the already adsorbed CH_4 that is collected from the production well. Mechanism of selective adsorption of CH_4 and CO_2 on the coal surface has been represented schematically in Figure 10.5.

10.2.2 The Dependency of India's Energy Scenario on Coal and Fossil Fuel

Being a developing country, rising economic growth demands additional energy consumption. Energy consumption is always associated with the growth of a nation. In the Indian scenario, most of the energy demand is currently being fulfilled by coal.

Since the first step toward development, coal is being used as the primary source of energy as it is found in abundance and easy to explore rather than any other source of energy. On the contrary, the use of coal results in a high rate of carbon emissions in the form of CO_2, which is one of the major constituents responsible for global warming. Figure 10.6 indicates the growth domestic product (GDP) along with the CO_2 emission of India in the past 50 years. High GDP growth has been supplemented with a high rate of CO_2 emissions.

Presently with 1.324 billion people, India is the second populated country after China, and for achieving the economic growth of such a vastly populated country, India turns to be a significant consumer of the energy sources. It is the world's third-biggest energy user after China and the USA.

Most of the energy demand in India is being achieved by the use of fossil fuels. Coal presently fulfils 60% of the total energy demand. Figure 10.7 represents year-wise energy consumption, which suggests that energy coming from the fossil fuel

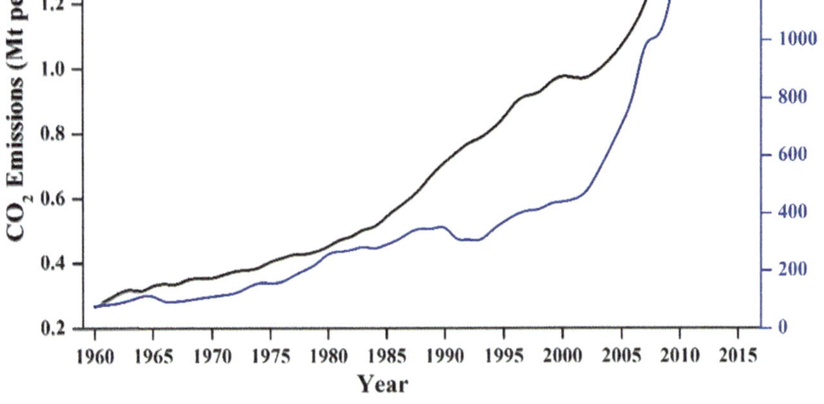

Fig. 10.6 A comparison between the per capita GDP and the CO_2 emission of India (The World Bank 2017)

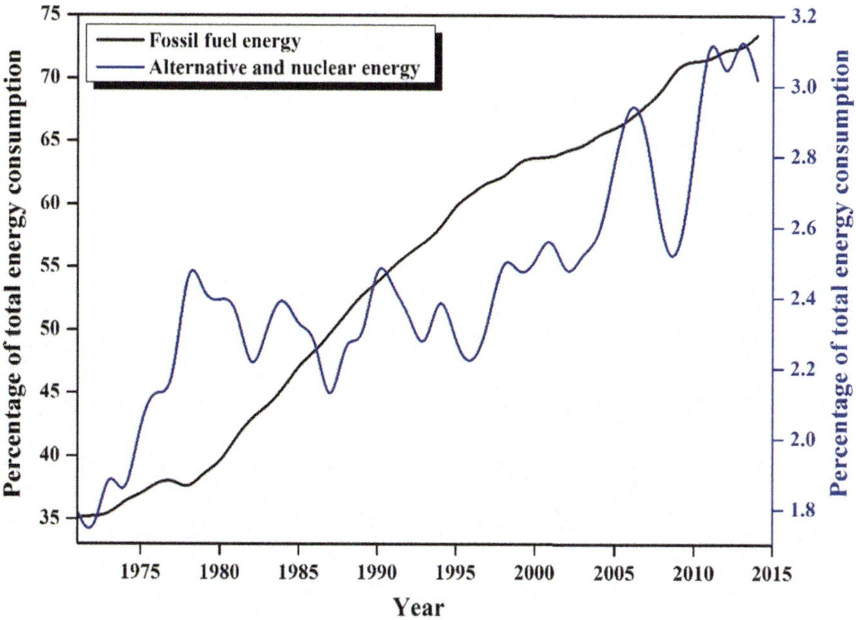

Fig. 10.7 Year-wise fossil fuel, alternative and nuclear energy consumption in India (The World Bank 2017)

source is the only dominant source over any other alternative or nuclear energy source.

Figure 10.8 represents the contribution of various sectors toward the CO_2 emission of India, which suggests a significant portion of the emitted CO_2 comes from the electricity and heat production sector. A major portion of the electricity produced in India comes from thermal power plants, which predominantly use coal as the source of energy (Fig. 10.9).

10.3 CO_2 Sequestration: Worldwide and Indian Scenario

10.3.1 World Scenario

At the present-day condition, fossil fuel provides 80% of the primary energy demand. The power generated as a result of coal fire bears the largest contribution of CO_2 emissions worldwide, and it fulfills around 40% of the total electricity output. Despite the various actions taken by the global communities to restrict the increment of temperature below 2 °C, the dependency on fossil fuel for world's primary energy will remain around 60% of the total energy demand by 2040, which is nearly 80% at present (Global CCS Institute 2015; IEA 2015). Carbon dioxide coming from

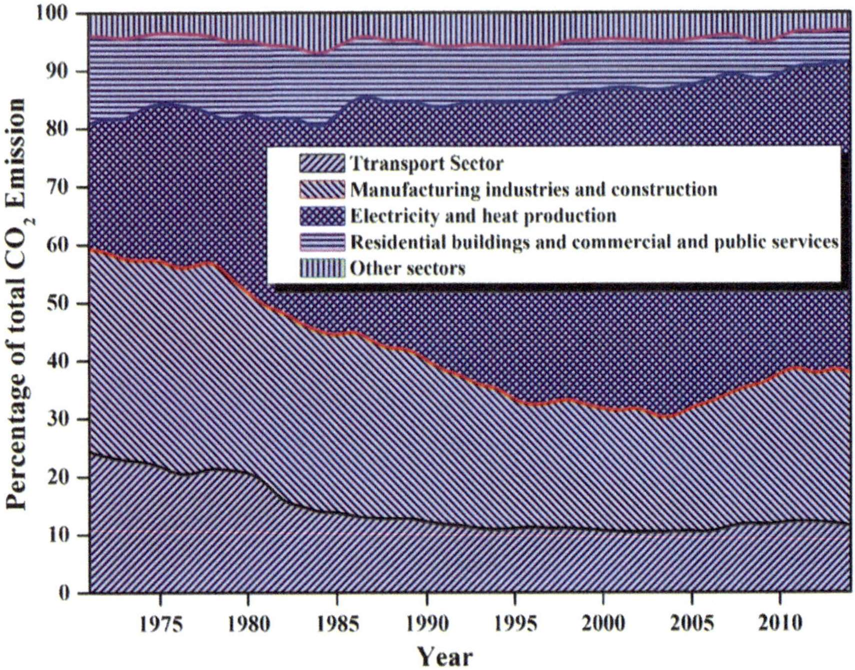

Fig. 10.8 Year-wise CO_2 emission from various sectors in India (The World Bank 2017)

industrial sources is considered the other major source, which holds around 25% of worldwide emission. Additionally, the by-product coming from various industries such as cement, steel, chemical adds significant CO_2 into the atmosphere. To have a better clean environment, IEA has implemented ***450 scenario***, which sets new energy pathway to reduce the release of GHG gases into the atmosphere and to limit the rise in global average temperature to 2 °C compared to the pre-industrial condition by limiting the atmospheric CO_2 concentration below 450 ppm.

Leung et al. (2014) summarized various approaches that could be possible methods for the mitigation of global climate change, and they are like better energy efficiency and conservation; promoting the usage of low-carbon fuels; implementation of renewable energy like solar, wind, bioenergy, and hydropower; afforestation; and CO_2 capture storage (CCS).

CCS is the only effective technique developed so far, which can result in a reduction of CO_2 emissions on a significant scale. However, along with the CCS technology, other sources of renewable energy and the level of low-carbon technology should be implemented for the mitigation of the global CO_2 level.

To date, in various parts of the world, 17 large-scale CCS projects are efficiently operational, and they are capable of capturing around 31.1 million tons of CO_2 per year (Mtpa). However, IEA's agenda to achieve 2 °C goal suggests that the capturing of CO_2 needs to be increased to 4000 million tons by 2040 and 6000 million tons in

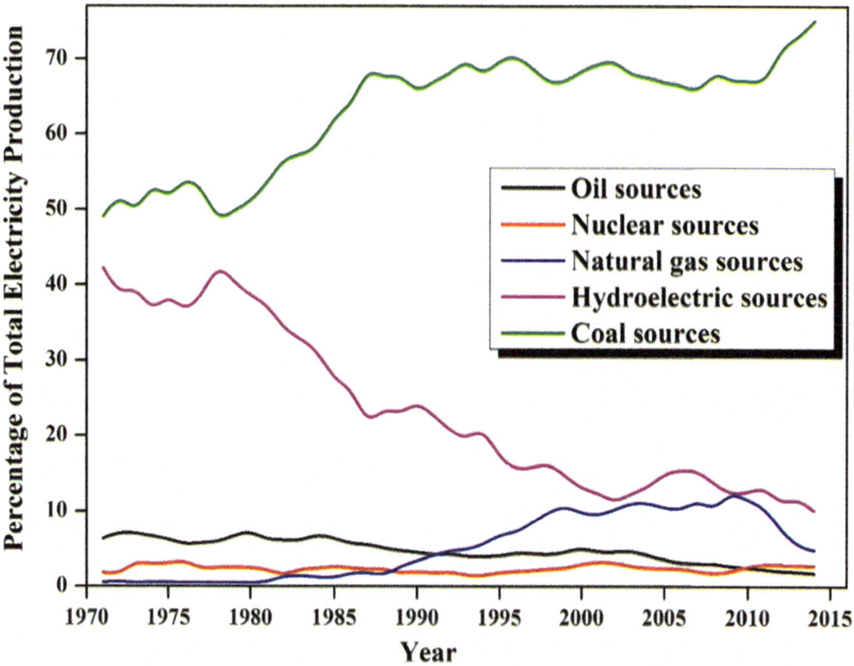

Fig. 10.9 Year-wise various source of electricity production in India (The World Bank 2017)

2050 (Global CCS Institute 2015, 2017). Table 10.2 represents the lists of 17 large-scale CCS projects in operation by 2017, which suggests most of the projects in operations are involved with all the developed countries. However, for achieving the objective of limiting the average global temperature by 2 °C compared to the pre-industrial time, all the developing countries should also pay strong attention to this problem. Although, the relatively high cost of the CCS turns to be a major barrier for its wider implication in the world (IPCC 2014).

10.3.2 Indian Scenario

Being a developing country, most of the energy demand in India is fulfilled by coal combustion, which contributes a major portion of the emitted atmospheric CO_2. The successful implementation of CCS technology primarily depends on factors such as the technological development, storage potential, cost reduction, social approval, GHG mitigation regimes, as well as global carbon price (Garg et al. 2017). With the progress of time, various research is being carried out on the Indian scenario to mitigate CO_2 production. In 2007, Department of Science and Technology (DST) started the National Program on Carbon Sequestration (NPCS) Research, under which research is being carried out in four different aspects, and these are microalgae bio-fixation techniques for the CO_2 sequestration; development of carbon capture

Table 10.2 Major large-scale CCS projects in operation to date (Global CCS Institute 2017)

Name	Location	Operation date	Capture capacity (Mtpa)	Primary storage type
Terrell natural gas processing plant	United States	1972	0.4–0.5	Enhanced oil recovery
Enid fertilizer	United States	1982	0.7	Enhanced oil recovery
Shute Creek gas processing plant	United States	1986	7.0	Enhanced oil recovery
Sleipner CO_2 storage	Norway	1996	1.0	Dedicated geological storage
Great Plains Synfuels plant and Weyburn-Midale	Canada	2000	3.0	Enhanced oil recovery
Snøhvit CO_2 storage	Norway	2008	0.7	Enhanced oil recovery
Century plant	United States	2010	8.4	Dedicated geological storage
Air products steam methane reformer	United States	2013	1.0	Enhanced oil recovery
Coffeyville gasification plant	United States	2013	1.0	Enhanced oil recovery
Lost cabin gas plant	United States	2013	0.9	Enhanced oil recovery
Petrobras Santos basin pre-salt oil field CCS	Brazil	2013	1.0	Enhanced oil recovery
Boundary dam carbon capture and storage	Canada	2014	1.0	Enhanced oil recovery
Quest	Canada	2015	1.0	Dedicated geological storage
Uthmaniyah CO_2-EOR demonstration	Saudi Arabia	2015	0.8	Enhanced oil recovery
Abu Dhabi CCS	United Arab Emirates	2016	0.8	Enhanced oil recovery
Illinois industrial carbon capture and storage	United States	2017	1.0	Dedicated geological storage
Petra Nova carbon capture	United States	2017	1.4 (31.1)	Enhanced oil recovery

Mtpa Million tons per annum

method; development of various policies; and terrestrial agro-forestry sequestration modeling (Global CCS Institute 2013). The DST also released the first ever roadmap for CCUS in India in June 2018 which recommends various pathways for its development and implementation in India.

The most crucial factor for the successful implementation of carbon sequestration is the determination of reliable sites having a good amount of storage capacity. Based on the globally integrated assessment model, a first-order estimation suggests a total capacity of 105 Gt of CO_2, mostly in saline aquifers and a successive

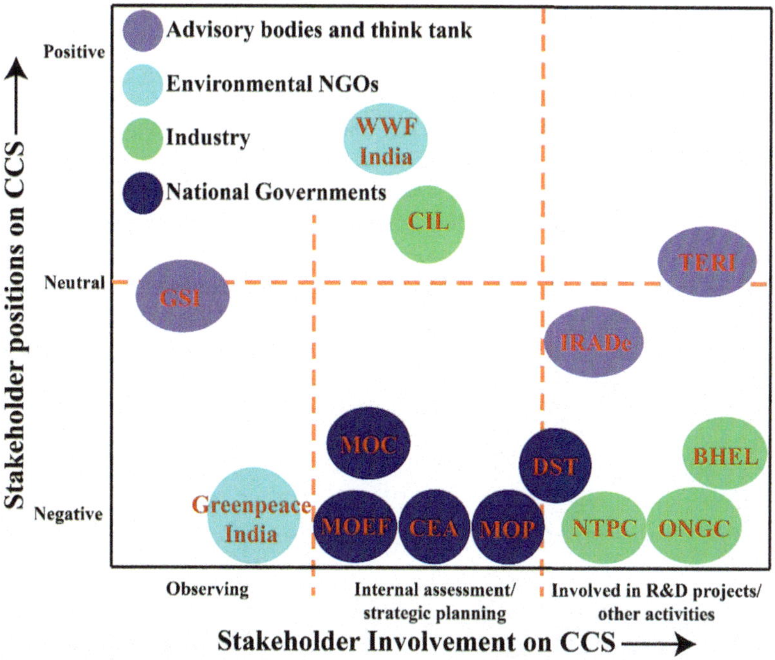

BHEL: Bharat Heavy Electricals Ltd.
CEA: Central Electricity Authority
CIL: Coal India Ltd.
DST: Department of Science and Technology
GSI: Geological Survey of India
WWF: World Wildlife Fund
IRADe: Integrated Research and Action for Developement
MOC: Ministry of Coal
MOEF: Ministry of Environment and Forests
MOP: Ministry of Power
NTPC: National Thermal Power Corporation
ONGC: Oil and Natural Gas Corporation Ltd.
TERI: The Energy and Resources Institute

Fig. 10.10 Major CCS stakeholders in India (modified after Viebahn et al. 2014)

assessment represents a figure of 360 Gt in saline aquifers and 200 Gt in basalt formation (Dooley et al. 2005; Singh et al. 2006; Viebahn et al. 2014). However, all the assessments are based on theoretical knowledge, and it requires the idea of practical significance or an in-depth assessment of the storage potential for a successful implementation of CCS in India. Viebahn et al. (2014) conducted an extensive study on the prospects of CCS in India's power sector. They opined that effective implementation of CCS in the Indian scenario mainly depends on better in-depth knowledge of the storage capacity for CCS, public support, as well as on the success of CCS in the industrialized nations (Viebahn et al. 2014). They also suggested that India's leading stakeholders can play a vital role in the successful implementation of CCS. Figure 10.10 represents the participation and progress of various leading stakeholders in India (Viebahn et al. 2014). However, in more recent years, there has been a change in stance of some of these organizations and a renewed stakeholder positions must be generated.

10.4 Potential Geological Sites for Carbon Capture and Storage (CCS) in India

Works by researchers such as Holloway et al. (2007, 2009) and Viebahn et al. (2014) have estimated the total CCS potential of India. Geological formations are investigated as potential sites for CO_2 sequestration for a considerable geological time (White et al. 2005; Bachu 2007; Holloway et al. 2009; Vishal et al. 2013). These geological formations include deep saline aquifers, depleted oil and gas fields, basalt formations, and the unmineable coal seams for anthropogenic carbon capture and storage. Among all the sink sites for CO_2 sequestration mentioned above, injection in the unmineable coal seams is quite convenient as CO_2-ECBM recovery not only reduces the concentration of GHG from the atmosphere but also recovers methane from the coalbeds, which can be used to meet the energy demand.

10.4.1 CO_2 Sequestration in Deep Saline Aquifers

In India, a significant amount of CO_2 can be stored especially in the saline water-bearing sedimentary rocks in various sedimentary basins. These deep saline aquifers can be traced around the margins of peninsular India, in various offshore basins, and sometimes in some onshore sedimentary basins that are located in the states of Gujarat and Rajasthan (Holloway et al. 2009). In addition to the saline aquifers as mentioned earlier, there are also few potentially less viable sink sites in states such as Assam, Tripura, and Mizoram. CO_2 storage in such saline reservoirs may prove expensive due to long-distance transport of CO_2 from the source (Vishal et al. 2013). Some previous researchers like Dooley et al. (2005), Singh et al. (2006), and Holloway et al. (2009) theoretically estimated the storage capacity in deep saline aquifers to be 100 Gt, 360 Gt, and 138 GT of CO_2, respectively. Considering such problems, to date, no such activities of CO_2 storage in saline aquifers in India have been initiated.

10.4.2 CO_2 Sequestration in Depleted Oil and Gas Fields

A recent study has shown that Indian basins such as Assam basins, Assam–Arakan fold belt, Mahanadi basins, Krishna–Godavari basins, Cauvery basins, Mumbai basins, Cambay basins, Barmer basins, and Jaisalmer basins are considered as having good potential for the CO_2 storage (IEA GHG 2008; Holloway et al. 2009). According to Viebahn et al. (2014), a hydrocarbon-producing reservoir is preferred over a nonproducing reservoir for CO_2 sequestration, as the producing reservoirs will ensure the presence of intact sealing rock that will enable holding the captured CO_2 for a considerably long time. The capacity of CO_2 storage in the

Table 10.3 Estimated CO_2 storage capacity in different Indian basins

Basin	Million tons
Assam Basin	186
Barmer Basin	24
Cambay Basin	156
Mumbai Basin	469

depleted oil and gas fields in India is estimated to be 4.6 Gt (Holloway et al. 2009), or 7 Gt of CO_2 (Singh et al. 2006), or 2 Gt of CO_2 (Dooley et al. 2005) as per various researchers. These oil and gas fields are the potential sites for carbon storage for and beyond the enhanced oil recovery (EOR) and enhanced gas recovery (EGR). However, for implementation of such techniques in the real-time, detailed site-specific scientific studies and cost analysis should be performed. Table 10.3 compiles the estimated CO_2 storage capacity of oil fields in different Indian basins (IEA GHG 2008).

The above-mentioned CO_2 storage capacity is estimated assuming that CO_2 can be injected into the pore spaces in these basins that are occupied by the oil and gas. However, not all oil and gas fields can be considered preferred sites for CO_2 storage as they may involve high transportation costs. Successful storage in these basins would depend on the quantity as well as the trapping mechanism.

10.4.3 CO_2 Sequestration in Basalt Formations

A good portion of India is covered with basalt formations. Deccan trap, one of the largest volcanic provinces in the world, contains major basaltic formations with more than 500,000 km^2 of the area covered in west-central India. The Rajmahal-traps also present with a spreading area of more than 200,000 km^2 in the eastern part of India. Based on the theoretical analysis, the storage capacity of basalt in India is estimated to be 200 Gt of CO_2 (Singh et al. 2006). The mechanism of CO_2 storage in basaltic formation is purely based on the reaction of aluminosilicate minerals with injected CO_2. In the present scenario, the technology for CO_2 storage in basalts is considered immature, and it is expected that in future it can be proved to have a good storage potential in India. Recent works on Deccan basalts using experimental analysis established that these rocks are potentially viable for CO_2 sequestration through the mineral trapping process (Prasad et al. 2012; Guha Roy et al. 2016). They explained that the secondary carbonate minerals over the surfaces and in the intergranular spaces result due to mineral carbonation reaction, and the CO_2 gets almost permanently stored.

10.4.4 CO_2 Sequestration Unmineable Coal Seams

Based on the period of formation, Indian coals are classified as Gondwana coal (Permo-Carboniferous) and tertiary coals, of which nearly 99% of coals are of Gondwana deposits. A brief description of the different categories of Indian coals is presented in ourearlier works (Vishal et al. 2013, 2015b).

Coal can be an effective sink for long-term CO_2 sequestration because of its highly porous nature and high carbon density that provides good surface area for gas adsorption (Vishal 2017a). Major portions of Indian coal are of bituminous to sub-bituminous type, which are of Permian age and are mostly situated in the Raniganj and Barakar coal measures that belong to Lower Gondwana Supergroup. Figure 10.11 illustrates a state-wise distribution of bituminous and sub-bituminous coal in India.

Out of the total estimated coal deposits, with the existing technologies, reserves up to a maximum depth of 300 m are successfully exploited. Due to the unavailability of modern technologies and mining limitations, beyond 300 m, the coal deposits do not appear to be feasible (Gupta 2004).

It has been estimated that around 3.2 billion tons of coal may present up to a depth range of 1200 m in the East Bokaro coalfields; and about 6.7 billion tons of Barakar coal and 0.97 billion tons of superior quality coal may present below mining depths in Jharia and Raniganj coalfields, respectively. Similarly, in South Karanpura and Sohagpur coalfields, coal reserves at a depth of 1200 m are expected (IEA GHG 2008). If these unexplored blocks of coal are proved suitable, they may be available for purposes such as underground coal gasification, carbon dioxide storage, and coal bed methane. The injected CO_2 gets adsorbed onto the porous surface of coal, and as a result of which the partial pressure of the methane is reduced, thereby facilitating its desorption from the coal matrix. For the storage of CO_2, the permeability of coal beds plays a key role; for the successful CO_2 injection and CBM production, the permeability of coal should be higher than 1 mD (Bachu et al. 2007). Permeability of the coal gradually decreases with depth because of the cleat closure with higher confinements at greater depths (Vishal et al. 2013). In general, the shallow depth coals have a permeability in the order of millidarcy (mD). In contrast the deeper coals have a permeability in the range of microdarcy (μD), which are not suitable for CO_2 injection and flow without reservoir fracturing. Worldwide, most of the CBM-producing wells are below 1000 m, and 1200–1300 m is considered the maximum suitable depth for successful production of CBM. The supercritical phase of CO_2 exists above the pressure and temperature condition of 7.38 MPa and 31.1 °C, respectively, which may be encountered at a subsurface depth of around 800 m (Vishal 2017b). So, at greater depths of nearly about 800–1000 m, the injection of CO_2 is likely.

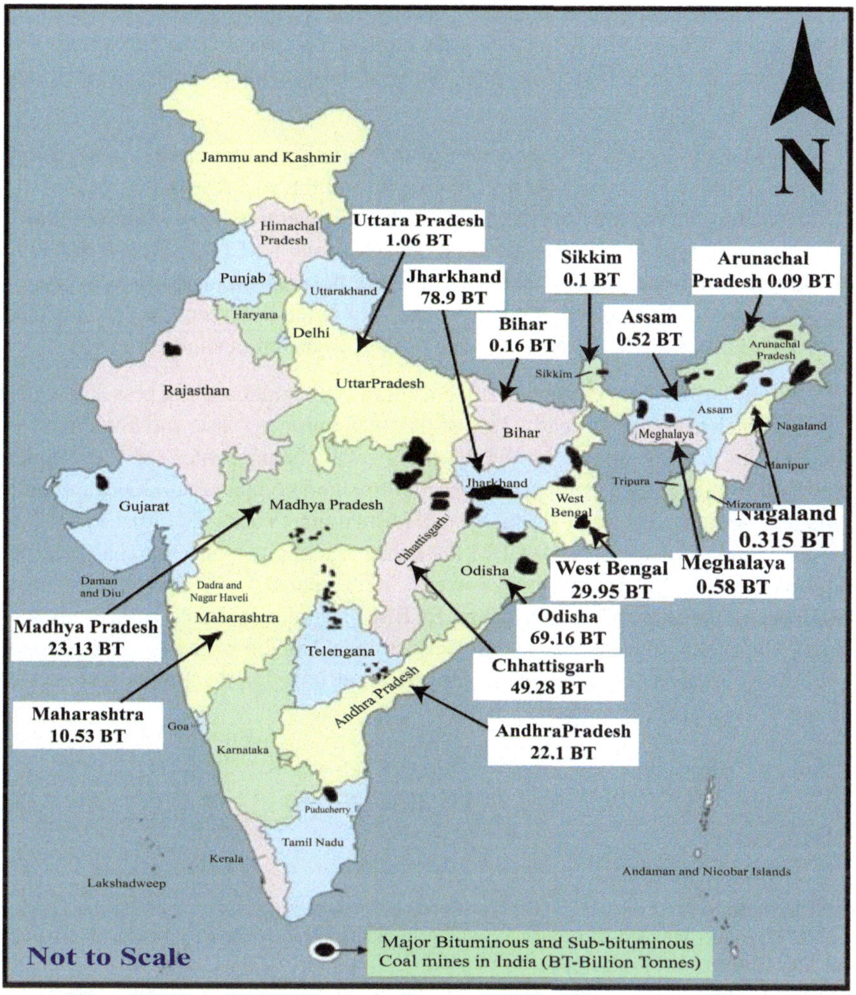

Fig. 10.11 State-wise bituminous and sub-bituminous coal in India (modified after Prabu and Mallick (2015))

10.5 Summary

A large amount of CO_2 may be sequestrated in coal seams, and the caprock of the coal seam, which is mostly impermeable, helps in holding the captured CO_2. Unless any serious sink formation damage is induced, the injected CO_2 can be stored in the coal seam for a long geological time. Both experimental and numerical approaches have been carried out by various researchers in the past few decades to understand the CO_2-ECBM recovery process. Experimental approach includes researchers such as Stevens et al. (1998), Zuber (1998), Dutta et al. (2011), Yang et al. (2011), Zhang et al. (2011), Ranathunga et al. (2015), Vishal et al. (2015a), Vishal (2017a, b) and Wang et al. (2016a). In the last few decades, many researchers have performed experiments on CH_4/CO_2 replacement. Some of the outcomes of these experimental works can be summarized as:

1. CO_2 injection results in a reduction of the partial pressure of CH_4, and thereby accelerating CH_4 desorption from the surface of the coal matrix.
2. Depending on the affinity, there is always a competitive adsorption mechanism between CH_4 and CO_2. As coal has a higher affinity for CO_2, CH_4 is replaced by CO_2 from the coal matrix, thereby enhancing the coalbed methane recovery.
3. As a result of CO_2 injection, the flow capacity of the coal seam can be enhanced, which will facilitate the easy flowage of CH_4 to the production wellbore.

Most of India's CO_2 emissions are a result of coal combustion in power and steel plants, which account for nearly 72% of the total emission (Garg and Shukla 2009). Considering the operational proximity and available storage sites for CO_2 sequestration in India, ECBM is the most preferred method compared to CO_2 trapping in offshore sedimentary basins, and basalt formations (Vishal 2017b). So far, the research and developments for CBM are concentrated on Gondwana coals, and the foremost work for CBM studies were implemented in the Raniganj basin. Addressing the implementation of the ECBM technique in India, some of India's energy analysts suggest the initial CO_2 storage can be opted-in areas already producing coalbed methane where the effective cost for the production can be lowered (Vishal et al. 2015a).

References

Akash AR, Rao AB, Chandel MK (2016) Prospects of implementing CO_2 capture and sequestration (CCS) in the proposed supercritical coal power plants in India. Energy Procedia 90:604–612. https://doi.org/10.1016/j.egypro.2016.11.229

Azmi AS, Yusup S, Muhamad S (2006) The influence of temperature on adsorption capacity of Malaysian coal. Chem Eng Process Process Intensif 45:392–396. https://doi.org/10.1016/j.cep.2005.10.006

Bachu S (2007) Carbon dioxide storage capacity in uneconomic coal beds in Alberta, Canada: methodology, potential and site identification. Int J Greenhouse Gas Control 1:374–385. https://doi.org/10.1016/S1750-5836(07)00070-9

Bachu S, Bonijoly D, Bradshaw J et al (2007) CO_2 storage capacity estimation : methodology and gaps. Greenhouse Gas Control 1:430–443. https://doi.org/10.1016/S1750-5836(07)00086-2

Baisheng N, Longkang W, Xiangchun L et al (2013) Simulation of the interaction of methane, carbon dioxide and coal. Int J Min Sci Technol 23:919–923. https://doi.org/10.1016/j.ijmst.2013.11.007

Boden T, Marland G, Andres R (2016) Global, regional, and national fossil-fuel CO_2 emissions

Butland CI, Moore TA (2008) Secondary biogenic coal seam gas reservoirs in New Zealand: a preliminary assessment of gas contents. Int J Coal Geol 76:151–165

Dooley JJ, Kim SH, Edmonds JA et al (2005) A first-order global geological CO_2-storage potential supply curve and its application in a global integrated assessment model. Greenhouse Gas Control Technol:573–581. https://doi.org/10.1016/B978-008044704-9/50058-6

Dutta P, Bhowmik S, Das S (2011) Methane and carbon dioxide sorption on a set of coals from India. Int J Coal Geol 85:289–299. https://doi.org/10.1016/j.coal.2010.12.004

EEA (2013) Observed trends in total global concentration of the Kyoto gases, 1850–2010. https://www.eea.europa.eu/data-and-maps/figures/observed-trends-in-the-kyoto-gases-1. Accessed 22 Oct 2017

Garg A, Shukla PR (2009) Coal and energy security for India: role of carbon dioxide (CO_2) capture and storage (CCS). Energy 34:1032–1041. https://doi.org/10.1016/j.energy.2009.01.005

Garg A, Shukla PR, Kapshe M (2006) The sectoral trends of multigas emissions inventory of India. Atmos Environ 40:4608–4620. https://doi.org/10.1016/j.atmosenv.2006.03.045

Garg A, Shukla PR, Parihar S et al (2017) Cost-effective architecture of carbon capture and storage (CCS) grid in India. Int J Greenhouse Gas Control 66:129–146. https://doi.org/10.1016/J.IJGGC.2017.09.012

Global Carbon Atlas (2016) CO_2 emission. http://www.globalcarbonatlas.org/en/CO2-emissions. Accessed 18 Sept 2017

Global CCS Institute (2013) India CCS scoping study: final report. https://hub.globalccsinstitute.com/publications/india-ccs-scoping-study-final-report/4-current-ccs-activity-india. Accessed 6 Sept 2017

Global CCS Institute (2015) The global status of CCS: 2015, summary report. Melbourne, Australia

Global CCS Institute (2017) Large-scale CCS facilities. https://www.globalccsinstitute.com/projects/large-scale-ccs-projects. Accessed 5 Sept 2017

Guha Roy D, Vishal V, Singh TN (2016) Effect of carbon dioxide sequestration on the mechanical properties of Deccan basalt. Environ Earth Sci 75:1–13. https://doi.org/10.1007/s12665-016-5587-4

Gupta SP (2004) India vision 2020: the report: report of the committee on India vision 2020, planning commission, government of India. Academic Foundation, New Delhi

Hamawand I, Yusaf T, Hamawand SG (2013) Coal seam gas and associated water: a review paper. Renew Sust Energ Rev 22:550–560. https://doi.org/10.1016/j.rser.2013.02.030

Hildenbrand A, Krooss BM, Busch A, Gaschnitz R (2006) Evolution of methane sorption capacity of coal seams as a function of burial history – a case study from the Campine Basin, NE Belgium. Int J Coal Geol 66:179–203

Holloway S, Pearce JM, Hards VL et al (2007) Natural emissions of CO_2 from the geosphere and their bearing on the geological storage of carbon dioxide. Energy 32:1194–1201. https://doi.org/10.1016/j.energy.2006.09.001

Holloway S, Garg A, Kapshe M et al (2009) An assessment of the CO_2 storage potential of the Indian subcontinent. Energy Procedia 1:2607–2613. https://doi.org/10.1016/j.egypro.2009.02.027

IEA (2012) Energy technology perspectives 2012: pathways to a clean energy system. International Energy Agency, Paris

IEA (2015) World energy outlook 2015. International Energy Agency, Paris

IEA GHG (2008) A regional assessment of the potential for CO_2 storage in the Indian subcontinent. IEA Greenhouse Gas R&D programm

IPCC (2007) Climate change 2007: IPCC fourth assessment report (AR4). Intergovernmental Panel on Climate Change, Geneva

IPCC (2014) Climate change 2014: mitigation of climate change. In: Edenhofer et al (eds) Contribution of working group III to the fifth assessment report of the Intergovernmental Panel on Climate Change. Cambridge University Press, Cambridge/New York

Kedzior S (2009) Accumulation of coal-bed methane in the south-west part of the Upper Silesian Coal Basin (southern Poland). Int J Coal Geol 80:20–34. https://doi.org/10.1016/j.coal.2009.08.003

Leung DYC, Caramanna G, Maroto-Valer MM (2014) An overview of current status of carbon dioxide capture and storage technologies. Renew Sust Energ Rev 39:426–443. https://doi.org/10.1016/j.rser.2014.07.093

Mazumder S, van Hemert P, Busch A et al (2006) Flue gas and pure CO_2 sorption properties of coal: a comparative study. Int J Coal Geol 67:267–279. https://doi.org/10.1016/j.coal.2005.12.001
Ministry of Coal (2016) Coal directory of India 2015-2016: coal statistics. Government of India, Ministry of Coal, Coal Controller's Organisation, Kolkata
Nair R, Shukla PR, Kapshe M et al (2003) Analysis of long-term energy and carbon emission Scenarios for India. Mitig Adapt Strateg Glob Chang 8:53–69
NOAA-ESRL (2017) Monthly CO2 data, Mauna Loa Observatory. https://www.co2.earth/monthly-co2. Accessed 22 Oct 2017
Prabu V, Mallick N (2015) Coalbed methane with CO_2 sequestration: an emerging clean coal technology in India. Renew Sust Energ Rev 50:229–244. https://doi.org/10.1016/j.rser.2015.05.010
Prasad PSR, Sarma DS, Charan SN (2012) Mineral trapping and sequestration of carbon-dioxide in Deccan basalts: SEM, FTIR and Raman spectroscopic studies on secondary carbonates. J Geol Soc India 80:546–552. https://doi.org/10.1007/s12594–012–0175-z
Ranathunga AS, Perera MSA, Ranjith PG et al (2015) A macro-scale experimental study of sub- and super-critical CO_2 flow behaviour in Victorian brown coal. Fuel 158:864–873. https://doi.org/10.1016/j.fuel.2015.06.047
Saini V, Gupta RP, Arora MK (2016) Environmental impact studies in coalfields in India: a case study from Jharia coal-field. Renew Sust Energ Rev 53:1222–1239. https://doi.org/10.1016/j.rser.2015.09.072
Singh AK, Mendhe VA, Garg A (2006) CO_2 sequestration potential of geologic formations in India. In: 8th International Conference on Greenhouse Gas Control Technologies (GHGT-8). Trondheim, Norway
Stevens SH, Spector D, Riemer P (1998) Enhanced coalbed methane recovery using CO2 injection: worldwide resource and CO2 sequestration potential. Soc Pet Eng:489–501. https://doi.org/10.2118/48881-MS
The World Bank (2017) World Bank Open Data. https://data.worldbank.org. Accessed 1 Aug 2017
Viebahn P, Vallentin D, Höller S (2014) Prospects of carbon capture and storage (CCS) in India's power sector – an integrated assessment. Appl Energy 117:62–75. https://doi.org/10.1016/j.apenergy.2013.11.054
Vishal V (2017a) Saturation time dependency of liquid and supercritical CO_2 permeability of bituminous coals: implications for carbon storage. Fuel 192:201–207. https://doi.org/10.1016/j.fuel.2016.12.017
Vishal V (2017b) In-situ disposal of CO_2: liquid and supercritical CO_2 permeability in coal at multiple down-hole stress conditions. J CO2 Util 17:235–242. https://doi.org/10.1016/j.jcou.2016.12.011
Vishal V, Singh L, Pradhan SP et al (2013) Numerical modeling of Gondwana coal seams in India as coalbed methane reservoirs substituted for carbon dioxide sequestration. Energy 49:384–394. https://doi.org/10.1016/j.energy.2012.09.045
Vishal V, Ranjith PG, Singh TN (2015a) An experimental investigation on of coal under fluid saturation, using acoustic emission. J Nat Gas Sci Eng 22:428–436. https://doi.org/10.1016/j.jngse.2014.12.020
Vishal V, Singh TN, Ranjith PG (2015b) Influence of sorption time in CO_2-ECBM process in Indian coals using coupled numerical simulation. Fuel 139:51–58. https://doi.org/10.1016/j.fuel.2014.08.009
Vishal V, Mahanta B, Pradhan SP et al (2018) Simulation of CO 2 enhanced coalbed methane recovery in Jharia coal fields, India. Energy 159:1185–1194. https://doi.org/10.1016/j.energy.2018.06.104
Wang K, Wang G, Ren T, Cheng Y (2014) Methane and CO_2 sorption hysteresis on coal: a critical review. Int J Coal Geol 132:60–80. https://doi.org/10.1016/j.coal.2014.08.004
Wang H, Ran Q, Liao X et al (2016a) Study of the CO_2 ECBM and sequestration in coalbed methane reservoirs with SRV. J Nat Gas Sci Eng 33:678–686. https://doi.org/10.1016/j.jngse.2016.06.007

Wang Q, Li W, Zhang D et al (2016b) Influence of high-pressure CO_2 exposure on adsorption kinetics of methane and CO_2 on coals. J Nat Gas Sci Eng 34:811–822. https://doi.org/10.1016/j.jngse.2016.07.042

White CM, Smith DH, Jones KL et al (2005) Sequestration of carbon dioxide in coal with enhanced coalbed methane recovery – a review. Energy Fuel 19:659–724. https://doi.org/10.1021/ef040047w

Yang K, Lu X, Lin Y, Neimark AV (2011) Effects of CO_2 adsorption on coal deformation during geological sequestration. J Geophys Res Solid Earth 116:1–11. https://doi.org/10.1029/2010JB008002

Zhang L, Aziz N, Ren TX, Wang Z (2011) Influence of temperature on coal sorption characteristics and the theory of coal surface free energy. Procedia Eng 26:1430–1439. https://doi.org/10.1016/j.proeng.2011.11.2320

Zuber MD (1998) Production characteristics and reservoir analysis of coalbed methane reservoirs. Int J Coal Geol 38:27–45. https://doi.org/10.1016/S0166-5162(98)00031-7

Part IV
Landslide: Monitoring

Chapter 11
Hydrological Behavior of Unsaturated Shallow Soils on a Slope and Its Failure Mechanism: A Case Study in Ren River Catchment, China

Xinsheng Wei, Wen Fan, Massimiliano Bordoni, and Claudia Meisina

Abstract Rainfall-induced shallow landslides have frequently occurred in the last decade, especially during or after long-lasting rainfall events in the summer rainy period, leading to the destruction of infrastructure and casualties in the Ren River catchment of Qinba Mountain region, at the center region of China. In this area, superficial soils of clayey silt texture with randomly distributed gravels overlapping the epimetamorphic stratums are generally affected by shallow landslides under the effect of rainfall infiltration. A long-term field hydrological monitoring research has been conducted in order to capture the variably saturated subsurface flow and to understand corresponding triggering mechanism of shallow landslides with similar geological features in the Ren River catchment, in which rainfall-induced shallow landslides have been perceived as a prevalent phenomenon in the last decades. Hydrological variables have been continuously monitored with regard to volumetric water content, matric suction, and positive pore water pressure derived from the ponding water table located at the interface between soil layers with differences in permeability, with the aim of identifying the intrinsic triggering mechanism of a rainfall-induced shallow landslide. Moreover, in situ measured soil water characteristic curves upon both wetting and drying paths were also identified, which facilitated to understand the hydro-mechanical properties of slope soil and can be utilized for the prediction of the infiltration process and the slope stability assessment through an analytical solution in response to seasonal meteorological factors. Preliminary results indicate that volumetric moisture content and matric suction were relative constant in drying period under a low rainfall and low evaporation rate, with a high saturation degree due to antecedent frequent rainfall conditions in shallow soils following the late autumn season. However, the hydrological parameters varied frequently under combined effects of alternative rainfall and pronounced

X. Wei · W. Fan (✉)
Chang'an University, Xi'an, China
e-mail: fanwen@chd.edu.cn

M. Bordoni · C. Meisina
University of Pavia, Pavia, Italy

M. De Maio, A. K. Tiwari (eds.), *Applied Geology*,
https://doi.org/10.1007/978-3-030-43953-8_11

evapotranspiration process. Furthermore, the matric suction increased in an exponential form and reached a peak value of 210 KPa before a long-term rainfall event occurred. The transient perched water table was identified after an extremely heavy rainstorm within a short period of time, which can be reasonably deduced as the triggering mechanism of shallow landslide. Preferential flow was also found to be a significant phenomenon, which led to a lag time between water content and pore water pressure response, as indicated by soil water retention curve.

Keywords Rainfall-induced shallow landslide · Monitoring · Volumetric water content · Matric suction · Pore water pressure · Triggering mechanism

11.1 Introduction

The Qinba Mountain region is located at the middle part of China where geo-hazards, consisting of landslides, debris flow, and rock fall, have frequently occurred in the last decades, particularly involving the shallow landslides triggered by long-lasting rainfall or extreme rainstorms in the rainy season each year. Although the impact of a single shallow landslide was considered to be far less than that of a catastrophic landslide because only a small-scale volume of sliding mass was involved, the extensive distribution and rapid triggering time of shallow landslides have made it possible to impose a threat to the local residence and infrastructure. Some previous researches have indicated that a decrease in matric suction and an increase in pore water pressure play an important role in the intrinsic triggering mechanism of a shallow landslide following the infiltration process into the soil mantle (Lim et al. 1996; Collins and Znidarcic 2004; Cascini et al. 2014; Bordoni et al. 2015). In the Qinba Mountain region, the soil covers overlying the bedrock were generally a production of the weathering parent bedrock, subject to recurrent leaching and erosion effects under humid environments. The macropores, fissures, and random distribution of gravels complicate the infiltration process, probably involving an additional preferential flow in the variably saturated conditions. Consequently, capturing the subsurface flow patterns and hydrological dynamics facilitate to understand the seasonal dynamics of hydrological parameters during different weather conditions and associated triggering mechanism of hydrological-driven shallow landslide.

With the development of several landslide monitoring techniques, high-quality and accurate sensors have been largely adopted for measuring the spatial–temporal variation of hydrological parameters, in terms of the volumetric water content, matric suction, and pore water pressure if the ponding water table formed, which have been testified to be impacted by antecedent hydrological conditions, soil thickness, geological features, and soil permeability (Godt et al. 2006; Smethurst et al. 2012; Springman et al. 2013).

In this research, a meteorological and hydrological monitoring station was established to better understand the interaction between weather conditions and hydrological dynamics within the shallow soil on a representative slope, with the further aim of applying the hydro-mechanical analysis for quantitatively evaluating the landslide triggering mechanism and determining the stability of a rainfall-induced shallow landslide. The monitoring station was located in the Ren River catchment of the southern part of Qinba Mountain region, where dozens of shallow landslide were triggered under long-lasting rainfall infiltration during the rainy season in each year. How hydrological dynamics develop within the superficial soil covers in response to seasonal climate regime and the associated hydro-mechanical characteristics remain unclear in this region, which was conceived as one of critical aspects for assessing the landslide and to facilitate the construction of early warning system for the shallow landslides in a similar geological environment.

11.2 Geological Backgrounds

The Ren River catchment, belonging to Ziyang County at the southern part of Qinba Mountain region, was characterized by complex geology and widely distributed shallow landslides (Fig. 11.1). Averaged annual rainfall amount ranges from 800 to 1200 mm, and most of the rainfall events concentrated in the rainy period from June to September each year. The primary land-use type in this area is cultivated land. From geological point of view, the study area is located at the contact zone between South-Qinling orogen and northern Yangtze platform where Daba arc-shaped fault zone formed extending from northwest to southeast, and resulting into steep slopes and fractured stratums with intersected cracks and joints. The epimetamorphic formations are prevalently distributed in South-Qinling orogen, mainly consisting of lower Ordovician Gaoqiao sandy slate with intermittently interbedded carbonaceous slate and upper Quanhekou argillaceous slate overlaid by Silurian Wuxiahe

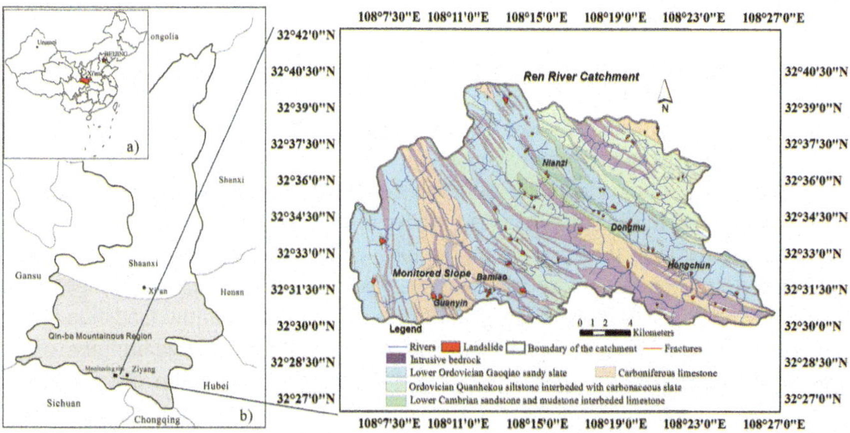

Fig. 11.1 Location and geology map of study area in Ren River Catchment

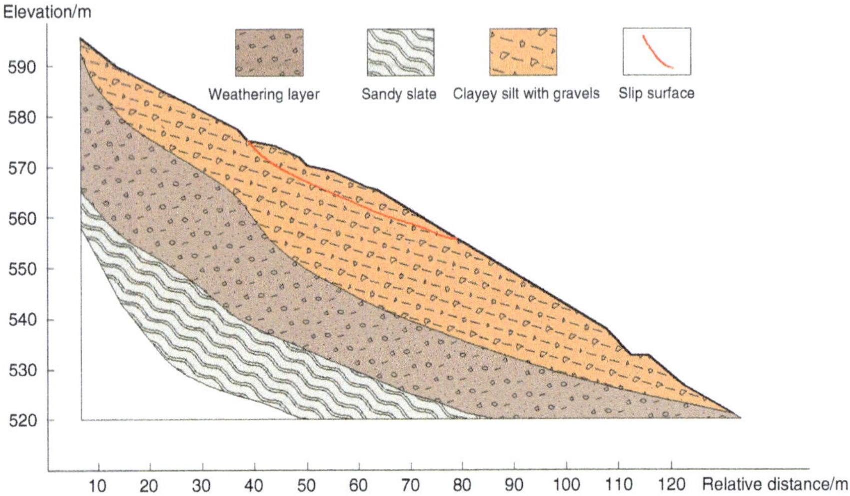

Fig. 11.2 Stratigraphic map of the monitoring slope

silty slate dominating in the northeast part of the catchment area. The formations developed in this period were featured by lower mechanical strength with developed cracks and fractured structure inside the rock mass due to the intense weathering process and erosion by water flow. In the Yangtze platform, the stratums exposed at the slope base mainly belong to sedimentary formation system, including Carboniferous limestone, lower Cambrian sandstone, and mudstone. All exposed formations extend from northwest toward southeast in the catchment scale. The quaternary colluvium mantling on slope was mainly derived from the weathering production of the parent bedrock, featured by unconsolidated soil texture with macropores and randomly distributed gravels. The thickness of soil mantle distribution was assumed to spatially vary in relation with slope morphology, ranging from about 10 m on gentle slope to only 2 m on steep slopes.

Field investigation, including trench pits and drilling exploration, has revealed the thickness of soil mantle and stratigraphic formations, validated by several geophysical profiles implemented during the investigation period. The cross section of the test-site slope was consequently reconstructed based on the obtained information (Fig. 11.2). The thickness of the soil was found to be largely related to the local morphology and the location of slope where thick soil layers of thickness 8–15 m were accumulated at the concave topography of lower part of slope with a gentle inclination and thin layers of thickness 2–3 m were located on steep slope at the upslope with a 30–40 slope degree. Failure surface of the shallow landslide developed in October 2017 was found to initiate at a morphological discontinuity, where the bedrock scarp was exposed. Clayey silt texture with random distribution of gravels was determined for the main soil type after soil particle distribution tests on samples were taken from the trench pits on the investigated and adjacent slopes in the study area. Undisturbed soil samples taken from the test-site were utilized to test the physical and mechanical properties of different soil layer in vertical profile of a slope, as shown in Table 11.1.

11.3 Hydrological Monitoring System

In order to capture seasonal hydrological behaviors of the slope soil and to provide field measured hydro-mechanical parameters for further variably saturated flow modeling, a hydrological monitoring system was established starting from the end of February in 2018. In this monitoring system, a rain gauge (7852 M, Jauntering Inc., Taiwan) and a hygrothermograph (111 N&222 N, Jauntering Inc., Taiwan) were installed to record the variation in atmospheric weather condition as the boundary conditions applied in subsurface hydrological analysis. Two arrays of paired moisture probes (ML3, Delta-T Devices Ltd., UK) and jet-fill tensiometers (SR, Irrometer Company, Inc., US) were buried at 20, 50, and 100 cm depths from the ground with the aim of understanding the variably saturated flow in response to rainfall and of the reconstruction of soil water retention curve (SWRC), considering the hysteresis of unsaturated soil under drying and wetting paths (Fig. 11.3). The vacuum manometer of a tensiometer was replaced with a pressure transducer (PDCR 4010, DRUCK Inc., UK) for automatically recording the monitoring data. Due to the limitation of measuring range of a jet-fill tensiometer (e.g., from −85 to 0 KPa) and potential higher matric suction particularly occurred in hot summer period because of the significant evapotranspiration effect, two equitensiometers (EQ3, Delta-T Devices Ltd., UK) were installed later at 20 cm depth in order to capture higher matric suction value. Then, piezometers (4500AL, Geokon Ltd., US) were both installed within shallow soil layers to capture the potential occurrence of ponding water, and at the top surface of impermeable fresh bedrock to track the variation in deep water table, which possibly affected the hydrological conditions in lower slope soil. The uprising water table could lead to an increase in positive pore water pressure, and subsequent decreasing effect stress and shear strength of the soil.

11.4 Seasonal Hydrological Features of the Slope Soil

The time series of volumetric water content and pore water pressure at different soil depths were collected and recorded every 10 min and transmitted to our laboratory wirelessly. Volumetric water content trends measured at each depth have revealed the soil water balance process involving both the rainfall infiltration and

Table 11.1 Averaged physical and mechanical parameters for shallow soil profile

Depth	Liquid limit	Plastic limit	Density	Dry density	Specific gravity	Porosity	Void ratio	Hydraulic conductivity	Shear strength	
D/cm	ωL/%	ωP/%	ρ/g/cm^3	?d/g/cm^3	Gs	p	e/%	k/10^{-3} cm/s	c/KPa	φ/Degree
20	31.7	23.3	1.63	1.32	2.0	0.756	0.732	5.251	\	\
50	30.4	23.8	1.76	1.28	2.53	0.749	0.691	3.265	\	\
80	33.8	21.7	1.77	1.39	2.65	0.733	0.637	\	\	\
100	32.1	22.5	1.81	1.56	2.7	0.731	0.659	2.407	10.48	2.4
200	\	\	1.84	1.43	\			\	\	\
300	32.6	23	1.91	1.54	\	\	\	3.582	7.24	4.6
400	33	23.4	2.06	1.6	2.706	0.691	0.654	\	9.22	5
500	33	23.2	2.16	1.59	2.712	0.706	0.658	\	6.52	4

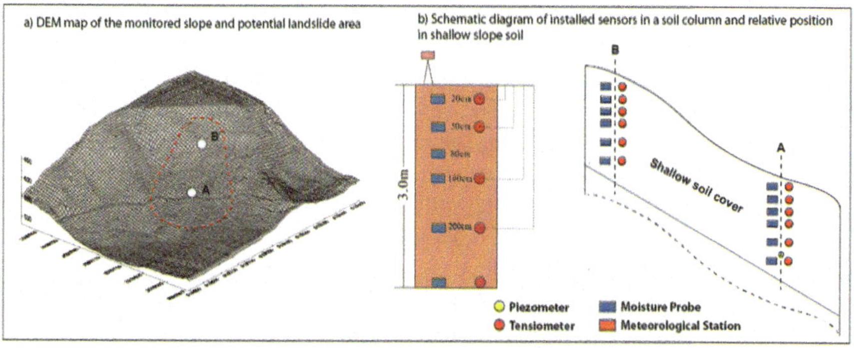

Fig. 11.3 Location of sensors arrays installed on monitoring slope

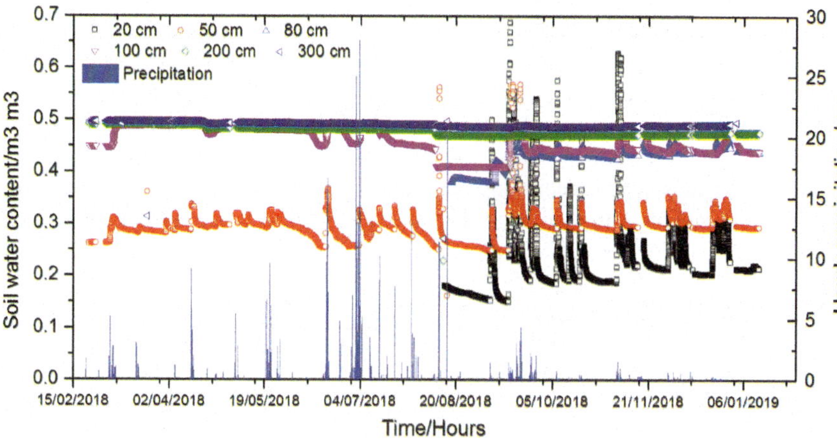

Fig. 11.4 Time series of water content at different soil depths

evapotranspiration effects and also have reflected the characteristics of downward wetting front movement. Results indicated some common hydrological behaviors of clayey silt soil cover during entire hydrological year (Fig. 11.4). In summer wetting periods from the end of May to the beginning of October, the shallowest soil layer typically has a wider fluctuation range and a much rapid variation of water content response compared with that within deeper soil depths. A typical case selected during the period from 14th to 21st September, 2018 (accumulated rainfall amount of 81 mm in 7 days) was used to illustrate the hydrological response. In this period, the water content value measured at 20 cm depth presented that the highest water content was 0.588, but it can only sustain for several minutes prior to a rapid downward gravitational drainage process. After a time lag of 0.41 h, another peak value experienced by the soil layer at 50 cm depth was attained. Compared with water content response to extreme rainfall events measured at 20 cm depth, the water content measured at the depth of 50 cm has a significant attenuation due to both effects of downward drainage, water retention capability of upper soil layer and surface evapotranspiration before another increase of water content induced by a rainfall event. In deeper soils, however, the effect of rainfall infiltration vanished, and each hydrological response was damped and limited with a longer response time, which was supposed to be the effect of decreasing saturated hydraulic conductivity and lower porosity with soil depths as a result of gradually higher consolidation pressure.

Notably, a rapid response to the infiltration and a wider fluctuation of water content monitored at 20 cm indicated the phenomenon of preferential flow leading to the fast development of water content.

The matric suction, also known as negative pore water pressure induced by capillarity effect at a soil particle scale (Lu and Likos 2006), was considered to make an additional contribution to the shear strength of unsaturated soil and consequent slope stability. Due to available sensors as the tensiometer and equitensiometer installed at 20 cm depth from the ground surface since the middle of August in 2018, the suction measurements began to be recorded since then. Available suction value recorded from the equitensiometer installed at 20 cm from the surface indicated that in the summer period, the value continuously increased in an exponential form at the interval period between two contiguous rainfall events due to an evapotranspiration effect near the soil surface. As shown in Fig. 11.5a, once the rainfall began to infiltrate into the soil surface, a sudden dissipation in the suction generally occurred in a few minutes near the surface till a steady condition was reached. The matric

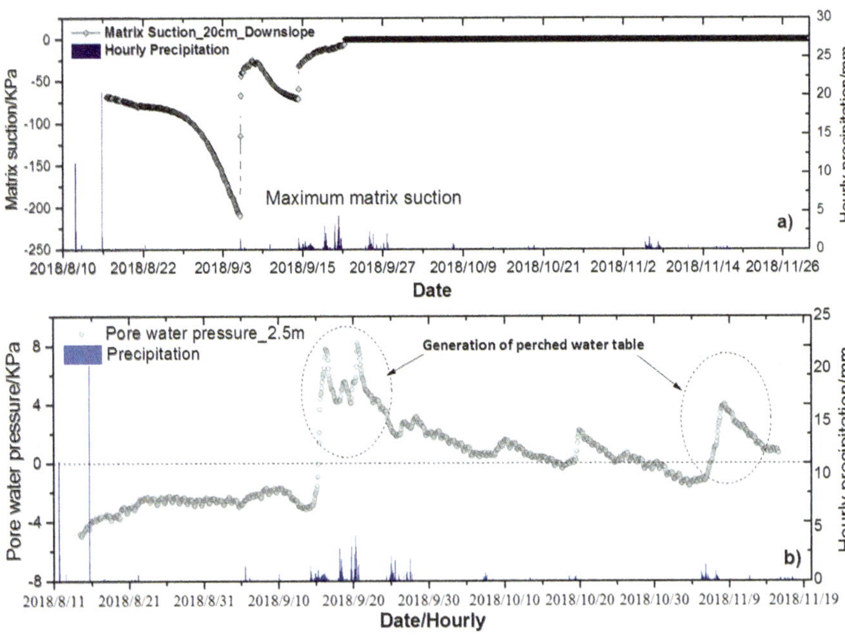

Fig. 11.5 Time series of pore water pressure measurements at depth of 20 cm by equitensiometer (**a**) and at depth of 250 cm by piezometer (**b**)

suction decreased to a certain value depending on the rainfall amount and the soil hydraulic properties. During summer wetting period, although frequent rainfall infiltration have induced the increase of soil water content, pronounced evapotranspiration effect in shallow soil layers due to high temperature and the water uptake of vegetation roots can result into soil water exfiltration upward, maintaining the soil in an unsaturated condition. During relatively drying period in winter and spring with less rainfall events (lower intensity and short duration), the hydrological condition of each soil depth remained quite stable due to lower temperature and less water infiltration. As shown by matric suction measured from the end of September in 2018, the suction value remained near zero in the period after a long-lasting rainfall event of accumulated amount of 110.2 mm after 11 days. Matric suction value measured from tensiometer installed at 100 cm indicated that deeper layer was featured by a near-zero pore pressure condition. Other installed tensiometers did not present reasonable suction due to a not well-coupled hydraulic connection between water in the ceramic cup and in the micropores of loose soil.

Correspondingly, positive pore water pressure was identified at a depth of 2.5 m to determine an ephemeral perched water table following the beginning of long-lasting rainfall event. As the rainfall intensity decreased till to zero, the pressure value also decreased and the ponding water was assumed to infiltrate downward progressively into a underlying less permeable soil layers.

Correspondingly, the field soil water retention curve (SWRC) was reconstructed by paired measurements of matric suction and water content at the same depths for clayey silt on both drying and wetting paths in order to understand the field slope soil hydro-mechanical properties that control the subsurface flow. Apparently, the hysteresis soil water characteristic was clearly identified because the two curves in drying and wetting paths, respectively, were not overlapped. Notably, a sudden increase of water content after rainfall infiltration into superficial soil indicated that the phenomenon of preferential flow might occur due to macropores and cracks in the unconsolidated soil layer (Fig. 11.6).

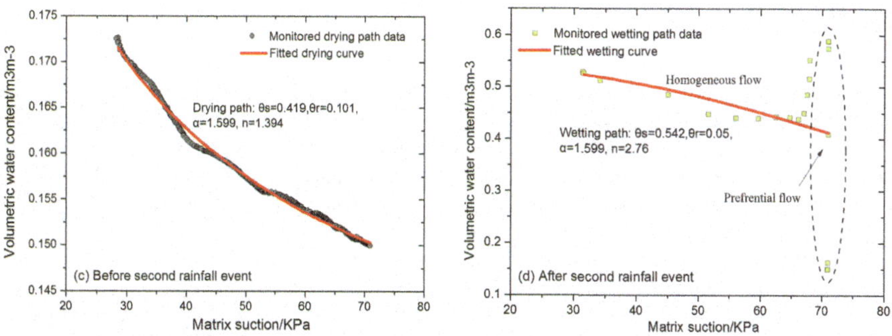

Fig. 11.6 Reconstruction of field soil water retention curve in drying and wetting paths

11.5 Discussion and Conclusion

Seasonal hydrological dynamics of the shallow slope soils during both wetting periods under frequent rainfall events and in drying periods were presented, with the further aim of quantifying the triggering mechanism of rainfall-induced shallow landslide in a hydro-mechanical framework. A seasonal unsaturated zone in shallow layers from the ground surface to about 100 cm deep and an underlying saturated zone were observed and identified based on seasonal hydrological measurements of water content and pore water pressure at different time. The lag time of hydrological response at different soil layers revealed that the progressively decreasing saturated hydraulic conductivities of soil and antecedent hydrological condition determines the feature of variably saturated flow. In drying seasons, due to relatively lower water content and corresponding higher matric suction, it takes longer time for wetting front to move toward a certain depth, leading shallow soil to get a higher saturation degree. However, in the late wetting season in autumn, even deep soil horizon have a much-pronounced response to the rainfall infiltration due to a higher antecedent water content along the vertical soil profile. Even more, a transient perched water table can develop during this period at the top of less permeable layer under long-lasting rainfall, or sometimes even if rainfall intensity was not high. Therefore, a shallow landslide have a higher probability to be initiated due to a rapid increase in the pore water pressure and resultant decreasing effective stress in the late summer period. Meanwhile, the phenomenon of preferential flow through macropores and cracks in the unconsolidated soil mantle have a deep influence on the hydrological behavior of slope, resulting a disequilibrium between water content and pore water pressure response in a transient time scale.

Combing a hydrological monitoring data and a variably hydro-mechanical framework would consequently facilitate to create an effective way to make hydrological-driven landslide evaluation work in the future.

References

Bordoni M, Meisina C, Valentino R, Lu N, Bittelli M, Chersich S (2015) Hydrological factors affecting rainfall-induced shallow landslides: from the field monitoring to a simplified slope stability analysis. Eng Geol 193:19–37. https://doi.org/10.1016/j.enggeo.2015.04.006

Cascini L, Sorbino G, Cuomo S, Ferlisi S (2014) Seasonal effects of rainfall on the shallow pyroclastic deposits of the Campania region (Southern Italy). Landslides 11(5):779–792. https://doi.org/10.1007/s10346-013-0395-3

Collins BD, Znidarcic D (2004) Stability analyses of rainfall induced landslides. J Geotech Geoenviron 130(4):362–372

Godt JW, Baum RL, Chleborad AF (2006) Rainfall characteristics for shallow landsliding in Seattle, Washington, USA. Earth Surf Process Land 31(1):97–110. https://doi.org/10.1002/esp.1237

Lim TT, Rahardjo H, Chang MF, Fredlund DG (1996) Effect of rainfall on matric suctions in a residual soil slope. Can Geotech J 33:618–628

Lu N, Likos WJ (2006) Suction stress characteristic curve for unsaturated soil. J Geotech Geoenviron 132(2):131–142. https://doi.org/10.1061/(asce)1090-0241(2006)132:2(131)

Smethurst JA, Clarke D, Powrie W (2012) Factors controlling the seasonal variation in soil water content and pore water pressures within a lightly vegetated clay slope. Geotechnique 62 (5):429–446

Springman SM, Thielen A, Kienzler P, Friedel S (2013) A long-term field study for the investigation of rainfall-induced landslides. Geotechnique 63(14):1177–1193

Chapter 12
First Steps for the Development of an Optical Fibre Strain Sensor for Shallow Landslide Stability Monitoring Through Laboratory Experiments

Monica Papini, Vladislav Ivov Ivanov, Davide Brambilla, Maddalena Ferrario, Marco Brunero, Gabriele Cazzulani, and Laura Longoni

Abstract Shallow landslides are land instability phenomena provoked by extreme precipitation events of either short duration and significant intensity or milder but longer lasting ones. Those slope failures generally involve limited volumes of terrain which could however induce significant amount of damage to property and cause casualties. In this work, we present an experimental study of the onset of shallow landslides under heavy precipitation in controlled laboratory conditions. The main objective of the work is to explore the applicability of an optical fibre strain sensor interrogated with an innovative low-cost technology. Thus, the system permits the exploitability of the otherwise costly and often unsuitable for field use interrogators. The optical fibre strain sensors were installed within the slope at different locations along the length of the plane. A second set of a more conventional optical fibre sensing system was used for a comparison. The results obtained indicate the ability of both devices to reflect the instabilities developing during the experiments and therefore encouraged the development of further investigation of the proposed sensors' applicability.

Keywords Shallow landslides · Downscaled simulation · Laboratory tests · Landslide monitoring · Coherent optical fibre sensing · Distributed optical fibre

---

M. Papini · V. I. Ivanov (✉) · D. Brambilla · L. Longoni
Department of Civil and Environmental Engineering, Politecnico di Milano, Milan, Italy
e-mail: vladislavivov.ivanov@polimi.it

M. Ferrario · M. Brunero
Department of Electronics, Information and Bioengineering, Politecnico di Milano, Milan, Italy

G. Cazzulani
Department of Mechanical Engineering, Politecnico di Milano, Milan, Italy

M. De Maio, A. K. Tiwari (eds.), *Applied Geology*,
https://doi.org/10.1007/978-3-030-43953-8_12

12.1 Introduction

Hydrogeological instabilities are clearly a great threat for the normal functioning of our modern society as an ever-increasing number of people and property is at risk of some sort of disaster that involves instable land masses. Mitigation of the risk posed by hydrogeological instabilities requires an up to date technology in order to be able to successfully manage a variety of calamitous scenarios, especially when it comes to landslides which may vary greatly in type and scale.

The use of optical fibre sensors for monitoring of hydrogeological instabilities has gained some strong interest during the past few decades. Numerous works could be found throughout the literature which explore the applicability of this technology as a landslide monitoring tool. Applications of optical fibre sensors include inclinometers (e. g. Yoshida et al. 2002; Ho et al. 2006; Li et al. 2010; Pei et al. 2011; Zeni et al. 2015), displacement sensors entrenched directly within the soil (e.g. Iten et al. 2011; Hauswirth et al. 2012; Lienhart, 2015), or various tools designed to incorporate the optical fibre technology in an attempt to ‘upgrade’ an already existing instrument such as a geophone (e.g. Zhang et al. 2015) or a piezometer (e.g. Huang et al. 2012). Some have also undertaken experimental campaigns at the laboratory level in order to explore the applicability of optical fibre sensors in controlled conditions and to confront the obtained readings with a range of monitoring tools which include tensiometers, TDR (Time Domain Reflectometry) probes, tensiometers, displacement sensors etc. (Olivares et al. 2009; Michlmayr et al. 2017; Schenato et al. 2017).

Several fibre optic systems defined as “distributed sensors” can provide a detailed monitoring of immense structures or territory with spatial resolution of meters (or even mm) along the entire sensing length. An overview of the broad range of applications of such sensors is proposed by Schenato (2017). On the other hand, a simpler and more cost-effective approach that can provide as much useful information, is represented by “integral coherent measurements”, in what they yield an integral of the signal response over the entire length of the sensor. This coherent approach proves to be particularly suitable to provide an overall indication of the state of the entire monitored area and yield fast indications of the onset of anomalous situations. Those innovative sensors are said to be a worthy competitor to what is already established and commercially available instruments. While many authors emphasize on the reduced costs and durability of fibre optic sensors, it has been seldom discussed that interrogators are in fact the exact opposite – costs extend up to several tens of thousands of euro and are generally designed to work in protected environment, such as a laboratory. This shortcoming plays a crucial role for the further development of optical fibre monitoring systems and especially when an early warning system is to be the final outcome of the work as constant presence of an interrogator on site would be required.

This work presents an experimental study on the application of fibre optic sensors for landslide stability monitoring. We propose a comparison between two sensing systems – an optical backscatter reflectometer (OBR), currently used in a series of applications as a strain and temperature measurement tool, and a recently developed integral coherent fibre optic sensing solution (Martinelli and Ferrario 2013). The goal of this work is to validate the applicability of the latter for landslide stability monitoring through a series of laboratory experiments. The manuscript is organized as follows: a description of the experimental setup introduces the procedure followed during the tests along with the instrumentation used. Particular attention is devoted to the optical fibre sensing systems. A description of the results and a discussion provides and overview of what was obtained after data processing, while a brief conclusion summarizes the outcome of this work.

12.2 Experimental Setup

Experimental activity was carried out in the laboratory of mountain hydraulics and applied geology situated in the Lecco campus of Politecnico di Milano. In order to accommodate the simulation of sliding soil masses, a steel structure facility was constructed (Fig. 12.1a). It is composed of two adjustable surfaces (Arosio et al. 2019). The upper one hosts the terrain under investigation. It has dimensions of 2 × 0.8 m and its inclination is adjustable up to 45°. The lateral sides of the channel are made of Plexiglass for better visual inspection, while the bottom is equipped with

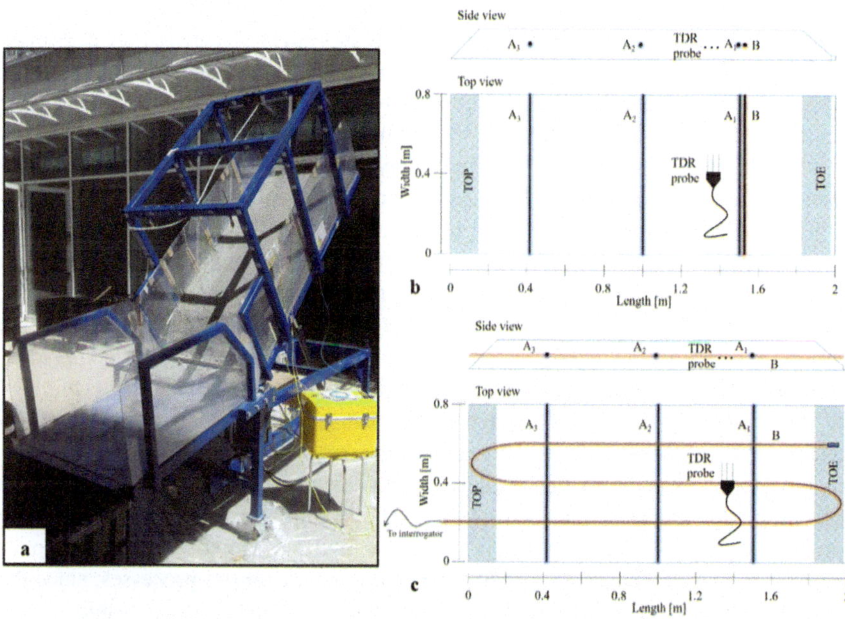

Fig. 12.1 Landslide simulator (**a**) Schematic representation of the experimental setups I (**b**) and II (**c**), indicating the position of the sensors within the slope. Sensors labelled A_i refer to the integral coherent optical fibre setup. Sensors labelled B refer to the OBR sensing system

a geogrid in order to ensure friction between the soil mass and the structure. The structure features a sprinkler system designed to simulate rainfall of various intensities as a main instability triggering factor. As the purpose of the experimental campaign was the functionality of the optical fibre sensors, generic homogeneous sand ($d_{50} = 0.35$ mm, $G_s = 2620$ kg/m^3) was used to create the soil body.

12.2.1 Monitoring

Saturation of the soil was monitored through a TDR probe at 1 min intervals. The probe was placed at mid-depth of the soil in order to avoid effects of the highly resistive boundaries at the bottom of the simulator and the air on the surface. The volumetric water content measured by the probe is an average value estimated along the probe length (15 cm). Action cameras were installed on the frame of the simulator (Scaioni et al. 2018) in order to document ongoing processes. Images were taken at 2 s interval so as to identify swiftly developing failure processes during the experiment. Two types of optical fibre sensors were put to the test within the scope of this laboratory work. The first, a commercially available optical distributed sensor interrogator, was employed to measure the strain exerted by the soil on a cable placed directly within the soil body. The LUNA ODiSI-B ® optical fibre measurement system is based on the use of a fibre optic cable as a distributed strain and temperature sensor, exploiting the Optical Backscatter Reflectometry (OBR). When light wave propagates in a medium, it provokes secondary ("scattered") electromagnetic waves due to its interaction with the medium. If the medium is not homogeneous in terms of density (thus, in terms of refraction index), as a real fibre is, the scattering provokes a back-scattered wave. Such phenomenon is called spontaneous Rayleigh scattering. Applying this phenomenon to optical sensors, when glass fibre is scanned through an Optical Frequency Domain Reflectometer (OFDR), a fluctuating intensity profile of the Rayleigh scattering along the fibre will be detected. Such profile is stable when the measurement is repeated under unaltered external conditions (no load applied), such that it represents the characteristic "*fingerprint*" for a specific fibre segment. If an external mechanical load is applied to the fibre (or its temperature changed), its fingerprint is spatially stretched or compressed (Fig. 12.2-left). This phenomenon is the basis for the Rayleigh sensor technology as the changes to the local Rayleigh pattern can be converted into local strain or temperature values. The potentiality represented by such fibre optic technology lays in the extremely high spatial resolution obtainable on a single fibre optic. With current commercially available systems, spatial resolution can be set to a minimum of approximately 1 mm with a fibre length up to 70 m. Moreover, the fibre optics employed by this technology needs no specific manufacturing processes, since the defects exploited to reflect back the laser signal are naturally available in any standard telecom fibre.

Two different configurations were tested with this sensing system, illustrated on Fig. 12.1b (I) and c (II) respectively. Within the scope of experimental configuration I, the optical fibre was placed transversally to the slope, while in

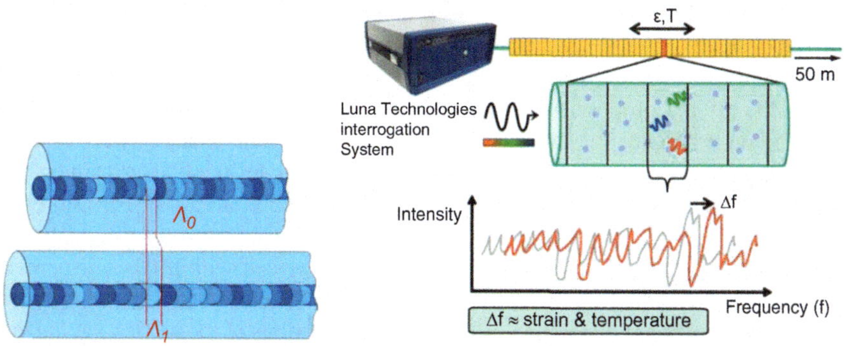

Fig. 12.2 Stretching of the fibre fingerprint (left) and consequent measured frequency response shift occurring on the fibre (right)

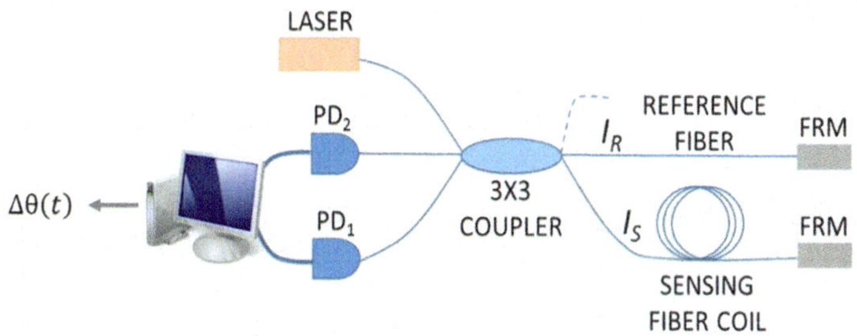

Fig. 12.3 Schematic representation of the fibre optic coherent sensing principle. (PD_i = Phase difference; FRM = Faraday Rotator Mirror)

configuration II the sensor was placed in parallel with the landslide direction. The sensor was long enough to pass three times through the landslide body and therefore, an '*S*' shape pattern was deemed suitable for the experiment. The output of the system is a direct measure of the strain undergone by the sensor within the soil. The sensing system operates at a sampling rate of 8 Hz with a spatial resolution of 1 mm.

An Interferometric Fibre Optic Sensing System (IFOSS) was designed to be tested together with the OBR sensor described above. The coherent fibre optic sensing solution exploited in this work relies on a novel phase-diversity coherent detection scheme (Martinelli and Ferrario 2013) which combines a 3×3 optical coupler with standard optical fibres terminated with Faraday rotator mirrors (FRMs), as shown in Fig. 12.3. The incident light wave is divided into two parts (sensing and reference) by a beam splitter on fibre optic coupler. In the sensing arm, the interaction of the measurand (i.e. the pressure exerted by the moving soil) with the sensing fibre, causes a strain ΔL of the fibre itself which, in turn, results in a perturbation of the phase $\Delta\theta(t)$ of the propagating optical signal. Thanks to the adopted phase-diversity scheme the phase behaviour, $\theta(t)$, can be recovered in a completely passive way without the need of an active feedback for quadrature point stabilization as in conventional fibre optic interferometry (Martinelli 1989). The proposed phase-diversity coherent detection scheme thus allows to overcome issues that until now have limited the actual in-field applicability, cost-effectiveness, broadband and high sensitivity featured by fibre optic interferometric sensors. The versatility of the proposed coherent sensing technique has already been widely proven, for example, for impact damage diagnostic in aeronautic structures, for cavitation onset detection in hydraulic pipes, as well as for predictive maintenance applications in industrial production process (Sbarufatti et al. 2016; Ferrario et al. 2016; Boffi et al. 2018).

Interferometric sensors used in this work were placed within the soil perpendicularly to the sliding direction in both experimental setups I and II, thus retrieving the integral phase difference, $\Delta\theta(t)$, at known sections of the model – at the toe, middle and top of the soil body. Variation of its value reflects the strain exerted on the cable by the moving soil mass. Sampling rate of the measurements could be set up to 20 kHz.

12.2.2 Experimental Conduct

Soil was placed in the flume in three layers of 5 cm each. Uniform compaction across the channel was ensured by fitting a known mass of soil to a predefined volume. Initial soil moisture content was set to 10% for both experiments by adding water and mixing it with the dry soil. A TDR probe was used to control this parameter until the desired value was reached. Optical fibre sensors were placed directly within the soil layers during their placement in the flume. Cables were led out of the soil on the side of the flume, where they were attached in order to limit their displacement and ensure that the sensing portion is situated within the limits of the mass that would eventually be moving. A TDR probe was installed during the soil preparation as well. After the terrain was settled, the upper platform was lifted to form an inclination of 40°. Rainfall was then initiated and controlled to maintain a constant value through a pressure control valve. Rainfall intensity was then estimated through the known relationship of pressure-sprinkler discharge, and the surface covered by the spray cones. Moisture content was then measured at 1 m intervals during the experiment, while photographic material was collected at 2 s intervals by the action cameras. Deformation induced on the optical fibre sensing cables was recorded by the respective interrogating system – the interferometric and OBR interrogation devices. The experiments lasted about 33 min each. The functionality of some cables was however undermined at an earlier instant due to a rupture of the sensor caused by collapsing soil. In the measurements carried out in this work the effect of environmental temperature fluctuations on the signals could be considered negligible with respect to deformations caused by landslide induced-strain and thus compensation techniques were not implemented (Table 12.1).

Table 12.1 Experimental setup parameters

Parameter	value
Initial volumetric water content [%]	10
Rainfall intensity [mm/h]	88
Soil porosity [−]	0.52
Inclination [°]	40

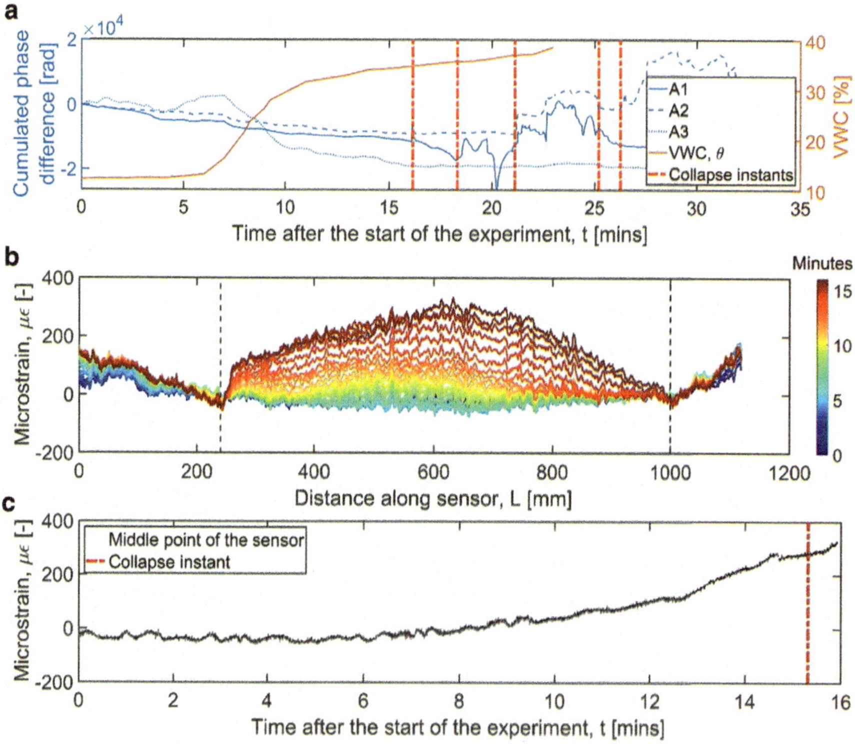

Fig. 12.4 Graphical representation of the data obtained from the first experimental setup. (**a**) Interferometric optical fibre sensing system. (**b**) OBR distributed optical fibre sensing system – strain along the cable length. Colorbar indicates the progression of strain with time at 30s intervals. Dashed lines indicate cable clamping positions. (**c**) Strain over time at L = 600 mm

12.3 Results and Discussion

The results of the experiments previously described are reported in a graphical manner on Figs. 12.4 and 12.5. Figure 12.4a illustrates the trends registered by the interferometric sensors A1, A2 and A3 respectively as well as the variation in volumetric water content recorded by the TDR probe. Instants corresponding to some notable collapses are additionally reported in order to provide reference points for data interpretation. The first 15 min of the experiment indicate a steady negative trend, due to the cable settling after their installation within the sand mass. Upon

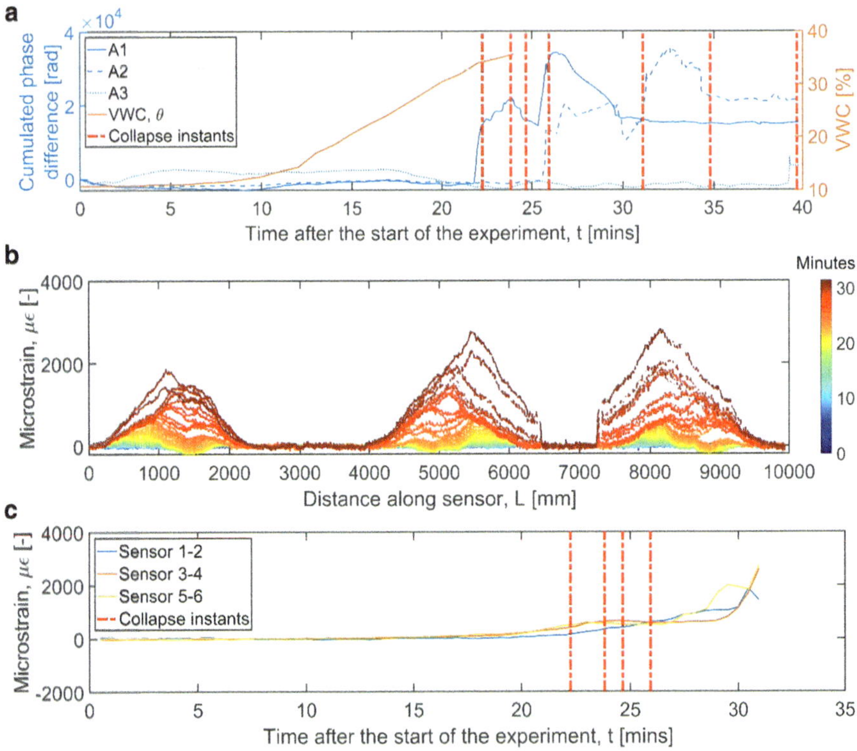

Fig. 12.5 Graphical representation of the data obtained from the second experimental setup. (**a**) Interferometric optical fibre sensing system. (**b**) OBR distributed optical fibre sensing system – strain along the cable length. Colorbar indicates the progression of strain with time at 30s intervals. (**c**) Strain over time at L = 1000 mm (sensor 1–2); L = 5500 mm (sensor 3–4); L = 8500 mm (sensor 5–6)

increase of soil saturation, some disturbance of the sensor can be observed, registered first by sensor A1, positioned at the toe of the slope, while sensors A2 and A3 continue to show a steady trend. After two registered collapses, sensor A1 remains exposed. The following collapses are well registered by sensor A2 which demonstrates jumps in the cumulated phase difference until it remains exposed as well. Sensor A3, positioned in the top of the landslide is hardly disturbed by the collapses of terrain which occur below. The TDR probe demonstrates the variation of water content within the soil with a sharp increase which indicates at the instant when rainfall water reaches the probe. This could provide an indication of the velocity with which the wetting front advances towards the bottom of the soil mass. After the failure of soil which covers the TDR probe, its measurements are no more viable and thus, monitoring of this parameter could only be maintained partially. Figures 12.5b and c illustrate the trends registered by the distributed OBR sensing system. As the interrogating device is able to locate the occurring strain, the process could be followed both with respect to time and along the length of the sensor.

Figure 12.4b represents the strain along the entire cable length. Different time instants are indicated by a colorbar, where lines are plotted at 30s intervals. Figure 12.4c illustrates the strain variation in time of the middle point of the sensor (which bears the highest strain). The extremely high sensitivity of the bare optical fibre allows for strain to be distinguished as early as 8 min after the start of the experiment, well before any visual indication of instability could be observed. Unfortunately, the first portion of collapsing soil was able to break the sensor and disrupt is functioning as optical fibre has a rather low ability to resist when strain is induced transversally to the cable, especially when the sensor remains unprotected. Instead, the interferometric sensors, which featured a PVC coating, did not undergo any physical damage. Their sensitivity, however, was decreased due to strain being sustained by the coating, as demonstrated by the trends presented below. This experimental setup allowed for the direct comparison of the two systems under investigation as sensors were placed in an immediate vicinity to each other.

A different approach was followed during the second experimental setup as previously described. While the arrangement of the interferometric sensing system was kept as in the previous experiment, the OBR distributed sensor was placed in an 'S'-shape configuration in order to capture the longitudinal strain induced on the cable by the soil mass. Figure 12.5a represents the trends of the cumulated phase difference yielded by the interferometric senor. As in the first landslide simulation, a steady trend can be observed in the beginning of the experiment, while collapsing terrain could be clearly distinguished by the sensors. Also here, strain is induced on the sensors mostly by the terrain collapsing in the immediate vicinity of the cable and thus initial collapses are detected by sensor A1 wile consecutive ones by sensor A2 respectively. Sensor A3 remains undisturbed for the entire duration of the experiment.

The configuration of the OBR distributed sensing used in this experiment proved to be more reliable with strain of up to 3000με (0.3%) or as much as 10 times more with respect to the first trial before a disruption of the sensor occurred. Similar values were obtained by Zeni et al. (2015) for a similar experimental setup, however using a heavily coated cable. Instead, Schenato et al. (2017) report cable strain of up to 1% with robustly designed corrugated cable configuration. All branches of the sensor were able to detect instabilities during the experiment, whereas some peaks could be distinguished upon the occurrence of sudden ruptures of the terrain. It could be noted however, that first signs of instability can be distinguished by both sensing systems at similar instants. Therefore, no advantage could be attributed to the distributed sensing system in this configuration in terms of detecting precursors. Even though the system is able to resist physical damage, its ability to detect instability is undermined as the direction of sliding is parallel to the direction of the sensor itself. A relative slip between the sensor and soil could be a reasonable explanation of this difference with respect to the first trial. The high sampling rate of the interferometric sensing system could be further exploited to extract information from events at the micro scale. Similarly to the work of Michlmayr et al. (2017), the applicability of the sensor for detection of elastic waves in the accosting range could be an interesting approach for various monitoring applications.

In summary, the OBR distributed optical fibre sensing system provides quite precise measurements of strain both in terms of time and along the length of the sensor. Placed transversally to the direction of sliding, micro displacements of the terrain are easily detected well before any visual sign of instability onset is present. However, the integrity of the sensor could be undermined due to excessive strain. On the other hand, the second experimental configuration was much more stable even though its ability of an early instability detection was compromised. It must be noted that the OBR distributed optical fibre sensing system could be a suitable tool for experimental work at the laboratory scale. However, the dimensions of the instrument, its sensitivity to unfavourable meteorological conditions and dust, as well as its elevated cost could render it rather impractical for onsite installation.

As a comparison, the interferometric optical fibre sensing system provided a robust detection of collapses during the experiments. All sensing cables remained intact, despite the large strain imposed by the soil. This stability is most likely due to the PVC coating of the cable which provides a certain degree of elasticity to the sensor. On the other hand, coating the cable decreases its sensitive to external disturbance. A disadvantage of the system is, at this stage, its inability to locate the occurring strain along the sensor length. The location of the measurand should therefore be known a priori for an effective result. This innovative technology however renders it simple, light and easy to apply at low cost.

12.4 Conclusions

Research into optical fibre technology allows for its use in various fields of expertise. Recently this technology was targeted as an alternative to conventional landslide stability monitoring which is notorious for elevated cost of operation and often low durability and therefore, reliability of the system. This study is aimed at the development of a sustainable fibre-optic sensing systems for landslide stability monitoring. Experimental works were undertaken as first steps for the verification of the functionality of such a monitoring system. The core of this work is a comparison between two optical fibre based interrogating systems, where one is an already well-established instrument that is however used with scopes that are much different from landslide stability monitoring. The other one is a recently developed tool that aims to provide the same or even better measurement performance at a much lower cost. These first tests indicate that the system is in fact able to effectively detect instabilities within the terrain which induce a physical deformation on the sensor. Several possible pitfalls merit further in-depth investigation, especially when a possible real scale application is concerned. In particular, temperature fluctuations could have an impact on sensor measurements due to the sensitivity of the optical fibre sensors to temperature gradients – an effect which was not considered relevant during the laboratory experiments due to the short temporal and spatial extents of the process. Thus, quantification of the temperature impact would be required. On-site applicability could be another critical aspect since various disturbances limited at the laboratory scale, could emerge and should be explored and taken into account during the design phase. Finally, limited strain resistance of the cables could emerge as a problematic aspect as stresses induced by failing land masses could lead to failure. Technical solutions for optimal cable placement are to be implemented to limit this possibility. Further tests are programmed to assess the applicability of the sensing system in different measurement modes. As those results provide a sound basis for further investigation, the interferometric optical fibre sensing system will be also improved to be able to locate the origin of the displacement as currently this is its main disadvantage with respect to similar instruments.

References

Arosio D, Hojat A, Ivanov VI, Loke MH, Longoni L, Papini M, Tresoldi G, Zanzi L (2019) A laboratory experience to assess the 3D effects on 2D ERT monitoring of river levees. Paper presented at the 24th European Meeting of Environmental and Engineering Geophysics, 9--13 September 2019, Porto, Portugal

Boffi P, Ferrarese G, Ferrario M, Malavasi S, Mastronardi, MV, Mattarei M (2018) Coherent optical fiber interferometric sensor for incipient cavitation index detection. Flow Meas Instrum J In press. https://doi.org/10.1016/j.flowmeasinst.2018.11.005

Ferrario M, Mattarei M, Boffi P, Martinelli M (2016) A software-defined coherent fiber optic sensor for manufacturing machine diagnostic. Paper presented at IEE Sensors Applications Symposium (SAS), 20–22 April, 2016, Catania, Italy

Hauswirth D, Iten M, Puzrin A (2012) Detection of ground movements using soil-embedded distributed fiber optic sensors. Paper presented at the Geotechnical and geophysical site characterization 4: Proceedings of the 4th International Conference on Site Characterization ISC'4, 18–21 September 2012, Galinhas, Pernambuco, Brazil

Ho Y-T, Huang A-B, Lee J-T (2006) Development of a fibre Bragg grating sensored ground movement monitoring system. Meas Sci Technol 17(7):1733. https://doi.org/10.1088/0957-0233/17/7/011

Huang AB, Lee JT, Ho YT, Chiu YF, Cheng SY (2012) Stability monitoring of rainfall-induced deep landslides through pore pressure profile measurements. Soils Found 52:737–747

Iten M, Hauswirtz D, Puzrin AM (2011) Distributed fiber optic sensor development, testing, and evaluation for geotechnical monitoring applications. Paper presented at the Smart Sensor Phenomena, Technology, Networks, and Systems, 7–9 March 2011, San Diego, California, United States

Li J, Correia R, Chehura E, Staines S, James SW, Tatam RP (2010) A fibre bragg grating-based inclinometer system for ground movement measurement. Paper presented at the Fourth European Workshop on Optical Fibre Sensors, 8–13 September 2010, Porto, Portugal

Lienhart W (2015) Case studies of high-sensitivity monitoring of natural and engineered slopes. J Rock Mech Geotech Eng 7(4):379–384. https://doi.org/10.1016/j.jrmge.2015.04.002

Martinelli M (1989) A universal compensator for polarization change induced by birefringence on a retracing beam. Opt Commun 72(6):341–344. https://doi.org/10.1016/0030-4018(89)90436-7

Martinelli M, Ferrario M (2013) Synoptic fiber optic sensor. Patent W02013179118A

Michlmayr G, Chalari A, Clarke A, Or D (2017) Fiber-optic high-resolution acoustic emission (AE) monitoring of slope failure. Landslides 14(3):1139–1146. https://doi.org/10.1007/s10346-016-0776-5

Olivares L, Damiano E, Greco R, Zeni L, Picarelli L, Minardo A, Bernini R (2009) An instrumented flume to investigate the mechanics of rainfall-induced landslides in unsaturated granular soils. Geotech Test J 32(2):108–118. https://doi.org/10.1520/GTJ101366

Pei H, Cui P, Yin J, Zhu H, Chen X, Pei L, Xu D (2011) Monitoring and warning of landslides and debris flows using an optical fiber sensor technology. J Mount Sci 8(5):728. https://doi.org/10.1007/s11629-011-2038-2

Sbarufatti C. Martinelli L, Mattarei M, Ferrario M, Giglio M (2016) Ultrasonic strain wave acquisition by a low-cost fiber optic coherent sensor for structural health monitoring applications. Paper presented at the 8th EWSHM, 5–8 July, Bilbao, Spain

Scaioni M, Crippa J, Yordanov V, Longoni L, Ivanov VI, Papini M (2018) Some tools to support teaching photogrammetry for slope stability assessment and monitoring. Int Arch Photogramm Remote Sens Spatial Inf Sci XLII-3/W4:453–460. https://doi.org/10.5194/isprs-archives-XLII-3-W4-453-2018

Schenato L (2017) A review of distributed fibre optic sensors for geo-hydrological applications. Appl Sci 7(9):896. https://doi.org/10.3390/app7090896

Schenato L, Palmieri L, Camporese M, Bersan S, Cola S, Pasuto A, Galtarossa A, Salandin P, Simonini P (2017) Distributed optical fibre sensing for early detection of shallow landslides triggering. Sci Rep 7(1):14686. http://10.1038/s41598-017-12610-1

Yoshida Y, Kashlwai Y, Murakami E, Ishida S, Hashiguchi N (2002) Development of the monitoring system for slope deformations with fiber Bragg grating arrays. In: Proceedings of SPIE – The International Society for Optical Engineering. Smart structures and Materials 2002: Smart sensor technology and measurement systems 4694:296–303. https://doi.org/10.1117/12.472632

Zeni L, Picarelli L, Avolio B, Coscetta A, Papa R, Zeni G, Di Maio C, Vassallo R, Minardo A (2015) Brillouin optical time-domain analysis for geotechnical monitoring. J Rock Mech Geotech Eng 7(4):458–462. https://doi.org/10.1016/j.jrmge.2015.01.008

Zhang D, Shi B, Sun Y, Tong H, Wang G (2015) Bank slope monitoring with integrated fiber optical sensing technology in three gorges reservoir area in Engineering Geology for Society and Territory, vol 2. Springer, pp 135–138

Chapter 13
The Giant Seymareh Landslide (Zagros Mts., Iran): A Lesson for Evaluating Multi-temporal Hazard Scenarios

Michele Delchiaro, Javad Rouhi, Marta Della Seta, Salvatore Martino, Reza Nozaem, and Maryam Dehbozorgi

Abstract The giant Seymareh Landslide (hereafter referred to as SL), located close to the border between Lorestan and Ilam provinces in the southwest of Iran, is one of the largest rock mass movements occurred on the Earth surface with an estimated volume of 44 Gm^3. This work focuses on geological and geomorphological constraints for the analysis of multi-temporal hazard related to the sequence of events and processes linked to the landslide occurrence, i.e., from the initial stage of deforming rock mass to the landslide debris cut by the Seymareh River. With this aim, the following four stages have been distinguished: (1) the mass rock creep (MRC) process evolving toward the paroxysmal slope failure; (2) the massive rock failure occurrence that caused the Seymareh rock avalanche and originated the natural dam responsible for the formation of a three-lake system (Seymareh, Jaidar, and Balmak lakes); (3) the lake infilling until the overflow of the landslide dam; (4) the cutting by the present Seymareh River of the natural dam, which caused the emptying of the lake system (except for the Balmak Lake) and the intense erosion due to the complex drainage system that originated after the cutting. A quantitative morphometric evaluation was also performed through the *Tu* index to predict the catchment-scale suspended sediment yield on the scar area and derive the erosion rate affecting the landslide after the cut of the landslide dam by the Seymareh River. Due to the very huge dimensions of the SL, the here depicted multistage scenario can be regarded as an "end-member" in view of future events that can occur in similar geological contexts.

M. Delchiaro (✉) · J. Rouhi · M. Della Seta · S. Martino
Department of Earth Sciences, Sapienza University of Rome, Rome, Italy
e-mail: michele.delchiaro@uniroma1.it

R. Nozaem
School of Geology, College of Science, University of Tehran, Tehran, Iran

M. Dehbozorgi
Department of Earth Sciences, Kharazmi University, Tehran, Iran

M. De Maio, A. K. Tiwari (eds.), *Applied Geology*,
https://doi.org/10.1007/978-3-030-43953-8_13

Keywords Risk scenario analysis · Landscape evolution model · Mass rock creep · Seymareh landslide · Zagros mountains

13.1 Introduction

Tectonically active landscapes are dynamic systems that record effects of geological processes acting inside and at surface on Earth. Threshold conditions on hillslopes are often reached, leading to gravity-induced phenomena with considerable implications for natural hazards (Montgomery and Brandon 2002; Binnie et al. 2007; Larsen and Montgomery 2012). Landslides affecting rock masses or soils are important landscape-shaping processes, releasing sediment/rock from hillslopes and delivering them to the fluvial system (e.g., Petley 2010). Landslide-driven erosion acts at different spatial and temporal scales, from short-term including shallow landslides with an annual or decennial recurrence to long-term such as gravity-induced slope deformations driven by time-dependent rheology, i.e., mass rock creep (MRC, Chigira 1992). The latter affects entire portions of slopes, giving rise to slope-scale deformational processes, i.e., deep-seated gravitational slope deformation (DSGSD) *Auct.*, evolving into local failures or general slope collapse, which originate catastrophic rock avalanches (Benko and Stead 1998; Casson et al. 2003; Cappa et al. 2004; Crosta et al. 2004; Eberhardt et al. 2004; Stead et al. 2006; Strom and Korup 2006; Evans et al. 2007; Marmoni et al. 2017). In such an evolutionary frame, transient forcing actions (i.e., earthquakes, teleseismic events, intense rainfalls, ground water changes, anthropic actions) can anticipate or delay the time to failure (Martino et al. 2004; Maffei et al. 2005; Lenti et al. 2015; Bozzano et al. 2016), since they could induce a restart of the MRC process (i.e., zeroing the creep time) or, on the contrary, they could be responsible for an increase in the creep strain rate. Although a wide literature exists on earthquake-induced landslides (Keefer 1994; Rodriguez et al. 1999; Delgado et al. 2011; Jibson and Harp 2012; Martino et al. 2014), some studies (Bozzano et al. 2016; Della Seta et al. 2017; Martino et al. 2017) have been focused on the role of landscape evolution rates on the development of gravity-induced slope deformations driven by time-dependent rheology. Anyway, a multi-modeling approach (Martino et al. 2017), incorporating contributions from morpho-evolutionary modeling, detailed engineering-geological modeling and time-dependent stress-strain numerical modeling is a useful methodology to analyze the space-time evolution of a slope, introducing the concept of "time-dependent risk," strictly related to a multi-temporal hazard scenario. The giant SL represents a exceptional case study to test the methodology, as one of the largest instability events among all continental landslides worldwide, whose relevance is also due to the multistage scenario, which can be depicted before and after the general slope collapse that can be regarded as "end-member" reference for future events occurring in similar geological contexts.

13.2 Geological Features

The SL is located close to the border between Lorestan and Ilam Provinces of Iran (33°03′N, 47°41′E), 460 km SW of Tehran and 300 km to the east of Baghdad, Iraq, in the western margin of the Zagros fold–thrust belt. The available geological maps (Setudehnia and Perry 1967; Macleod 1970; Takin et al. 1970) have been used as reference for the geological setting of the study area (Fig. 13.1), in addition to data collected during geological field surveys. Figure 13.1 shows the study area, which includes the entire Seymareh River valley from the artificial dam to the SE termination of the Kabir-kuh fold, which is characterized by a series of smooth and gentle folding with a NW–SE trending (Aghanabati 2004): the Maleh-kuh, Halush-kuh, Chenareh-kuh, and Kabir-kuh anticlines appearing as box fold structures because of its whale back morphology. The geological formations outcropping in the anticline areas refer to the Middle Cretaceous up to the Lower Miocene, whereas more recent deposits, from Miocene to Present, constitute the geological bedrock of the valleys.

The most ancient geological formation is represented by the Sarvak Formation, a thick carbonate unit including shale and sandstone with high erodibility (Elyasi et al. 2014). The Ilam-Surgah Formation is superimposed on the Sarvak and consists of grainstone, packstone, and wackestone (Elyasi et al. 2014). At the top of Ilam-Surgah Formation, the Gurpi Formation is composed of marly limestone, marl, and shale (Elyasi and Goshtasbi 2015). In this area, the lower Gurpi Formation laterally changes in a considerably more calcareous member called Emam Hassan. The Pabdeh Formation overlies the Gurpi Formation and can be lithologically distinguished in three members, according to Delchiaro et al. 2019: (1) the lower Pabdeh member (marls), (2) the Taleh Zang member (limestone), and (3) the upper Pabdeh member (calcareous marls). The Asmari Formation overlies the passive margin succession creating a carbonatic carapace originally covering the top of the Kabir-kuh fold (Khoshboresh 2013). The most recent geological formations of the local stratigraphy include the following: the Gachsaran Formation, consisting of salt, anhydrite, colored marl, and gypsum; the Agha Jari Formation, composed of marl and red sandstone; and the Bakhtiari Formation, consisting of siltstone and conglomerate. Moreover, according to Delchiaro et al. (2019), conglomerates ascribable to inactive, terraced alluvial fans and alluvial deposits are preserved in the middle reach of the Seymareh River valley. They form a suite of fill terraces entrenched in the terraced lacustrine deposit of Seymareh and Jaidar lakes upstream of the landslide, in the area where the lakes originated by the natural damming could be extended. A second set of terraced alluvial fans and alluvial deposits are part of a suite of fill terraces preserved downstream of the SL, whose debris hosts a suite of strath terraces and a flood plain shaped along the Seymareh River gorge. The landscape evolution of the Seymareh River valley before and after the failure occurrence was studied by several Authors (Harrison and Falcon 1938; Watson and Wright 1969; Roberts 2008; Roberts and Evans 2013; Shoaei 2014; Delchiaro et al. 2019; Rouhi et al. 2019). The SL detached from the NE flank of Kabir-kuh fold involving about 44 Gm^3 of debris and covering an area of about 220 km^2 with a

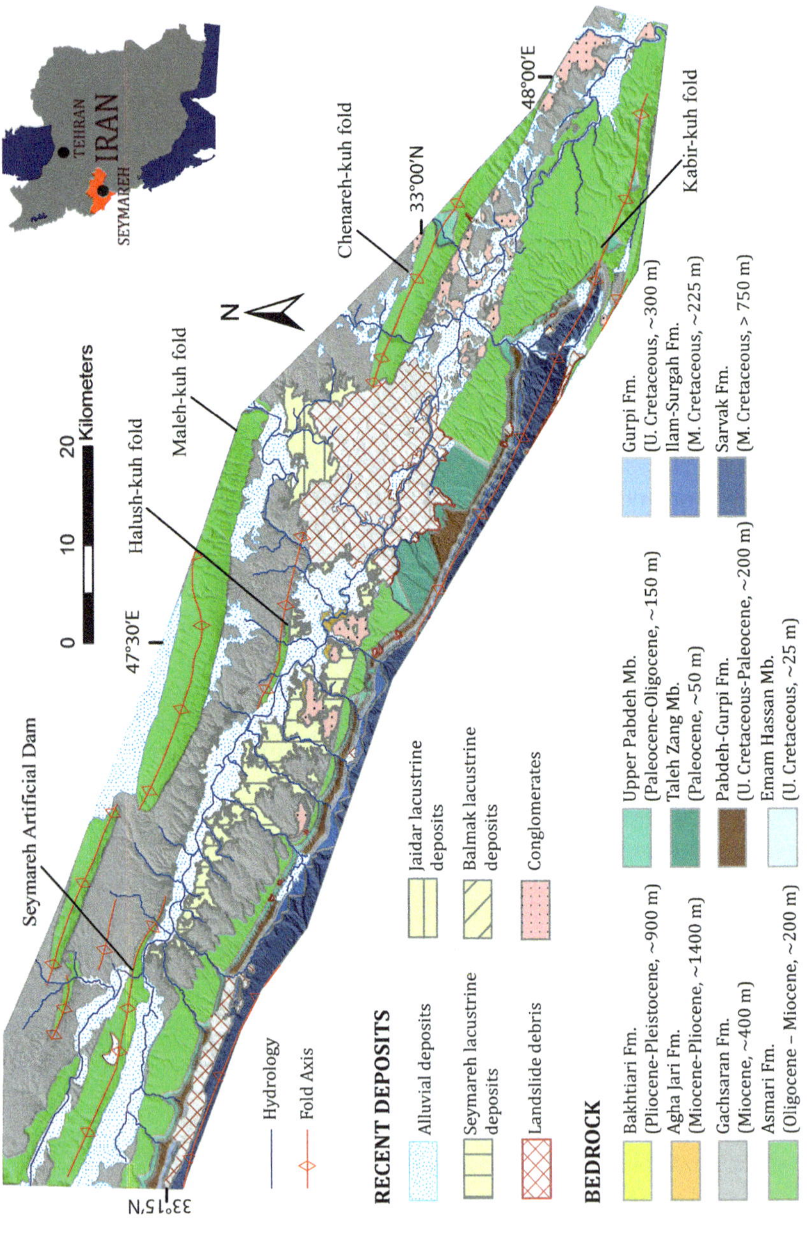

Fig. 13.1 Geological Map of the Seymareh River valley

run-out of 16 km and an apparent friction angle of 3.6 degrees (Roberts and Evans 2013). The landslide debris dammed the course of the Seymareh and Kashgan rivers generating the three-lake system, which includes the Seymareh, Jaidar, and Balmak lakes. Roberts and Evans (2013) obtained C^{14} ages from the organic matter sampled at the bottom of the lacustrine deposits, thus fixing the time occurrence of the landslide at almost 10 ka. According to Harrison and Falcon (1938) and Shoaei (2014), the Jaidar Lake was drained into the largest Seymareh Lake causing the overflooding and the consequent natural cutting of the dam (Delchiaro et al. 2019). According to Delchiaro et al. (2019), the OSL age of the highest strath terrace formed on the SL debris, along the river gorge, constrains the beginning of the lake emptying phase at ~6.6 ka, with a persistence time of the lake of ~3500 years. The scientific community is still debating on the triggering of the landslide that could be due either to the achievement of the tertiary phase of the MRC process or to a trigger occurrence such an earthquake. In the second case, the trigger could have anticipated the time to failure expected for the ongoing tertiary creep.

13.3 Geomorphological Features

Tectonic and stratigraphic setting strongly control the physiography of the Zagros fold-thrust belt (Borne and Twidale 2011) since its drainage network clearly adapted to the growth of the thrust-fold structures, also in relation with the erodibility of the outcropping formations. Folds are typically symmetrical to slightly asymmetrical and are commonly crossed by rivers according to a normally distributed pattern (i.e., transverse drainage according to Oberlander 1965). Most anticlines show well-preserved "hogback" landforms due the resistant Asmari limestone carapace. Folds with breached carapaces show exposed cores of alternating layers of resistant and erodible units forming in some cases small perched basins. Where breaching reaches a lower, highly resistant unit (e.g., Sarvak Formation), the anticlines form a pair of parallel ridges composed of stratigraphically higher and resistant lithologies, which are often incised to form cols. The two ridges are often separated by a minor ridge constituted of the lower resistant unit in the hinge region of the anticline. Synclines host the Miocene-Pleistocene units overlying the Asmari, as well as the most recent deposits. The highly mobile Gachsaran Formation predominates in many valley bottoms where it is commonly cut by rills to form badlands (Roberts 2008). In relation with the erodibility of the outcropping formations and active tectonics, the action of fluvial and runoff erosion is very evident in the area. Water and wind gaps (both visible along the Chenareh-kuh fold) are typical evidences of drainage evolution in the present-day landscape. According to Ramsey et al. (2008), the river courses are modified because of laterally growing anticlines and fan-shaped tributary patterns on fold flanks and other geomorphological features (such as wind gaps and water gaps) are interpreted as markers of the fold propagation. Runoff erosional landforms, like rills and gullies, as well as fluvial erosion, play an important role in the release of the Asmari Formation carapace, causing the initiation of the MRC process in the rock mass (Delchiaro et al. 2019). Shoaei (2014) found that the particular hydro-morphological characteristic of the landslide source area, together with the enormous eroding energy, resulted from the joining of the Kashkan and Seymareh rivers to originate the Karkheh River, which resulted in an unusual condition of high discharge and strong eroding force in the basin, changing its direction toward the northern flank of Kabir-kuh anticline and caused severe erosion at the toe of the rock slab prone to sliding, so activating the MRC process.

13.4 Methods

To depict the multi-temporal hazard related to the sequence of events linked to the SL occurrence, a multidisciplinary approach was applied by incorporating contribution from landscape evolution and engineering-geological modeling, as well as by resuming the results of previous work (Harrison and Falcon 1938; Watson and Wright 1969; Roberts 2008; Roberts and Evans 2013; Shoaei 2014; Delchiaro et al. 2019; Rouhi et al. 2019). Geomorphological analysis and remote surveying (aerial photos and Google Earth satellite optical images) were first carried out; at this aim, the 30 m resolution DEM of the Shuttle Radar Topographic Mission (SRTM; NASA JPL 2013) was used to map landforms and derive a geomorphological map of both the scar and the debris areas. In this regard, an integrated geomorphological and engineering-geological survey allowed to provide the main constraints to the assumed multi-hazard scenario. The geomorphic index *Tu* was also computed to predict the catchment-scale suspended sediment yield (Della Seta et al. 2009 and references therein) for estimating the erosion rate affecting the landslide scar area after the cut of the landslide dam due to the Seymareh River engraving.

13.5 Results

The geomorphological and engineering-geological field surveying, supported by a remote surveying based on satellite images, allowed to point out the evidence of runoff and gravity-induced features, such as gullies, blocks, ridges, topographic obstacles, surface evidence of MRC (such as buckling and bulldozer deformation affecting the rock mass), as well as the evidence of the geomorphic response to the emplacement of the landslide dam on landslide scar and debris and in its neighborhood, such as lacustrine deposits, springs and minor landslides, strath terraces, and terraced flood plains.

These landforms were mapped in order to analyze their spatial distribution (Fig. 13.2) providing relative time constraints to the evolution in different stages of the Kabir-kuh slope, as well as to the evaluation of the multi-hazard scenarios and to reconstruct the landslide mechanism. Figure 13.3 shows the evidence used to reconstruct the temporal sequence of events and to define the multi-hazard. Ridges

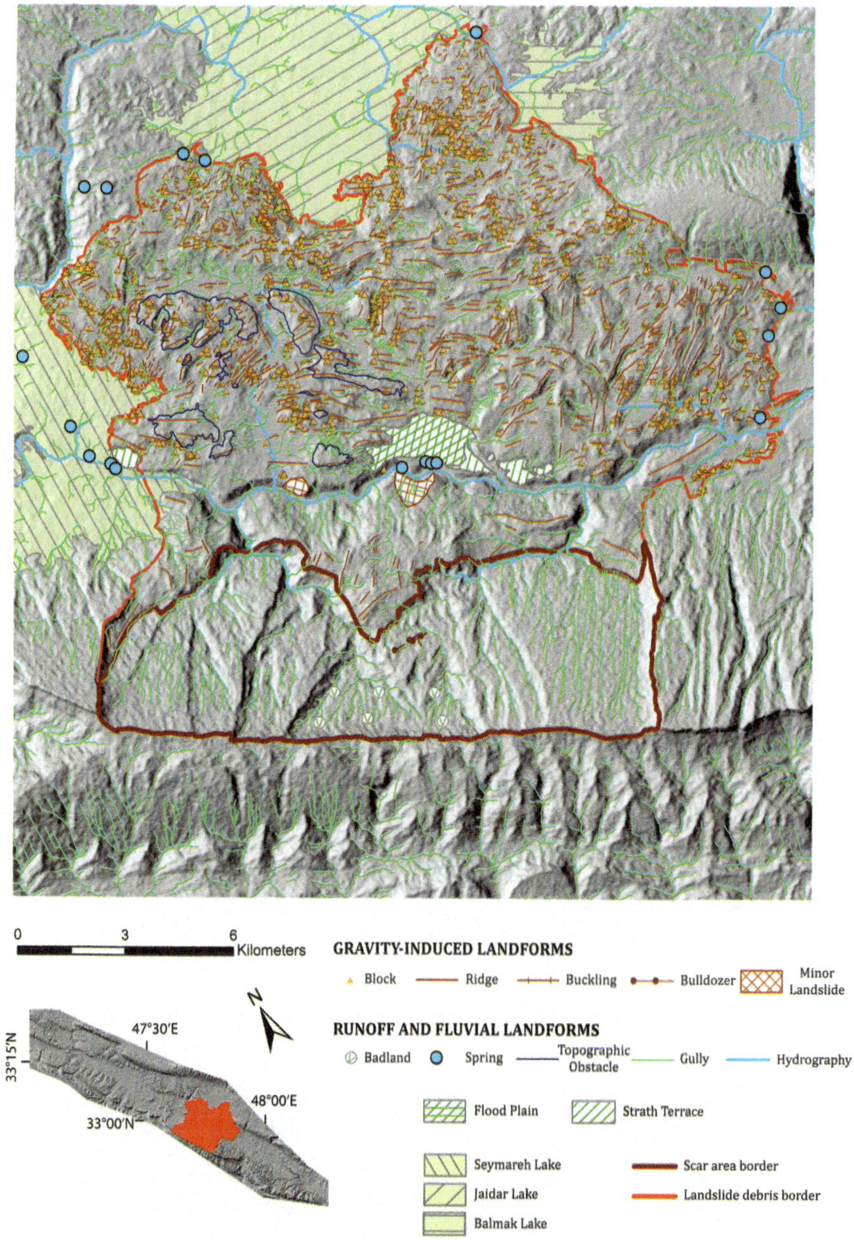

Fig. 13.2 Geomorphological map reporting the main landforms surveyed within the debris and the scar area of the SL

Fig. 13.3 Gravity-induced (G1-G5), runoff, and lacustrine (R1-R5) landforms collected to constrain the multi-hazard scenario. G1) Ridges; G2) Blocks; G3) Buckling evidence; G4) Bulldozing evidence; G5) Minor landslide involving landslide debris; R1) Topographic obstacles; R2) Lacustrine deposits; R3) Strath terrace and terraced flood plain; R4) Gullies; R5) Springs located at the bottom of the landslide debris

(G1), blocks (G2), and evidence of MRC (such as G3 and G4) were fundamental to reconstruct the landslide mechanism and the precursor of the massive slope failure. Specifically, the spatial distribution of blocks (larger than 10 m) and ridges with their orientation provide information on the emplacement mechanism of the SL and allow us to confirm what has already been stated in the previous studies (Roberts 2008; Roberts and Evans 2013; Shoaei 2014) that parts of Chenareh-kuh and Halush-kuh anticlines are buried under the landslide debris. Indeed, a preliminary analysis of ridges directions shows that during the single landslide event, debris run out occurred along two main directions: NE and NW toward Chenareh-kuh and Halush-kuh anticlines, as clearly demonstrated by the predominant orientation of the gravity-induced landforms, respectively SW-NE and SE-NW in the NW and NE parts of the debris. Moreover, the study of the curvature of the ridges, whose concavities provide evidence of the versus of the gravitational movement, highlights that the presence of the anticlines caused that a part of landslide debris overpassed the topographic obstacles (concavities of a part of the ridges toward S-SE and S-SW, respectively, in the NW and NE parts of the debris) and a part was swept back into SE and SW directions (concavities of the other part of the ridges toward N-NW and N-NE, respectively, in the NW and NE parts of the debris). Other evidence, such as buckling (G3) and bulldozing (G4) of downslope dipping strata, surveyed along the scar of the SL confirm that MRC can be regarded as responsible for the ultimate massive slope failure. These features could be associated to different rheological behaviors of the litho-technical units, which are part of the succession outcropping in the northeastern slope of the Kabir-kuh (stiffness contrast between the upper member of the Pabdeh Formation and the overlying Asmari Formation). Such a different behavior can justify the MRC process evolving toward the ultimate massive rock failure, which generated the SL (Shoaei 2014; Delchiaro et al. 2019). Moreover, the small-size landslides (G5) detected along the river gorge allowed to assign a residual hazard for the Seymareh area due to a possible new river damming. A strath terrace and a terraced flood plain (R3) sculpted on the landslide deposit demonstrate that the cut of the natural dam occurred due to overflow, which caused the progressive lake emptying (Delchiaro et al. 2019). To reconstruct the maximum hydrometric height reached by the Seymareh Lake before an emissary was generated due to the disequilibrium in the hydrogeological balance between the drainage in the natural dam and that of the tributary, the topographic obstacles (R1) in the landslide debris were taken into account. Also, the springs (R5) recognized during the field survey were mapped in order to hypothesize the presence of an underground water circulation in the multi-hazard reconstruction.

In response to the natural dam, three lakes (Seymareh, Jaidar and Balmak; Figs. 13.1 and 13.2) were generated. The persistence of the natural dam is constrained by the age of the strath terrace dated 6.59 ± 0.49 ka (Delchiaro et al. 2019), since it marks the cut of the natural dam due to overflow. After the overflow, a dense drainage system set on the scar area, thanks to the high erodibility and low permeability of the less competent lithotypes of the Kabir-kuh NE slope stratigraphic succession (Pabdeh-Gurpi Formation), generated the badland landforms mapped in Fig. 13.2. This interpretation implies a partial revision of the one by Roberts and

Evans (2013), which included the aforementioned formation in the source rock mass involved in the landslide phenomenon, and it is strengthened by the computing of the *Tu* index value (Della Seta et al. 2009 and references therein) for the badland catchments developed on the scar area. This index provides the catchment-scale predicted suspended sediment yield (Della Seta et al. 2009 and references therein) in badlands areas, as a function of specific drainage network parameters.

The obtained suspended sediment yield can be easily traduced into erosion rates, which we used to support the hypothesis that the Pabdeh-Gurpi Formation was not originally involved in the collapse of the rock mass, but it was eroded later. The *Tu* index was calculated using the following experimental Eq. (13.1):

$$\log Tu = a + b \log \mathrm{D} + c\,\Delta a \tag{13.1}$$

where $a = 1.060$, $b = 2.797$ and $c = 0.140$, which had been obtained empirically from statistical correlations among measured suspended sediment yield and geomorphic parameters calculated for catchments with badlands developed on marly/clayey lithological units (Della Seta et al. 2009 and references therein). *D* is the drainage density $= \sum L_i/A$ ($\sum L_i$ is the length sum of all the drainage segments in the drainage area; A is the drainage area) (Della Seta et al. 2009 and references therein); Δa is the hierarchical anomaly index $= G_a/N_1$ (G_a is hierarchical anomaly number, i.e., the minimum number of Strahler's first-order channels that would be necessary to make the drainage perfectly hierarchized; N_1 is the number of Strahler's first-order channels present in the drainage network). The above drainage network hierarchical parameters (Della Seta et al. 2009 and references therein) have been calculated for each catchment unit area. The *Tu* index was computed for both the two basins of the scar area identified as badlands areas and characterized by the outcrop of the Pabdeh-Gurpi Formation (NW and SE basins in Fig. 13.4).

The resulting *Tu* value was multiplied by the time elapsed since the cut of the Seymareh Lake dam (~6.59 ky) and by the density of the eroded lithological unit (ρ of the Pabdeh-Gurpi Formation $= 2600$ kg m^{-3}) in order to estimate the volume eroded by the drainage system since the beginning of the overflowing. The obtained volume was then compared to the calculated volume of the missing Pabdeh-Gurpi Formation, which was estimated in ArcGIS® using the *Trend* tool to reconstruct the top of the Pabdeh-Gurpi Formation and the *Cut-Fill* tool to obtain the missing volume given the present topography (Wilson and Gallant 2000). Table 13.1 shows the obtained results: the missing volume of Pabdeh-Gurpi Formation is comparable with that possibly eroded over the considered time span by the dense channel network, as computed through the *Tu* index.

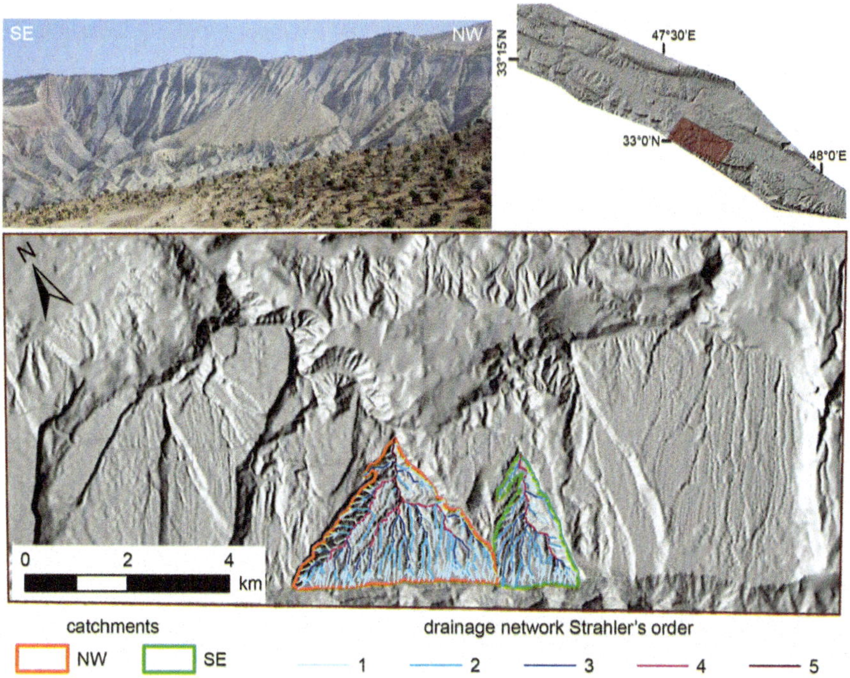

Fig. 13.4 Drainage network of the badland catchments developed on the Pabdeh-Gurpi Formation, for which the *Tu* index was computed

Table 13.1 Eroded volume in the scar area due to badlands development as obtained through the *Tu* index and compared to the computed missing volume of the Pabdeh-Gurpi Formation

Basin						
NW	T_u *(ton* $km^{-2}y^{-1}$*)*	$logT_u$	*D (km* km^{-2}*)*	Δa	*area* *(*km^2*)*	$\sum L$ *(km)*
	17,559.95	4.24	12.22	1.03	5.83	71.23
	ton by area in 6.59 ky	*ρ Pabdeh-Gurpi (ton/*km^3*)*	*Estimated Removed Volume (*km^3*)*	*Estimated Removed Volume (*m^3*)*	*Computed Volume (*m^3*)*	***% of Estimated Volume to the Computed Volume***
	$6.75\ 10^8$	$2.6\ 10^9$	0.26	$2.59\ 10^8$	$2.75\ 10^8$	**94.47**
SE	T_u *(ton* $km^{-2}y^{-1}$*)*	$logT_u$	*D (km* km^{-2}*)*	Δa	*area* *(*km^2*)*	$\sum L$ *(km)*
	16,520.73	4.22	12.26	0.81	2.45	30.04
	ton by area in 6.59 ky	*ρ Pabdeh-Gurpi (ton/*km^3*)*	*Estimated Removed Volume (*km^3*)*	*Estimated Removed Volume (*m^3*)*	*Computed Volume (*m^3*)*	***% of Estimated Volume to the Computed Volume***
	$2.66\ 10^8$	$2.6\ 10^9$	0.10	$1.03\ 10^8$	$1.04\ 10^8$	**98.51**

13.6 Discussion

According to Delchiaro et al. (2019) and based on the constraint provided by *Tu* computation, the multi-stage evolution of the SL includes the following four stages:

1. The MRC process responsible for the gravity-driven slope deformations of the Kabir-kuh fold (before 10 ka);
2. The landslide dam, which generated three lakes, named Seymareh, Jaidar, and Balmak, respectively (~10 ka – ~7 ka);
3. The lake infilling until the overflow of the landslide dam (~10 ka – ~7 ka);
4. The cutting of the natural dam, which caused the lakes emptying (except for Balmak Lake) and the intense erosion due to a complex drainage system that originated after the landslide dam cutting (~7 ka – Present).

The multistage sequence of events associated to the SL occurrence is analyzed in the following in terms of potential, effective, residual, and inherited hazard (Fig.13.5).

Potential hazard refers to any possible hazardous phenomena, and it can be regarded as the largest involvement scenario. Effective hazard is associated with events that really occurred and therefore represents the real occurred scenario. Residual hazard is related to persistent hazardous conditions after the event occurrence, not necessarily detectable nowadays. Inherited hazard refers to the persistence, up to present, of the hazardous conditions that derive from residual hazards. Based on the multistage sequence introduced before, step 1 is related to the massive slope failure as paroxysmal evolution of a creep process driven and controlled by the geo-structural setting of the Kabir-kuh fold (Roberts and Evans 2013), even if is not possible to exclude the occurrence of triggering actions anticipating the ultimate failure. At this stage, the potential hazard was defined taking into account the possible zonation of the detachment area related to the geo-structural and geomorphological setting of the NE slope of Kabir-kuh anticline. To support the landslide reconstructions, the directions of ridges (G1) and the location of blocks (G2) allowed to deduce that, in a potential scenario, the landslide debris during a single mass movement event could have buried part of the two front anticlines (Chenareh-kuh and Halush-kuh anticlines). According to literature (Costa and Shuster 1988; Nicoletti and Sorriso-Valvo 1991) in this landslide potential scenario, the rock mass could have barred by the topographic obstacle and settle in the valley for 2/3 of its downstream area and for 1/3 upstream . The effective hazard scenario refers to the really occurred event, during which debris traveled to two main directions of NE and NW toward Chenareh-kuh and Halush-kuh anticlines whose presence caused that a part of landslide debris overpassed the preexisting landforms and a part swept back into SE and SW directions. The residual hazard regards the possibility that the scar area could give rise to other gravity-induced instabilities as demonstrated by buckling and bulldozing evidence (G3 and G4) on the sliding surface, possibility inherited up to the present. Step 2 is related to the damming produced ~10 ka ago by 44 Gm^3 of debris and to the subsequent geomorphic response of the Seymareh valley. Based on the elevations of the topographic obstacles on the landslide debris

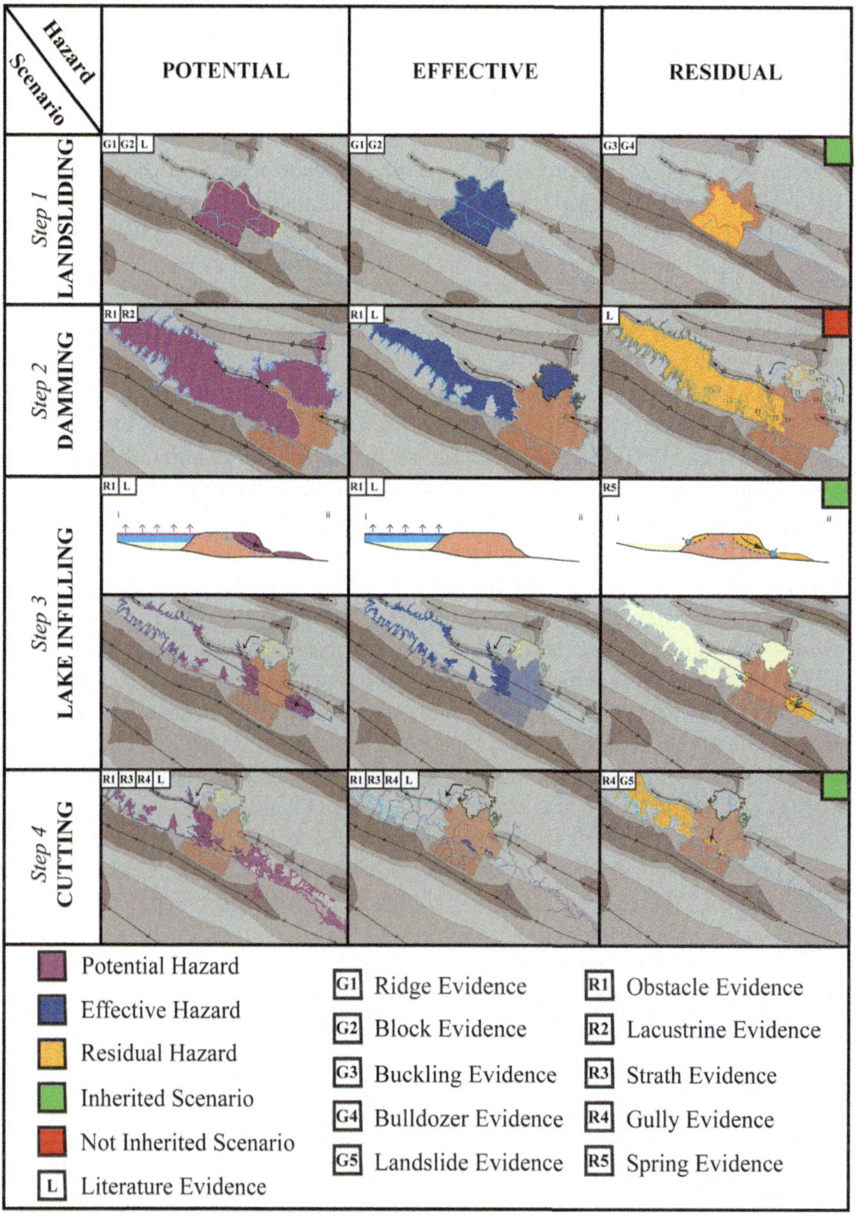

Fig. 13.5 Multi-temporal hazard scenario sketch obtained based on the landform evidences for each time step of the evolution of the valley-slope system. Potential, effective, residual, and inherited hazards were estimated

(R1) derived from the DEM and on the field evidence of the lacustrine deposits (R2), it is possible to assume that a complex lake system, including the Seymareh, Jaidar, and Balmak lakes, was originated by the landslide damming, which constrains the potential hazard zoning. The effective hazard was evaluated based on literature data (Harrison and Falcon 1938, Roberts and Evans 2013; Shoaei 2014; Delchiaro et al. 2019) and on the field evidence of the lacustrine deposits of the Seymareh Lake (R2). The residual hazard scenario can be related to the drainage of the Balmak Lake into the Jaidar Lake, due to the intense discharge of the Kashkan River, a paleo-tributary of the Seymareh River. No inherited hazard can be assumed as the lakes have been naturally dried. Step 3 represents the lake infilling and is substantially based on the same evidences of the Step 2 (R1, R2 and L), except for the springs (R5). The potential hazard was referred to the overflowed of the Jaidar Lake into the Seymareh River (Harrison and Falcon 1938; Shoaei 2014), inducing a complex ground water circulation within the landslide dam responsible for local landslides involving the dam itself. The effective hazard scenario represents similar conditions but without considering the landslide involving the dam since they cannot be documented. As the overflowing and drainage of the three lakes have occurred, the residual hazard, which can be considered inherited up to present, includes minor landslides from the debris. Step 4 implies the natural cut of the landslide dam and the consequent emptying of the Seymareh and Jaidar Lakes occurred ~7 ka ago according to the reconstruction of Delchiaro et al. (2019) on the basis of the OSL age of a strath terrace along the Seymareh River gorge, as well as on the *Tu* index calculation. The intense erosion due to a complex drainage system, which originated after the cutting justifies the potential hazard which has been zoned based on topographic obstacles on the landslide debris, strath terraces, spatial distribution of gullies (R1, R3, and R4), as well as on literature constraints (Shoaei 2014; Delchiaro et al. 2019). The collected evidence explains the cutting of the landslide dam through the drainage of the Jaidar Lake into the Seymareh Lake and the overflow of the Seymareh Lake downstream to the dam in the preexisting Seymareh River valley. The effective hazard scenario represents similar conditions but considering a new river system that gradually involved the landslide dam since flooding evidences downstream of the landslide debris cannot be documented. The residual and the inherited hazard related to the Step 4 express the possibility that minor landslides (G5) could occur along the gorge causing a new damming of the Seymareh River and generating a new lake.

13.7 Conclusion

The multistage scenario resulting from the occurrence of the giant SL in northwestern Zagros Mts. (Iran) provides a lesson that can be regarded as an extreme reference for future events in similar geological contexts. The SL represents an exemplary case of study to test multi-modeling approaches, as well as to analyze slope evolution as an effect of time-dependent behavior evaluating the time-space hazard change over time, from potential to effective, to residual up to inherited, allowing to experience the concept of "time-dependent risk." The results obtained here provide main geological and geomorphological constraints for the analysis of the multi-temporal hazard related to the sequence of events and processes linked to the landslide occurrence in terms of potential, effective, residual, and inherited hazard. This study focused on four temporal steps based on the landscape evolution deduced for the Seymareh River valley by Delchiaro et al. (2019), constrained by OSL dating of soil samples, as well as by *Tu* index calculation. The aforementioned sequence of events has been distinguished: (1) the MRC process responsible for the evident slope deformations of the Kabir-kuh fold; (2) the landslide dam that generated three lakes (Seymareh, Jaidar, and Balmak lakes); (3) the lake infilling until the overflow of the landslide dam; (4) the cutting by the present Seymareh River of the natural dam, which caused the emptying of the lake system (except for the Balmak Lake), and the intense erosion due to the complex drainage system which originated after the cutting. Grounding on such a sequence of events, an engineering-geological model of the landslide debris and scar will allow in future to point out a quantitative characterization of the landslide mechanism and kinematics, longevity, and hydrogeological role of the natural dam.

References

Aghanabati A (2004) Geology of Iran. Geological Survey of Iran Press, Tehran, 586 pages

Benko B, Stead D (1998) The frank slide: a reexamination of the failure mechanism. Can Geotech J 35(2):299–311. https://doi.org/10.1139/t98–005

Binnie SA, Phillips WM, Summerfield MA, Fifield LK (2007) Tectonic uplift, threshold hillslopes, and denudation rates in a developing mountain range. Geology 35(8):743–746. https://doi.org/10.1130/G23641A.1

Bourne JA, Twidale CR (2011) Neglected and cryptostructural effects in drainage development. Cadernos do Laboratorio Xeolóxico de Laxe 36:41–60

Bozzano F, Della Seta M, Martino S (2016) Time-dependent evolution of rock slopes by a multi-modelling approach. Geomorphology 263:113–131. https://doi.org/10.1016/j.geomorph.2016.03.031

Cappa F, Guglielmi Y, Soukatchoff VM, Mudry J, Bertrand C, Charmoille A (2004) Hydromechanical modeling of a large moving rock slope inferred from slope levelling coupled to spring long-term hydrochemical monitoring: example of the La Clapiere landslide (Southern Alps, France). J Hydrol 291(1–2):67–90. https://doi.org/10.1016/j.jhydrol.2003.12.013

Casson B, Delacourt C, Baratoux D, Allemand P (2003) Seventeen years of the "La Clapiere" landslide evolution analysed from ortho-rectified aerial photographs. Eng Geol 68 (1–2):123–139. https://doi.org/10.1016/S0013–7952(02)00201–6

Chigira M (1992) Long-term gravitational deformation of rocks by mass rock creep. Eng Geol 32 (3):157–184. https://doi.org/10.1016/0013–7952(92)90043-X

Costa JE, Schuster RL (1988) The formation and failure of natural dams. Geol Soc Am Bull 100 (7):1054–1068. https://doi.org/10.1130/0016–7606(1988)100<1054:TFAFON>2.3.CO;2

Crosta GB, Chen H, Lee CF (2004) Replay of the 1987 Val Pola landslide, Italian alps. Geomorphology 60(1–2):127–146. https://doi.org/10.1016/j.geomorph.2003.07.015

Delchiaro M, Della Seta M, Martino S, Dehbozorgi M, Nozaem R (2019) Reconstruction of river valley evolution before and after the emplacement of the giant Seymareh rock avalanche (Zagros Mts., Iran). Earth Surf Dynam 7:929–947. https://doi.org/10.5194/esurf-7-929-2019

Delgado J, Garrido J, López-Casado MS, Peláez JA (2011) On far field occurrence of seismically induced landslides. Eng Geol 123:204–213. https://doi.org/10.1016/j.enggeo.2011.08.002

Della Seta M, Del Monte M, Fredi P, Lupia Palmieri E (2009) Space-time variability of denudation rates at the catchment and hillslope scales on the Tyrrhenian side of Central Italy. Geomorphology 107(3–4):161–177. https://doi.org/10.1016/j.geomorph.2008.12.004

Della Seta M, Esposito C, Marmoni GM, Martino S, Scarascia MG, Troiani F (2017) Morpho-structural evolution of the valley-slope systems and related implications on slope-scale gravitational processes: new results from the Mt. Genzana case history (Central Apennines, Italy). Geomorphology 289:60–77. https://doi.org/10.1016/j.geomorph.2016.07.003

Eberhardt E, Stead D, Coggan JS (2004) Numerical analysis of initiation and progressive failure in natural rock slopes – the 1991 Randa rockslide. Int J Rock Mech Min 41(1):69–87. https://doi.org/10.1016/S1365–1609(03)00076–5

Elyasi A, Goshtasbi K (2015) Using different rock failure criteria in wellbore stability analysis. Geomech Energy Environ 2:15–21. https://doi.org/10.1016/j.gete.2015.04.001

Elyasi A, Goshtasbi K, Saeidi O, Torabi SR (2014) Stress determination and geomechanical stability analysis of an oil well of Iran, Sadhana. Acad Proc Eng Sci 39(1):207–220. https://doi.org/10.1007/s12046–013–0224–3

Evans SG, Mugnozza GS, Strom A and Hermanns RL (2007). Landslides from massive rock slope failure. Springer, Drodrecht (Vol. 49)

Harrison JV, Falcon NL (1938) An ancient landslip at Saidmarreh in southwestern Iran. J Geol 46:296–309

Jibson RW, Harp EL (2012) Extraordinary distance limits of landslides triggered by the 2011 mineral, Virginia, earthquake. B Seismol Soc Am 102(6):2368–2377. https://doi.org/10.1785/0120120055

Keefer DK (1994) The importance of earthquake-induced landslides to long-term slope erosion and slope-failure hazards in seismically active regions. Geomorphol Nat Haz 10:265–284. https://doi.org/10.1016/B978–0–444–82012–9.50022–0

Khoshboresh AR (2013) A study on deformation of tunnels excavated in fractured rocks, 1st edn. Universitè Laval, Quebec

Larsen IJ, Montgomery DR (2012) Landslide erosion coupled to tectonics and river incision. Nat Geosci 5(7):468. https://doi.org/10.1038/ngeo1479

Lenti L, Martino S, Paciello A, Prestininzi A, Rivellino S (2015) Recorded displacement in a landslide slope due to regional and teleseismic earthquakes. Geophys J Int 201(3):1335. https://doi.org/10.1093/gji/ggv063

Macleod JH (1970) Kabir Kuh, 1: 100 000 Geological Map. Iran Oil Operating Companies, Geological Exploration Division, Tehran, Iran

Maffei A, Martino S, Prestininzi A (2005) From the geological to the numerical model in the analysis of gravity-induced slope deformations: an example from the Central appennines (Italy). Eng Geol 78(3):215–236. https://doi.org/10.1016/j.enggeo.2004.12.009

Marmoni GM, Martino S, Heap MJ, Reuschlé T (2017) Gravitational slope-deformation of a resurgent caldera: new insights from the mechanical behaviour of Mt. Nuovo tuffs (Ischia Island, Italy). J Volcanol Geoth Res 345:1–20. https://doi.org/10.1016/j.jvolgeores.2017.07.019

Martino S, Prestininzi A, Scarascia Mugnozza G (2004) Geological-evolutionary model of a gravity-induced slope deformation in the carbonate Central Appennines (Italy). Q J Eng Geol Hydroge 37(1):31–47. https://doi.org/10.1144/1470–9236/03–030

Martino S, Prestinizi A, Romeo RW (2014) Earthquake-induced ground failures in Italy from a reviewed database. Nat Hazard Earth Sys 14:799–814. https://doi.org/10.5194/nhess-14–799–2014

Martino S, Della Seta M and Esposito C (2017) Back-analysis of rock landslides to infer rheological parameters. In: Rock Mechanics and Engineering, Vol. 3: Analysis, Modeling and Design 237–269

Montgomery DR, Brandon MT (2002) Topographic controls on erosion rates in tectonically active mountain ranges. Earth Planet Sc Lett 201(3–4):481–489. https://doi.org/10.1016/S0012–821X(02)00725–2

NASA JPL (2013) NASA shuttle radar topography Mission global 1 arc second [Data set]. NASA EOSDIS Land Processes DAAC. https://doi.org/10.5067/MEaSUREs/SRTM/SRTMGL1.003

Nicoletti PG, Sorriso-Valvo M (1991) Geomorphic controls of the shape and mobility of rock avalanches. Geol Soc Am Bull 103(10):1365–1373. https://doi.org/10.1130/0016–7606(1991)103<1365:GCOTSA>2.3.CO;2

Oberlander TM (1965) The Zagros streams: a new interpretation of transverse drainage in an orogenic zone. In: Syracuse Geographical Series, Syracuse University Press, New York

Petley DN (2010) Landslide hazards. In: Alcantara-Ayala I, Goudie A (eds) Geomorphological hazards and disaster prevention. Cambridge University Press, Cambridge, UK, pp 63–74

Ramsey LA, Walker RT, Jackson J (2008) Fold evolution and drainage development in the Zagros mountains of Fars province, SE Iran. Basin Res 20(1):23–48. https://doi.org/10.1111/j.1365–2117.2007.00342.x

Roberts NJ (2008) Structural and geologic controls on gigantic (> 1 Gm^3) landslides in carbonate sequences: case studies from the Zagros Mountains, Iran and Rocky Mountains, Canada (Master Thesis)

Roberts NJ, Evans SG (2013) The gigantic Seymareh (Saidmarreh) rock avalanche, Zagros Fold–Thrust Belt, Iran. J Geol Soc Lond 170(4):685–700. https://doi.org/10.1144/jgs2012–090

Rodrıguez CE, Bommer JJ, Chandler RJ (1999) Earthquake-induced landslides: 1980–1997. Soil Dyn Earthq Eng 18(5):325–346. https://doi.org/10.1016/S0267–7261(99)00012–3

Rouhi J, Delchiaro M, Della Seta M, Martino S (2019) Emplacement kinematics of the Seymareh rock-avalanche debris (Iran) inferred by field and remote surveying. Italian J Eng Geol Environ. https://doi.org/10.4408/IJEGE.2019-01.S-16

Setudehnia A, Perry JTOB (1967) Dal Parri. 1: 100 000 Geological Map. Iran Oil Operating Companies, Geological Exploration Division, Tehran, Iran

Shoaei Z (2014) Mechanism of the giant Seimareh Landslide, Iran, and the longevity of its landslide dams. Environ Earth Sci 72(7):2411–2422. https://doi.org/10.1007/s12665–014–3150–8

Stead D, Eberhardt E, Coggan JS (2006) Developments in the characterization of complex rock slope deformation and failure using numerical modelling techniques. Eng Geol 83(1–3):217–235. https://doi.org/10.1016/j.enggeo.2005.06.033

Strom AL, Korup O (2006) Extremely large rockslides and rock avalanches in the Tien Shan Mountains, Kyrgyzstan. Landslides 3(2):125–136. https://doi.org/10.1007/s10346–005–0027–7

Takin M, Akbari Y, Macleod JH (1970) Pul-E Dukhtar. 1: 100 000 Geological Map. Iran Oil Operating Companies, Geological Exploration Division, Tehran, Iran

Watson RA, Wright HE Jr (1969) The Saidmarreh landslide, Iran. In: Schumm SA, Bradley WC (eds) United States contributions to quaternary research, Geological Society of America, Special Papers, vol 123, pp 115–139

Wilson JP, Gallant JC (2000) Terrain analysis: principles and applications. Wiley, New York

Chapter 14
Toward Real-time Geodetic Monitoring of Landslides with GNSS Mass-market Devices

Paolo Dabove, Ambrogio M. Manzino, Alberto Cina, Marco Piras, and Iosif H. Bendea

Abstract The use of GNSS for landslide monitoring is not a novelty. In most cases, the GNSS receivers and antennas are only a subset of instruments usually considered and used in a complex monitoring network, where more often a continuous monitoring is required. In the present work, the proposed GNSS solution is composed by a dual frequency (L1 + L2) receiver, with a solar power and with a radio connection to a ground station, where the rover's measurements are collected and processed. The most critical aspect is the management of the collected data because the approach, which is normally used, assumes a fixed position of the GNSS antenna during the acquisition time window. Methodology for data acquisition and positioning (real-time or post-processing) and its duration, type of receivers, and antenna used (single or multi-constellation, single or dual frequency, mass-market or geodetic), data processing strategies (i.e., single epoch, static, kinematic), and GNSS network services are fundamental factors, which may favor one or another solution, according to time, economy, and infrastructure readiness in the field. Starting from the previous experience of the research group, this work investigates the possibility of employment of GNSS mass-market receivers and antenna for Network Real-Time Kinematic positioning for displacement detection. Using a dedicated slide which allows to define a micrometric displacement, several tests have been performed in a test-site at Politecnico of Torino (Italy), where different combinations of receivers and antennas (from geodetic to mass market) and displacement strategies have been considered. After this phase, some instruments are settled on a "real" landslide in the Verbano-Cusio-Ossola province (NW of Italy) in order to verify the feasibility of the employment of these devices and whose accuracies can be reached. Furthermore, data processing has been realized by means of different software (commercial and free and open source) and different kinds of solution. In this way, it could be possible

P. Dabove (✉) · A. M. Manzino · A. Cina · M. Piras · I. H. Bendea
Department of Environment, Land, and Infrastructure Engineering, Turin, Italy
e-mail: paolo.dabove@polito.it

M. De Maio, A. K. Tiwari (eds.), *Applied Geology*,
https://doi.org/10.1007/978-3-030-43953-8_14

to reduce costs for monitoring activities, improving the quality of the solutions, and to allow a "smart" use of GNSS technologies for monitoring.

Keywords GNSS · Monitoring · Low-cost · Landslides · Real-time

14.1 Introduction

Landslide is defined as "the movement of a mass of rock, debris, or earth down a slope" (Cruden 1991). Many disasters occurred in recent years, not only in Italy but also in Brazil, Pakistan, Philippines, Indonesia, and have killed thousands of people, destroyed infrastructures, creating injuries and heavy economic losses (Reid et al. 2012). The increasing number of these terrible natural events has increased the demands for improving new emerging technologies in order to prevent landslides, as well as to analyze data in real time for monitoring activities and prevention actions.

Many geomatics techniques and instruments have been considered for these kind of activities, starting from typical traditional instruments such as theodolites or geodetic GPS receivers, up to new more sophisticated instruments developed in recent years, such as robotic total stations, laser scanners, or Unmanned Aerial Vehicles (UAVs). In addition, the development of new Global Navigation Satellite System (GNSS) constellations (GLONASS, Galileo, BeiDou) coupled with the size reduction of the GNSS chipset allowed the diffusion of GNSS mass-market devices not only for mapping activities but also for engineering applications, including landslides monitoring. On the one hand, geodetic GPS/GNSS receivers have been used coupled with other instruments such as Electronic Distance Measurement (EDM), levels, total station (Rizzo 2002), inclinometers (Calcaterra et al. 2012), and wire extensometers (Bertachini et al. 2009; Coe et al. 2003; Gili et al. 2000; Malet et al. 2002; Moss 2000; Tagliavini et al. 2007). On the other hand, new GNSS mass-market devices have been employed for landslide-monitoring activities both in real time (Janssen and Rizos 2003; Bellone et al. 2016) and post-processing (Cina and Piras 2015).

Starting from these considerations, the aim of this study is to analyze and illustrate the use of GNSS mass-market instrumentation and its limitations in order to monitor real-time geological instability events such as slow-moving landslides by means of the employment of GNSS mass-market receivers and antennas. Firstly, a dedicated infrastructure is considered for analyze the sensitivity of these instruments: providing micrometric displacement manually, several tests have been carried out at Politecnico of Torino, considering different receivers-antenna combinations (from geodetic to mass market) and displacement velocities. Then, some instruments have been installed on a "real" landslide in the Verbano-Cusio-Ossola province (NW of Italy) in order to verify the feasibility of the employment of these devices and to analyze what kind of accuracies can be reached. All collected data have been processed considering different software (commercial, free and open source – FOSS) and different kinds of solutions.

Thus, this work aims to provide an overview about the state-of-the art of the available technology and to show an innovative method for real-time monitoring activities considering low-cost GNSS devices, in order to allow a "smart" use of GNSS technologies for monitoring.

14.2 Landslide Monitoring Using GNSS Receivers

In the past, total stations have been usually adopted for landslide monitoring but starting from the XXI century, GPS (before) and GNSS (now) techniques have replaced them and it is recently the most used monitoring system for different applications (Gili et al. 2000; Heunecke 2011; Wang 2013; Gassner et al. 2002).

At the beginning, the main restriction of the use of GPS for monitoring was due to the limited number of satellites (GPS only), the cost of the receivers (dual frequencies only), and the possibility to analyze a wide area, because the monitoring was only made in the point where the receiver is installed. The total station is not abandoned, but it is still used when it is necessary to monitor several number of points (e.g., dams, rock façades, etc.) or when the velocity of the landslide is not compatible with the GNSS precision or accuracy (e.g., < mm/years).

Considering the availability of new constellations (e.g., GLONASS, BeiDou, Galileo) and GNSS infrastructures (e.g., CORSs, Networks, etc.), the use of GNSS receivers is considerably increased because the number of the satellites and their signal quality are improved, and there are new products (e.g., Virtual RINEX, Dabove et al. 2016) and PPP services (Martín et al. 2015), which can be used even for monitoring applications. Moreover, considering a network of permanent stations (see next section), low-cost systems can be used and even single-frequency devices can be adopted (Piras et al. 2011; Cina and Piras 2015).

Traditionally, landslide monitoring with GNSS is realized installing several receivers on the field. The typical schema (Fig. 14.1) is composed by at least two receivers installed outside of the area affected by the landslide deformation, in order to create a "stable and reference" network of known points. However, other receivers are instead installed inside the area where some displacements occur, considering a

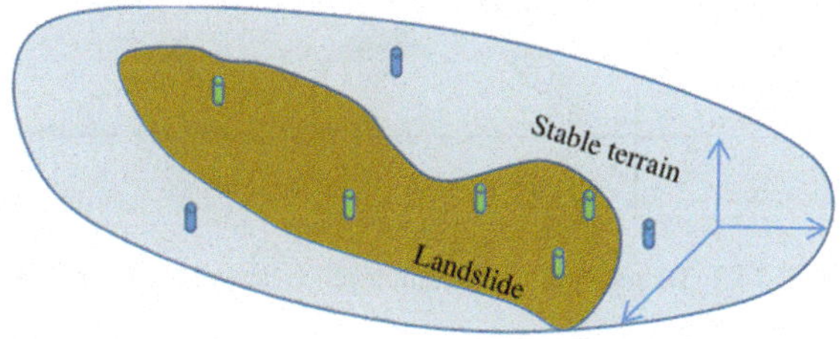

Fig. 14.1 Example of schema GNSS monitoring (blue = references, green = monitoring stations)

sparse distribution, with the purpose of identifying the main behavior of the phenomenon.

Each receiver is connected to a control station, which is devoted to collect data and making the data processing. The communication is performed by radiolink or GSM devices, which are powered by solar panels (in remote site) or power supply, where available. Adopting the following schema, several independent baselines are estimated using double differences approaches (Hofmann-Wellenhof et al. 2007), in order to estimate the coordinates of the single "check" point. These coordinates are compared with the past catalog, with the purpose of obtaining a time series analysis and to estimate the deformations. This procedure is very important and needs specific competences to be carried out, because it is fundamental to avoid false alarms.

As mentioned before, in some cases, the "reference stations" are replaced with some "virtual stations," which are generated by the Continuous Operating Reference Stations (CORS) network; in this way, the monitoring cost is lower and the stability of the external points is implicit, because these points are not physical and are not affected by local deformations.

As will be explained later, the low-cost technology is today a real approach and it allows to increase the number of active sensors installed on the body of landslides, improving the quality of the monitoring.

Nowadays, there are many COTS systems (*Commercial off-the-shelf*), with different costs, performances, and sizes, which allow to customize and design different monitoring networks, adopting different strategies.

In the following, a brief overview of some available GNSS receivers for monitoring issues is shown (Fig. 14.2), where costs and performances are described.

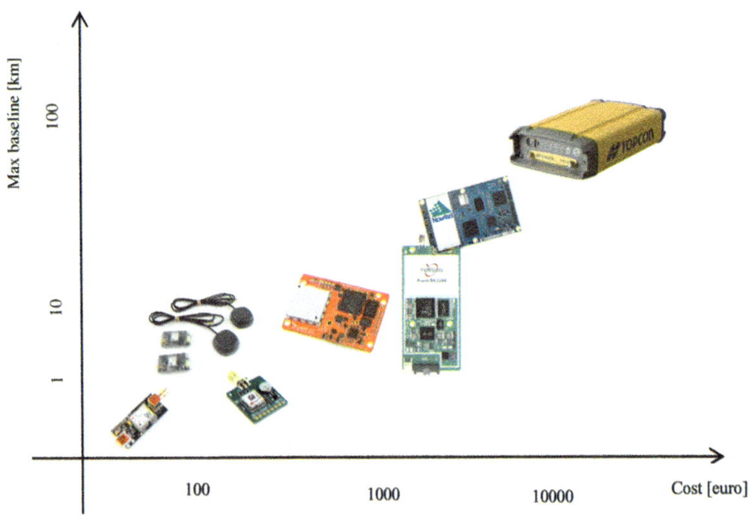

Fig. 14.2 Diagram of maximum distance between master and rover stations (baseline) vs cost

14.2.1 The Simulation of Obtainable Results

As mentioned before, the new generation of low-cost GNSS extended the possibility of being adopted for landslide monitoring, but it is necessary to analyze the limit and the performance achievable.

A good approach is to test these sensors in a controlled condition, using a geodetic GNSS receiver as reference, in order to compare the results.

A dedicated test was carried out, using a particular sliding system where controlled deformation was applied with a level of accuracy of a few millimeters. In this case, two Virtual RINEX have been used, in order to analyze the performances of these products. A dedicated mechanism (Fig. 14.3) for slide rectification has permitted to reach a high precision of movement definition along the path: it has been verified that the precision of the slide movement estimation is always better than 1 mm, which can be considered as the "scale resolution" of this device. The dimensions of this slide allow us to provide movements up to 1.30 m and 0.30 m in the horizontal and vertical components, respectively.

Three main situations, in terms of instruments, have been considered:

- Geodetic receiver and geodetic antenna
- Mass-market receiver (L1 - u-blox) and geodetic antenna
- Mass-market receiver (L1 - u-blox) patch antenna

First, displacements of about 2.5 cm and 5 cm were provided, without changing the height component. Then, when the end of the bar was reached, the return was realized, increasing the height component by 5 cm for each step. GNSS raw measurements were processed, comparing the estimated coordinates with the

Fig. 14.3 Micrometric slide for controlled deformations

known ones. As expected with a geodetic receiver, ambiguity fixing was always achieved and the coordinate differences were limited to a few millimeters, which are considered as "reference" solutions.

In particular, working with a mass-market solution (u-blox M8T receiver), the coordinate differences are those shown in Fig. 14.4.

In this case, the residuals are greater than the values obtained using the geodetic receiver, while the ambiguity is also fixed as integer numbers. This aspect is particularly highlighted in the vertical component, where the differences can exceed 3 cm and appear rather scattered.

In another test, the u-blox receiver was connected to a geodetic antenna already used with the geodetic receiver with a professional splitter. Using the same measurement protocol, the differences between the estimated coordinates and those imposed were newly estimated. Figure 14.5 shows the results obtained.

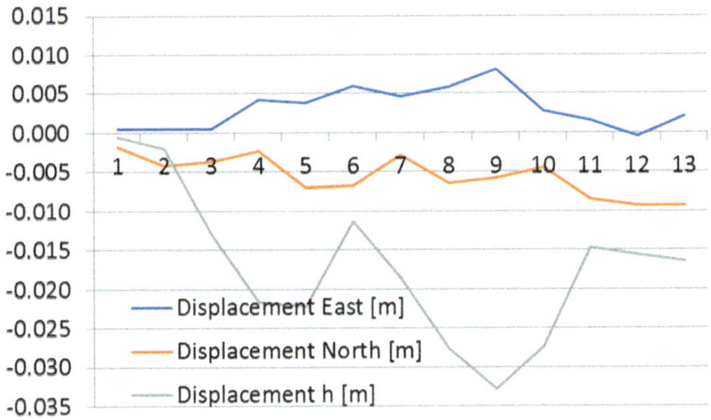

Fig. 14.4 Coordinate differences [m] vs test – case study: mass-market receiver and patch antenna

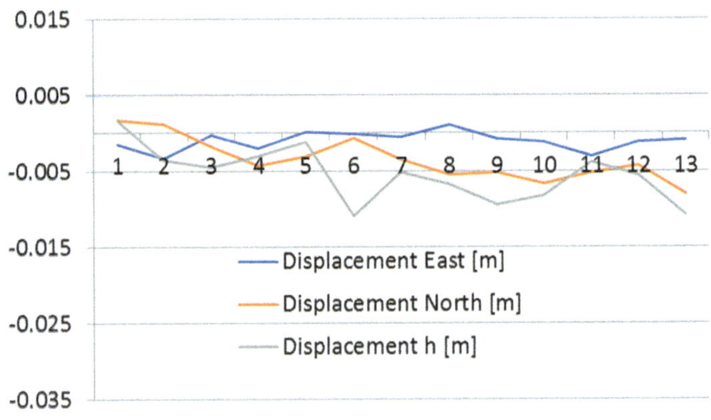

Fig. 14.5 Coordinate differences [m] vs test: case study mass-market receiver and geodetic antenna

Table 14.1 Summary of residuals [mm] obtained considering different configurations

Receivers	Residual East [mm]	Residual North [mm]	Residual h [mm]
(1) Geodetic antenna and receiver	−2 ± 1	−5 ± 3	2 ± 5
(2) u-blox+ u-blox antenna	3 ± 3	−6 ± 3	−17 ± 9
(3) u-blox +geodetic antenna	−1 ± 1	−3 ± 3	−5 ± 4

As expected, the use of a high-performance antenna can influence the positioning accuracy: the residuals obtained with this configuration are comparable and not worse than those obtained with the geodetic receiver (Table 14.1). Obviously, deformation monitoring with a geodetic antenna is not a low-cost solution. However, there are several antennas on the market that may represent an intermediate proposal between the geodetic and low-cost solutions, with prices that are comparable to mass-market receivers.

A final consideration concerns the possibility of reducing the acquisition time.

The tests conducted show that the accuracy achieved in post-processing with a mass-market single-frequency receiver allows this class of device to be used for monitoring or surveying. This class of receiver costs 200–300 € and is able to acquire only the carrier phase on L1.

It is comparable to the more expensive geodetic receivers in the following particular conditions:

- Acquisition time of at least 10 min
- Base-rover distance within 1 km
- Use of an external antenna according to the characteristics of the accuracy required

The end user will need to evaluate the device customization or their interfaces: in this paper, only the performance in terms of accuracy and precision is considered.

14.3 Toward the Use of Mass-Market GNSS Receivers in a CORSs Network

Today, thinking about the GNSS real-time positioning, one of the most used techniques is the Network Real Time Kinematic (NRTK) approach. When discussing about GNSS NRTK positioning, it is common to consider a typical network of permanent stations composed by multi-frequency and multi-constellation receivers (Dabove et al. 2012). The development of new satellite constellations (such as BeiDou and Galileo) allows to track more than 20 satellites mostly everywhere. This has permitted the use of low-cost receivers for real-time positioning as rover, reaching accuracies of about a few centimeters. One of the new developments for NRTK positioning, especially for monitoring purposes, is to consider these kind of devices as master stations in a CORS network (Nykiel and Szolucha 2014). The reason is mainly their cost, one order of magnitude less than the geodetic receivers, coupled with their performances, which can be compared in case of proper use, as shown in the previous section. One of the main problems is to verify if software that manage the CORS network (defined also as network software) are able to consider these kind of receivers as master, since the signals' quality (the noise of the measurements, for both pseudoranges and carrier-phase) in this case is higher than those obtainable with geodetic receivers.

Today, there are two main network software available for managing CORS networks: the Leica GNSS Spider, provided by Leica Geosystems®, and the GNSMART, provided by Geo++®. Thus, we have tried to consider both in order to verify the feasibility of considering mass-market GNSS receivers as master stations in a CORSs network. A dedicated network has been considered, composed by all geodetic receivers with a mean inter-station distance of about 50 km, comparable to the typical distances of CORSs available in Italy today.

In addition, a mass-market L1 GNSS receiver (called TEST in Fig. 14.6) has been added: this station, about 19 km far from the closest permanent station (NOVR), is composed by a single-frequency and multi-constellation (GPS + GLONASS) mass-

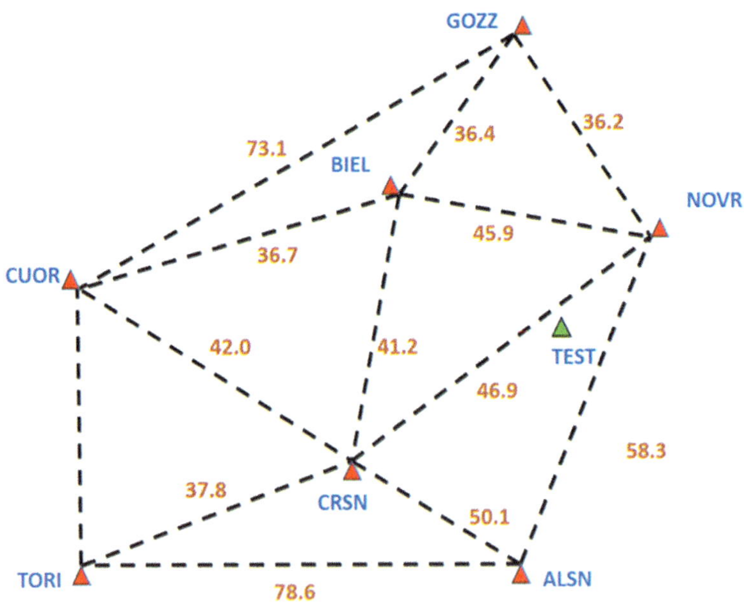

Fig. 14.6 The network considered: geodetic (red triangle) and mass-market (in green) GNSS receivers. All distances are expressed in kilometers

market receiver (u-blox EVK-6 T) coupled with a low-cost Garmin GA38 antenna. This station has been connected to an AC socket for power supply and to a Wi-Fi modem for internet connection, in order to be able to provide real-time data to the network software. For this task, the STRSRV tool of RTKLIB software (Takasu and Yasuda 2006) is used because it is able to convert the .ubx binary stream into an RTCM3 message, which can be used as input in all network software. Both the coordinates of TEST and all reference stations were adjusted with the Bernese 5.2 software (Dach et al. 2015), considering a static session of 2 days with a sampling rate of 1 s also in this case.

The first impressive result obtained is that the two network software provides completely different results: while considering GNSMART, it is possible to have a "mixed" network. This does not happen with Spidernet. This means that with Spidernet it is not possible to consider a single-frequency receiver as a permanent station: this software can manage pseudorange and carrier phase measurements but these data cannot be used for the network calculation. A different behavior is instead shown by GNSMART: only this software can be considered useful for this new positioning method because both single- and dual-frequency receivers can be used as master stations. It is important to underline that the GNSMART software needs about 8 mins to fix the phase ambiguities for all stations of the network (it is the same time if a network with only geodetic receivers is considered).

The next step is to verify if a generic GNSS receiver (mass-market or geodetic) can be considered as rover in a mixed network, reaching the fixing of phase ambiguities in order to obtain the typical accuracies available with the NRTK positioning technique.

Considering the mixed network and GNSMART as the network software, some tests were made to determine the performances of NRTK positioning with geodetic and mass-market instruments in a mixed network. Test sites 2.5 km, 5 km, 10 km, and 15 km far from TEST are considered in order to verify the precision and accuracy obtainable (Fig. 14.7).

For these tests, the same instruments installed on the TEST site are considered for the rover; different network products, such as VRS and Nearest (NRT) corrections, are used with the RTKNAVI tool for real-time positioning of the mass-marker rover receiver, with an updating rate equal to 1 s. Particular results are obtained: even if the phase ambiguities are fixed for the networks stations, no FIX positions are available for the rover receiver considering all differential corrections: only float solutions can be obtained. This behavior is the same even if a geodetic (Leica 1230 + GNSS) receiver is considered as rover. On the other hand, the float solutions are very good even if the low-cost instrument is considered: the difference between the estimated and the reference positions are about 10 cm if the distance with respect to the nearest station is less than 2.5 km, as it is possible to see from Table 14.2. In this context, these results are not useful for landslide monitoring because the accuracies reachable today are insufficient for these purposes.

Up to now, this technique is useful only for other applications, such as for traffic control: even if this new methodology has shown impressive performances, this technology is not ready today. It could be interesting to repeat these tests in the

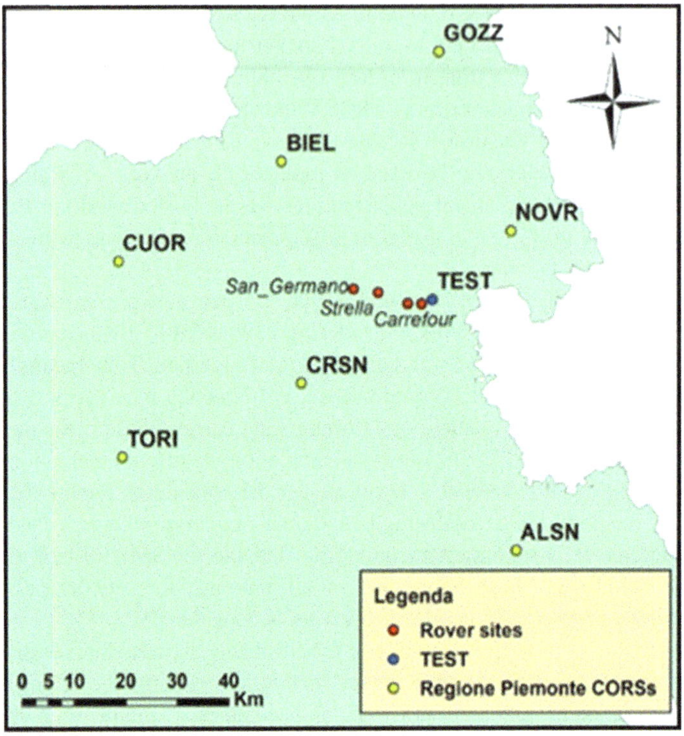

Fig. 14.7 Rover test sites (in red), CORSs (in yellow), and TEST station (in blue)

Table 14.2 Results of mass-market receivers in a mixed network: no FIX solutions can be obtained, only float

			Mass-market	Geodetic
Test site name	*d* from test	Correction type	*3D* accuracy at 95%	
Carrefour	2.5 km	VRS	0.09 m	0.05 m
		NRT	0.07 m	0.07 m
Cascine Strà	5 km	VRS	0.18 m	0.15 m
		NRT	0.22 m	0.18 m
Strella	10 km	VRS	0.53 m	0.44 m
		NRT	0.60 m	0.51 m
San Germano	15 km	VRS	0.56 m	0.50 m
		NRT	0.61 m	0.55 m

future, when the development of new positioning algorithms in network software could permit the fixing of phase ambiguities as integer values, in order to reach a positioning accuracy in real time of about a few centimeters.

In synthesis, the obtained results exclude the possibility to perform NRTK positioning in mixed networks because no software allows this kind of survey as of today. Despite that, a new possible solution can be adopted, for example, for landslide monitoring: the idea is to consider all stations (mass-market and geodetic) as master and analyze the variation of coordinates in order to decide if a displacement occurs. This is possible thanks to GNSMART: the a priori coordinates can be weighted in different ways by the network manager. So it is possible to use a more constraint weight for geodetic receivers and a more loose weight for mass-market ones. At the same time, this software can determine the difference between estimated and a priori coordinates in real time. This is useful because, coupled with some statistical techniques, this approach gives the possibility to analyze if a difference is due to a noise of measurements or is a real displacement.

14.4 A Case-study of Landslide Monitoring Using GNSS Low-cost Receivers

In this section, the goal is to show the GNSS monitoring results performed on five landslides located in the north of Piedmont (Italy), near the borders with Switzerland.

These landslides are subject to slow deformation phenomena, so the most convenient operation is post-processing monitoring. Nevertheless, for the sake of completeness, and to extend the methodology to future applications, also the real-time monitoring procedures are considered. In both cases, the u-blox M8T receiver has been used.

First of all, having to use mass-market antennas (the Garmin GA38 GNSS antenna in this case), it was necessary to precisely determine the position of the phase center of the antenna (also known by the PCV acronym). The procedure by which the position of the phase center has been determined is called "antenna swap," and has provided millimeter precision in all three coordinates.

14.4.1 Post-processing Approach

This operation was designed and built completely "home-made," as the mass-market receivers used provide their data only through an RTCM flow, which must be transmitted to a control center.

All the received flows arriving at the control center at a 1 s rate. The control center converted the RTCM flows in the standard RINEX formats for the hourly, daily, and real-time positioning procedures, which were carried out with the routines of the RTKLIB package.

Hourly and daily scheduled procedures were built in a Windows environment, while the statistical inference procedures on the results were performed in the Matlab environment.

The five mass-market stations are respectively located as follows: two on the landslide of Madonna del Sasso (SASA and SASB), two on the landslide of Loreglia (LORA and LORB), and one on the landslide of Crodo (CROD). Data processing involved the use of the NRTK permanent stations network of the Piedmont and Lombardy Regions called SPIN GNSS (https://www.spingnss.it/spiderweb/frmIndex.aspx).

Hourly positioning was performed both for the deformation monitoring of the two stations of Madonna del Sasso between them, and for the positioning of the stations of Madonna del Sasso starting from the Gozzano station which is only 6 km away. Only the L1 frequency on u-blox receivers was considered as well as only CDMA constellations (i.e., GPS, Galileo, and BeiDou) were tracked.

The movements and deformation analysis of the five stations of Loreglia and Madonna del Sasso, for a total of seven result files, are then read by a software that analyzes the significance of the results and constructs appropriate graphs for this purpose.

More precisely, for these seven series and for each of the three components, two types of control are performed:

(a) A check on a possible rapid movement on the current day
(b) A check of the significance of the speed movements

(a) The rapid movements

Based on the time series of each site and for each component, it was decided to analyze only the solutions with a minimum ratio greater than or equal to 4 (Dabove et al. 2016).

Obviously, the "float" solutions (i.e., solutions where the phase ambiguities are estimated as real numbers) are not analyzed and considered. On this subset of data, a robust estimation of the MAD (median absolute deviation, Jearkpaporn et al. 2005) type was performed for the estimation of the regression line for the three components.

This analysis takes place if the data of the day just passed are valid, but the parameters of the straight line are calculated without using these last data. With the estimation of the regression line in fact, the measurements of the coordinates of the last epoch are predicted, that is of the last measurement day. The comparison between the expected measure and the actual measure serves as a risk assessment parameter.

The robust estimate of the regression line also evaluates the mean square deviation of the data. The coordinates obtained from the latest GNSS measurements are considered "without alarms" if the last measure is included in the range of twice the average standard deviation around the value set by the regression line at the last day.

(b) Control of velocity movement

It may happen that the values foreseen for the three components are perfectly in line with the values of the three coordinates calculated on that day. This does not mean that there is no danger, but it only means that the trend line, whose derivative

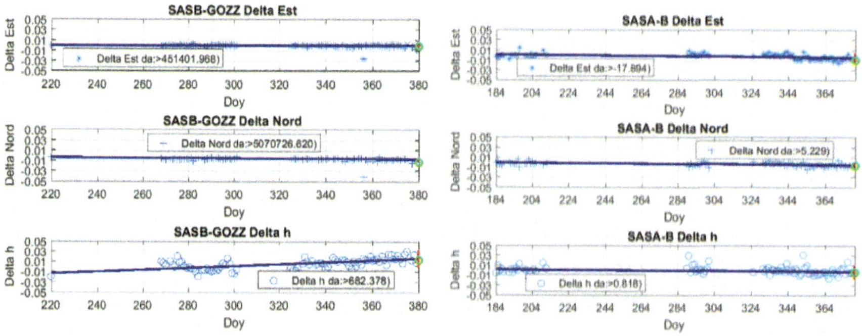

Fig. 14.8 Displacements and deformations [m] of Madonna del Sasso (Italy)

indicates the speed, is precise and also the daily measurements are precise. However, it is necessary to provide an alarm if this velocity (i.e., the slope of the straight line) is significantly different from zero.

This check is carried out every day, regardless of data; there are no phase ambiguities fixed with a high "ratio."

Also, in this case, a vector of three components is written for each baseline, indicating the non-significance of the velocity with the number zero.

A non-zero value, with the sign, indicates the speed in mm / 100 days of that speed component.

In any case, any speed that, in absolute terms, does not exceed 3 mm / 100 days is considered "not significant." This parameter and the probability value of significance are both parametric indices that can be modified as desired.

Ultimately, a summary report is generated every day, sent by email, and located in the shared Google Drive directory. We can define this relationship as a "traffic light" even if only in the first columns, it contains traffic light indexes and in the second part, it contains some deformation velocities.

Figure 14.8 contains an example of two graphs obtained for the displacements of one of the two stations of Madonna del Sasso (SASB) on the left and the deformations between the two stations both on the landslide of Madonna del Sasso (SASA-B). All diagrams are also located daily in the shared Google Drive directory.

14.4.2 Real-time Approach

RTK positioning with the use of single-frequency receivers has greater limitations than in the case of use of multi-frequency instruments, which allows rapid fixing of phase ambiguity by exploiting phase combinations.

Furthermore, due to the ionospheric refraction, the length of the baseline using single-frequency receivers must be limited to 10–15 km.

In these cases, the Virtual Reference Station (VRS) correction can also be used effectively.

The RTK solution has been analyzed for the five points where the multi-sensor control units have been installed but here, for reasons of space, only the most significant results are provided. Therefore, only comparisons of the GNSS and GPS positions of the two stations of Madonna del Sasso and the comparisons of the NRST and VRS positions of Madonna del Sasso station, which is 6 km from the nearest station but is outside the SPIN NRTK network, are shown.

The analysis has been conducted for 24 h of RTCM data at a 1 s rate, capturing the streams of the stations and processing them in real time with the RTKLIB software and with the real or virtual stations of the SPIN GNSS network.

Analyzing the statistics of 24 h measurement, it is possible to obtain the following considerations:

- The phase ambiguity fixing is mainly lost when the constellation changes, for example, if a new satellite arises.
- If the ratio value exceeds the threshold (i.e., the ambiguities are declared as "fixed," which implies a FIX solution), the correct estimation of the phase ambiguities is not guaranteed.
- The solution with centimeter accuracy is however also reachable with distances of 15 km from the nearest station. The use of GNSS (Galileo and GPS constellation) does not lead to an improvement in the percentage of fixings in the RTKNAVI software; however, considering only the FIX results, the GNSS solutions have a better accuracy with respect to the GPS only ones, even if the percentage is less.

SASA-B GNSS (about 20 m)	SASA-B GPS (about 20 m)
n°epoch FIX FLT perc.FIX	n°epoch FIX FLT perc.FIX
65,126 20,052 45,045 30.8%	65,285 46,355 18,892 71.1%
FIX:residual 3D %	FIX:residual 3D %
< 1 cm 2 cm 5 cm >5 cm	< 1 cm 2 cm 5 cm >5 cm
30% 86% 96% 4%	23% 70% 86% 14%
gozz-sasa (about 6 km)	vrs-sasa
n°epoch FIX FLT perc.FIX	n°epoch FIX FLT perc.FIX
64,147 47,915 16,179 74.8%	64,128 51,771 12,324 80.8%
FIX:residual 3D	FIX:residual 3D
<1 cm 2 cm 5 cm >5 cm	< 1 cm 2 cm 5 cm >5 cm
38% 77% 90% 10%	19% 60% 93% 7%

Considering the comparison between the solution obtained from the real station (nearest station: NRT) and the virtual one, this last one provides slightly better performances, with a greater percentage of epochs where phase ambiguities are declared as fixed and where the differences between the estimated results with respect to the reference one are a little bit lower.

However, the VRS solution does not always solve the problems of distance from the GNSS base stations: this solution is still feasible if the rover is inside the network or close to the boundaries but become unfeasible if it is outside.

In the case of the landslide of Crodo, 13 km outside the perimeter of the SPIN GNSS network, the VRS solution provides only 35% of solutions where the phase ambiguities are declared as fixed in a correct way.

14.5 Conclusions

In this work, we have provided an overview about the state-of-the art of the available technology and to show an innovative method for real-time monitoring activities considering low-cost GNSS devices, in order to allow a "smart" use of GNSS technologies for monitoring. In particular, attention has been focused on analyzing and illustrating the use of GNSS mass-market instrumentation and its limitations for monitoring geological instability events, such as slow-moving landslides, in real time by means of the employment of GNSS mass-market receivers and antennas.

First, a sensitivity analysis of these kinds of instruments has been provided considering a dedicated infrastructure carried out in the Topography Lab at DIATI - Politecnico di Torino: with these instruments, it has been possible to understand the accuracy and precision obtainable, comparing the obtained results of displacements, thanks to micrometric screws. Then, some instruments have been installed on a "real" landslide in the Verbano-Cusio-Ossola province (NW of Italy) in order to verify the feasibility of the employment of these devices and to analyze what kind of accuracies can be reached. All collected data have been processed considering different software (commercial and free and open source - FOSS) and different kinds of solutions. Considering the post-processing approach, it has been demonstrated that it is possible to obtain a millimeter level of accuracy and precision if a daily dataset is processed. About the real time, the level of accuracy and precision is one order of magnitude greater with respect of the previous case, even if medium baseline (15 km) are considered. Also the percentage of FIX solutions is interesting and it leaves opened interesting perspectives in the near future. A possible solution could be to consider a network of mass-market GNSS devices in order to improve the redundancy of observations and to increase the reliability of the obtained solutions.

References

Bellone T, Dabove P, Manzino AM, Taglioretti C (2016) Real-time monitoring for fast deformations using GNSS low-cost receivers. Geomat Nat Haz Risk 7(2):458–470

Bertachini E, Capitani A, Capra A, Castagnetti C, Corsini A, Dubbini M, Ronchetti F (2009) Integrated surveying system for landslide monitoring, Valoria landslide (Appennines of Modena, Italy). In: Proceedings of FIG working week 2009, Eilat, Israel
Calcaterra S, Cesi C, Di Maio C, Gambino P, Merli K, Vallario M, Vassallo R (2012) Surface displacements of two landslides evaluated by GPS and inclinometer systems: a case study in Southern Apennines, Italy. Nat Hazards 61(1):257–266
Cina A, Piras M (2015) Performance of low-cost GNSS receiver for landslides monitoring: test and results. Geomat Nat Haz Risk 6(5–7):497–514
Coe JA, Ellis WL, Godt JW, Savage WZ, Savage JE, Michael JA, Kibler JD, Powers PS, Lidke DJ, Debray S (2003) Seasonal movement of the Slumgullion landslide determined from Global Positioning System surveys and field instrumentation, July 1998-March 2002. Eng Geol 68 (1–2):67–101
Cruden DM (1991) A simple definition of a landslide. Bulletin of the International Association of Engineering Geology/Bulletin de l'Association Internationale de Géologie de l'Ingénieur 43 (1):27–29
Dabove P, De Agostino M, Manzino A. 2012. Achievable positioning accuracies in a network of GNSS reference stations. In: Global Navigation Satellite Systems: Signal, Theory and Applications / Jin S.G. InTech, pp 189–214. ISBN 9789533078434
Dabove P, Cina A, Manzino AM (2016) How reliable is a Virtual RINEX? In: 2016 IEEE/ION Position, Location and Navigation Symposium (PLANS), IEEE, pp 255–262
Dach R, Lutz S, Walser P, Fridez P (2015) Bernese GNSS Software Version 5.2. User manual, Astronomical Institute, University of Bern, Bern Open Publishing. ISBN: 978–3–906813-05-9
Gassner G, Wieser A, Brunner FK (2002) GPS software development for monitoring of landslides. In Proceedings of FIG XXII Congress (CD-ROM)
Gili JA, Corominas J, Rius J (2000) Using Global Positioning System techniques in landslide monitoring. Eng Geol 55(3):167–192
Heunecke O, Glabsch J, Schuhbäck S (2011) Landslide monitoring using low cost GNSS equipment n experiences from two alpine testing sites. Journal of civil engineering and architecture 5 (8)
Hofmann-Wellenhof B, Lichtenegger H, Wasle E (2007) GNSS–Global Navigation Satellite Systems: GPS, GLONASS, Galileo, and more. Springer, Wien
Janssen V, Rizos C (2003) A mixed-mode GPS network processing approach for deformation monitoring applications. Surv Rev 37(287):2–19
Jearkpaporn D, Montgomery DC, Runger GC, Borror CM (2005) Model-based process monitoring using robust generalized linear models. Int J Prod Res 43(7):1337–1354
Malet J-P, Maquaire O, Calais E (2002) The use of global positioning system techniques for the continuous monitoring of landslides: application to the super-sauze earth flow (Alpes-de-Haute-Province, France). Geomorphology 43:33–54
Martín A, Anquela AB, Dimas-Pagés A, Cos-Gayón F (2015) Validation of performance of real-time kinematic PPP. A possible tool for deformation monitoring. Measurement 69:95–108
Moss JL (2000) Using the global positioning system to monitor dynamic ground deformation networks on potentially active landslides. Int J Appl Earth Obs Geoinf 2(1):24–32
Nykiel G, Szolucha M (2014) GNSS reference network real-time service control techniques. In: Environmental Engineering. Proceedings of the International Conference on Environmental Engineering. ICEE, Vol. 9. Vilnius Gediminas Technical University, Department of Construction Economics & Property, p 1
Piras M, Marucco G, Cina A (2011) Mass-market receiver for static positioning: tests and statistical analyses. Coordinates 7(1):16–18
Reid ME, LaHusen RG, Baum RL, Kean JW, Schulz WH, Highland LM (2012) Real-time monitoring of landslides. USGS Annual report; Fact sheets 2012–3008
Rizzo V (2002) GPS monitoring and new data on slope movements in the Maratea Valley (Potenza, Basilicata). Phys Chem Earth, Parts A/B/C 27(36):1535–1544

Tagliavini F, Mantovani M, Marcato G, Pasuto A, Silvano S (2007) Validation of landslide hazard assessment by means of GPS monitoring technique – a case study in the Dolomites (Eastern Alps, Italy). Nat Hazards Earth Syst Sci 7:185–193

Takasu T, Yasuda A (2006) Evaluation of RTK-GPS Performance with Low-cost Single-Frequency GPS Receivers. In Proceedings of International Symposium on GPS/GNSS, Tokyo, Japan, 11–14 November 2006; pp 852–861

Wang GQ (2013) Millimeter-accuracy GPS landslide monitoring using Precise Point Positioning with Single Receiver Phase Ambiguity (PPP-SRPA) resolution: a case study in Puerto Rico. J Geogr Sci 3(1):22–31

Part V
Landslide: Climate Change

Chapter 15
Advances in Rainfall Thresholds for Landslide Triggering in Italy

Stefano Luigi Gariano, Samuele Segoni, and Luca Piciullo

Abstract We reviewed the Italian scientific literature published in the period 2008–2018 on the topic of rainfall thresholds for the landslide triggering, with the aim of analyzing the most significant advances and the main open issues. In the international literature, Italy occupies a relevant position from both a quantitative and a qualitative viewpoint: 65 out of the 163 thresholds published worldwide in the considered period are defined in Italy. The main improvements can be ascribed to rigorous cataloguing of landslides; definition of standard and objective methods for thresholds analysis; quantitative validation of the results and evaluation of the performance of related warning systems; attempts to improve the spatial resolution of the forecasts. However, some shortcomings still limit the research on landslide rainfall thresholds and some open issues recently emerged as priorities to be further investigated: the effects of climatic and environmental changes on the thresholds; their implementation into hazard management procedures and early warning systems; the adoption of combined approaches to account for the hydrological conditions of the slopes; the quantification of diverse uncertainties. This review disseminates the best practices among scientists and stakeholders involved in landslide hazard management, and it draws a national framework of procedures for defining reliable rainfall thresholds, in particular for early warning purposes.

Keywords Landslides · Debris flows · Rainfall threshold · Validation · Early warning · Forecasting · Review

---

S. L. Gariano
CNR-IRPI, Perugia, Italy
e-mail: stefano.luigi.gariano@irpi.cnr.it

S. Segoni (✉)
Università degli Studi di Firenze, Dipartimento di Scienze della Terra, Firenze, Italy
e-mail: samuele.segoni@unifi.it

L. Piciullo
Norwegian Geotechnical Institute, Oslo, Norway
e-mail: luca.piciullo@ngi.no

M. De Maio, A. K. Tiwari (eds.), *Applied Geology*,
https://doi.org/10.1007/978-3-030-43953-8_15

15.1 Introduction

Landslides are widespread and frequent natural phenomena responsible for casualties and damages worldwide; Italy is acknowledged as a country highly exposed to landslide hazard and risk (Salvati et al. 2014; Haque et al. 2016, 2019). The Italian geoscience community is particularly involved in landslide prediction and in the mitigation of the related risks. Since extremely intense or exceptionally prolonged rainfall is the main triggering cause of landslides in Italy and in the world (Froude and Petley 2018), rainfall thresholds for landslide initiation have become a common and well-established approach to study landslide occurrence at all spatial scales and to set up landslide early warning systems (LEWSs).

A rainfall threshold defines, often with a mathematical law, the rainfall amount needed for one or more landslides to occur in a given study area (Segoni et al. 2018a and references therein). Rainfall thresholds are usually based on a very straightforward empirical correlation between cause (rainfall) and effect (landsliding). Their analysis require only few input parameters (rainfall data and information on landslide occurrences) and can be easily used as a scientific base for LEWSs at different scales given that they are conceptually simple, well understandable, and because rainfall is the only needed monitoring parameter.

After some pioneering works (Endo 1969; Onodera et al. 1974; Campbell 1975; Caine 1980), the rainfall threshold approach became very popular in landslide prediction studies, producing dozens of applications worldwide at all geographical scales, considering many different landslide types and using diverse rainfall parameters with several approaches (Guzzetti et al. 2007, 2008).

In the last years, the success of the rainfall threshold approach has not diminished; on the contrary, it received a growing consideration: several improvements were introduced to increase the scientific soundness and the forecasting reliability and effectiveness of the methodology (Segoni et al. 2018a).

In this work, we focus on the most recent advances brought by rainfall thresholds defined in Italy, by Italian geoscientists. We analyze all peer-reviewed, indexed scientific articles published in the period 2008–2018 and provide an overview of the most important improvements and of the main issues that still need to be investigated and debated, in the hope of providing a contribution to the dissemination of best practices among geoscientists and stakeholders involved in landslide hazard management.

15.2 Materials and Methods

To identify the main recent advances in the definition and application of rainfall thresholds in Italy, we investigated the *Scopus* database, searching for the keys "rainfall" and "threshold" in the title, abstract, or keywords of items included therein. We limited the results to peer-reviewed articles written in English and published in the period 2008–2018. Furthermore, all the outcomes were screened to extract only works related to rainfall thresholds developed for study areas in Italy.

To summarize significant information from these works, all articles were accurately examined, and a list of relevant details was extracted and organized according to the following scheme.

- General Information and Operational Details
  - Publication details (authors, journal, year of publication, doi)
  - Geographical information (name, location, and extension of the study area)
  - Spatial scale (slope, local, basin, regional, national)
  - Implementation stage (threshold implemented in an operative LEWS, in a prototype version of LEWS, threshold not intended for early warning purpose)
  - Number of warning levels of the possible LEWS
- Dataset Details
  - Period of analysis
  - Number of events and number of single landslides considered in the study
  - Types of landslide considered in the study
  - Information source for landslide catalogue
  - Source of rainfall data
  - Number and density of rainfall measuring points
  - Rainfall time scale
  - Other monitoring instruments (if any)
- Methodological Details
  - Method to relate landslides and rainfall measurements
  - Method to extract rainfall parameters from rainfall records
  - Method to define the threshold
  - Threshold parameters
  - Validation and performance evaluation

The collected information was analyzed looking for research trends and main methodological advances proposed; finally some basic statistics were examined.

15.3 Results

The bibliographic search and the subsequent screening activities returned 163 articles published worldwide, concerning rainfall thresholds in the 2008–2018 period. Among these, 63 articles (39%) regarding case studies in Italy were further investigated. For the sake of clarity, we maintain that 40 works are related to the 2008–2016 period and were already included in the review by Segoni et al. (2018a), while 23 are related to the 2017–2018 period and were reviewed in this work for the first time.

We found that rainfall thresholds were defined using information on landslides that occurred in all Italian regions (Fig. 15.1a), and, concerning different geographical scales of analysis: slope, if a single landslide is investigated; basin, where the threshold definition considers all landslides occurred in a hydrographic catchment; local, from a few to some hundreds of square kilometers; regional, intended as study areas equal to one or more administrative subdivisions. Finally, we found also that some thresholds were defined for the whole national territory (Fig. 15.2a; e.g. Brunetti et al. 2010; Peruccacci et al. 2017). In particular, the regional-scale applications are the most represented (30 out of 63) in the surveyed literature, with some administrative regions (*Campania*, *Emilia-Romagna*, *Piemonte*, *Toscana*, and *Umbria*) interested by a relevant number of works at several scales.

The trend of published articles almost constantly increased in the last decade, and, especially starting from 2012, a relevant number of works were published every year (Fig. 15.1b). Most of the articles were published on the journals *Natural Hazards and Earth System Sciences* (13), *Landslides* (12), and *Geomorphology* (10).

Almost half of the published thresholds are strictly related to operative landslide prediction and early warning, describing applications to prototypal or operational LEWSs, or accounting for possible implementations and improvements to the said LEWSs (Fig. 15.1b). The remaining thresholds are not directly related to a warning system and are used either to characterize an area prone to landslides or to explore the effects that methodological aspects or physical constraints can play on landslide prediction.

One of the most important input data for threshold definition is a reliable, accurate, and complete catalogue of landslide information (Fig. 15.2c). The effectiveness of the thresholds relies on the quality, number, and distribution of empirical data (Peruccacci et al. 2012). A wide range of sources are employed to build or update a landslide dataset with a sufficient detail regarding the location and time of occurrence: technical reports from public administration (mainly municipalities, civil protection, and fire brigades), newspapers, pre-compiled inventories, and internet news. An overview of the landslide types included in the threshold analysis is portrayed in Fig. 15.2d. Shallow landslides are the most investigated types (30 cases), followed by debris flows (8). A few examples have been published also related to earth flows (Floris et al. 2012; Greco et al. 2013) and deep-seated landslides (Floris and Bozzano 2008). A relevant number of thresholds (22) consider all landslide types without distinction. Regarding the periods of data collection and analysis, they range from some months (Floris et al. 2012; Papa et al. 2013) to several decades (e.g., Giannecchini et al. 2012; Frattini et al. 2009).

Reliable rainfall measurements constitute the other essential input for threshold analyses. The large majority of thresholds is based on rainfall data retrieved by rain gauges, which almost everywhere are organized in regional automated networks with a good spatial density (on average 1 station every 100 km^2). Only a few recent studies are based on continuous rainfall fields estimated by satellite (Nikolopoulos et al. 2017; Rossi et al. 2017; Brunetti et al. 2018) or radar (Nikolopoulos et al. 2015; Marra et al. 2016, 2017). Lastly, we found thresholds defined without using real

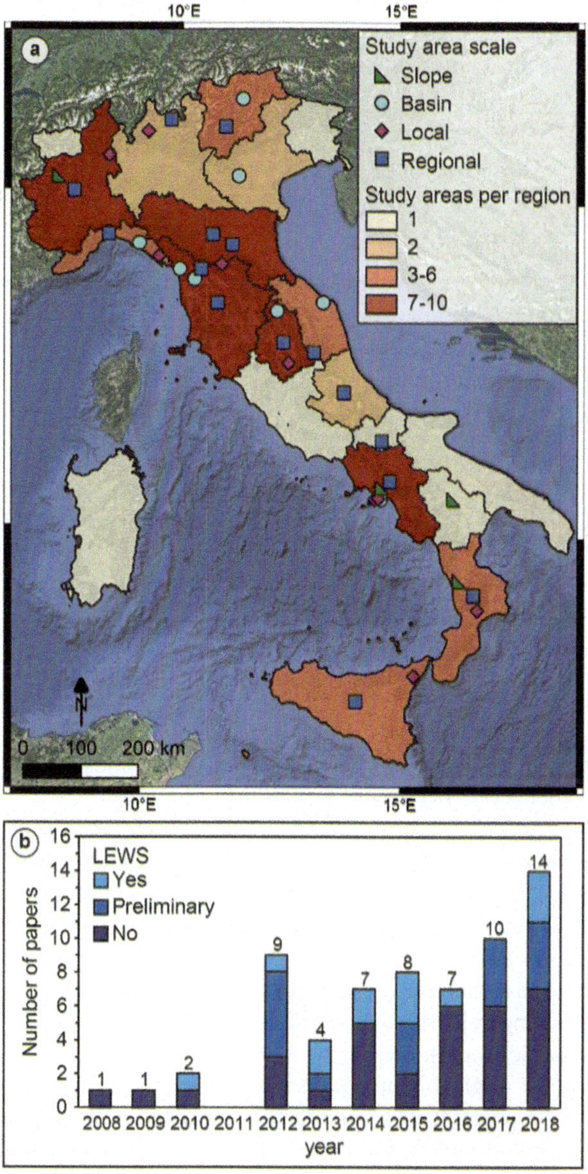

Fig. 15.1 (**a**) Geographical locations of the study areas considered in the articles dealing with rainfall thresholds, classified according to the spatial scale. The 20 Italian administrative regions are colored based on the number of study areas. (**b**) Bar chart displaying the number of analyzed papers published in scientific journals from 2008 to 2018. Bars show, for every year, how many thresholds were implemented in an operative or preliminary LEWS and how many thresholds were not conceived for future LEWS implementation

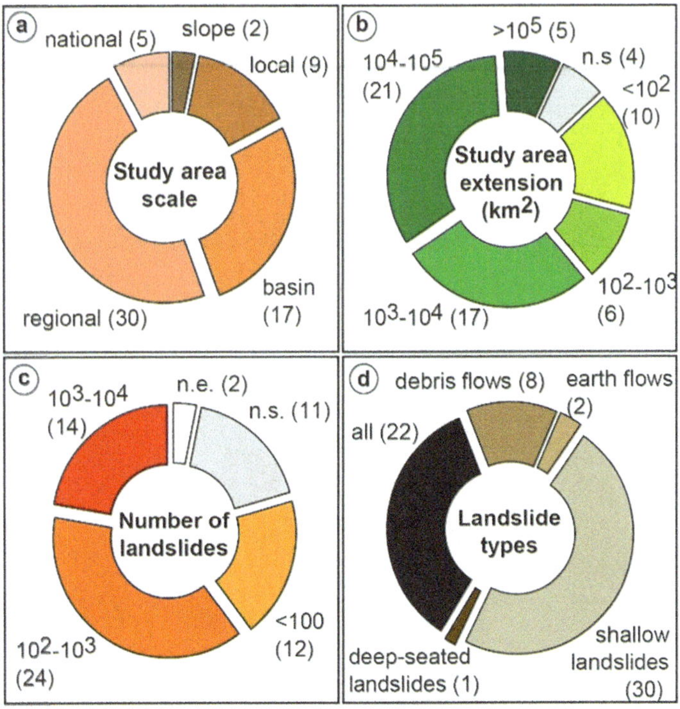

Fig. 15.2 Doughnut charts showing the number of analyzed papers per (**a**) spatial scale (see text for explanation); (**b**) study area extension, in classes; (**c**) number, in classes; and (**d**) types of landslides used to define the thresholds. Key: *n.s.* not specified, *n.e.* not expected

rainfall data: four of them are thresholds defined with theoretical calculations (Bovolo and Bathurst 2012; Salciarini et al. 2012; Papa et al. 2013; Alvioli et al. 2014), while two works report numerical experiments carried out on synthetic rainfall data (Peres and Cancelliere 2014; Peres et al. 2018) (Fig. 15.3a).

Rain gauges and radars recently installed in Italy are capable of measuring rainfall with a very fine temporal resolution (i.e. sub-hourly). Nevertheless, for almost all the investigated thresholds (53), rainfall measures are based on hourly time steps either because this is the original temporal resolution of the instruments, or because the measures have been aggregated (Fig. 15.3b). Six cases are based on daily rainfall measures (e.g. Frattini et al. 2009) and one is based on tri-hour rainfall data (Alvioli et al. 2018).

Given that rain gauges are the most used instrumental source of rainfall measurements, typically the first step of the threshold analysis consists in associating each landslide to a rain gauge, to further estimate the triggering rainfall. There is no consensus on which is the most suitable approach and, consequently, different methods are used (Fig. 15.3c). In eight cases, a single rain gauge was a-priori selected as representative for the landslide(s) in the study area (e.g. Lagomarsino et al. 2013; Terranova et al. 2015; Napolitano et al. 2016; Ciervo et al. 2017). Ten

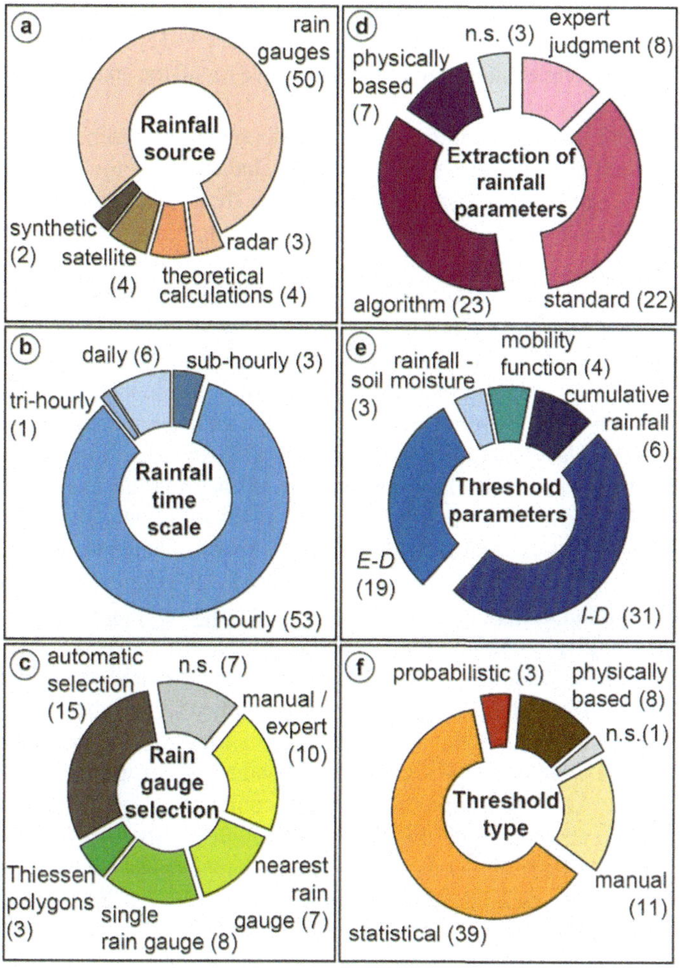

Fig. 15.3 Doughnut charts showing the number of analyzed papers classified per (**a**) sources of rainfall data used to define thresholds; (**b**) rainfall time scales adopted in the analyses; (**c**) methods employed for rain gauge selection; (**d**) methods adopted for the extraction of rainfall parameters; (**e**) main parameters adopted for defining the thresholds; (**f**) types of thresholds, i.e., methods used for drawing or defining the thresholds. Key: *n.s.* not specified

thresholds rely on expert judgment to manually identify the rain gauge that is deemed more suitable. The disadvantage of this approach is the high degree of subjectivity, which pushed the research community to develop algorithms and other standardized methods to automatically define the rain gauge representative for each landslide (15 occurrences). Usually, the automatic selection relies on both geographical features and the characteristics of the rainfall events (Segoni et al. 2014a; Melillo et al. 2016, 2018). In some other cases, straightforward methods are used, such as the selection of the nearest rain gauge (7 cases, e.g. Tiranti and Rabuffetti 2010;

Nikolopoulos et al. 2014; Gioia et al. 2015; Caracciolo et al. 2017), the use of Thiessen polygons (3 cases, Berti et al. 2012; Rosi et al. 2012; Tiranti et al. 2018), or the selection of a single rain gauge as representative for all the landslides occurred in the same area.

A similar outcome can be found when considering the approaches used to extract from the rainfall records the rainfall parameters characterizing the triggering conditions. On one end, eight thresholds are based on the visual inspection of rainfall charts and manual identification of the triggering parameters according to pre-defined criteria (e.g. Rosi et al. 2012; Vennari et al. 2014). On the other end, 23 more recent works rely on complex algorithms developed to operate a completely objective and standardized procedure (e.g. Segoni et al. 2014a; Vessia et al. 2014, 2016; Melillo et al. 2015, 2016, 2018; Giannecchini et al. 2016). As a compromise between simplicity and objectiveness, 22 thresholds are defined making use of a standardized definition of rainfall parameters such as using a standard duration to define the accumulated rainfall (e.g. Ciabatta et al. 2016) or setting a standard time span without rain to define the start and end of each rainfall event (Tiranti and Rabuffetti 2010; Roccati et al. 2018). Lastly, seven thresholds avoid this step of analysis because they are defined with physically based calculations (e.g. Papa et al. 2013; Napolitano et al. 2016). The most used combination of parameters for rainfall threshold definition in Italy is composed of the intensity (in mm/h) and the duration (in h) of the triggering rainfall events, *I*-*D* (Fig. 15.3e). This approach dates back to the work of Caine (1980), but relies on different methods and procedures. As an instance, Brunetti et al. (2010) uses the mean intensity of the rainfall events associated with the landslides, while other works (Segoni et al. 2014a, b) try to identify the *I*-*D* combination associated to the higher return period of the events. Another widely used approach relies in the combination of cumulated rainfall (in mm) and duration (in h) of the triggering rainfall events, *E*-*D* (e.g. Peruccacci et al. 2012, 2017; Palladino et al. 2018). The two approaches could be considered equivalent when I stays for the mean intensity, since it is $I = E/D$. In these cases, it would be preferable to use the *E*-*D* approach in order to work with two independent variables. Contrarily to the global threshold review (Segoni et al. 2018a), only few works dealt with antecedent rainfall conditions in Italy (e.g. Martelloni et al. 2012). We also found four cases in which temperature data are used in conjunction with rainfall, mainly to account for snow melting and accumulation effects (Martelloni et al. 2013) or for a better assessment of soil moisture condition (Ponziani et al. 2012; Ciabatta et al. 2016; Segoni et al. 2018c). Finally, few works presented thresholds based on a mobility function representing the hydrological response of the slope (s) to the rainfall (e.g. Greco et al. 2013; Terranova et al. 2015, 2018; De Luca and Versace 2017a, b).

Concerning the method used to define the threshold equation, Fig. 15.3f shows that a manual definition of the curve that low-bounds the triggering conditions was

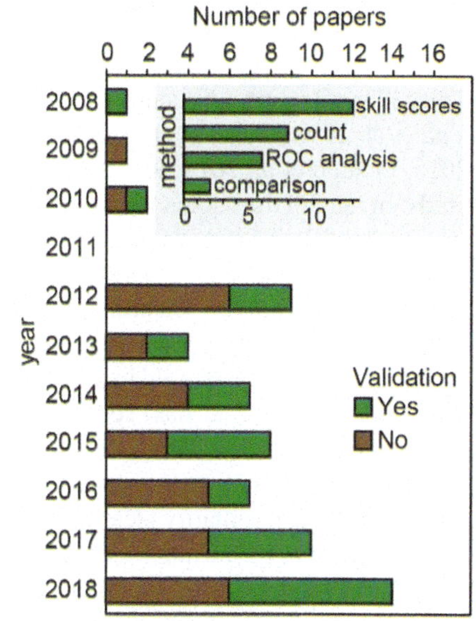

Fig. 15.4 Bar chart plotting the number of papers per year for which a validation procedure was (in green) and was not (in brown) conducted. The inset shows a bar chart reporting the main methods used for validation

still used (11 occurrences, e.g. Tiranti and Rabuffetti 2010; Giannecchini et al. 2012), even if it is largely overcome by statistical techniques (39 occurrences), among which is the frequentist approach proposed by Brunetti et al. (2010) and modified by Peruccacci et al. (2012). Three thresholds are defined using Bayesian probabilistic approaches (well described in Berti et al. 2012) and eight are defined according to physically based approaches.

A proper validation of the predictive capability of the thresholds is mandatory, in particular if they are supposed to be implemented in an operational LEWS. Figure 15.4 reports the number of thresholds per year that underwent a validation procedure. Regrettably, only 29 works (46% of the total) present a proper validation of the thresholds. However, an increasing trend in the application of validation procedures seems to be present in the last 4 years. Despite a common validation method not being established yet, most of the thresholds were validated using contingency matrices and related skill scores (e.g. Segoni et al. 2014a, b; Battistini et al. 2017; De Luca and Versace 2017a, b; Piciullo et al. 2017). In some cases, the skill scores are used to compare different threshold models to identify the threshold approach with the highest forecasting effectiveness in a given area (Lagomarsino et al. 2015). In other cases, skill scores are used to test different model configurations and to quantitatively assess the effectiveness of a methodological improvement before implementation in a LEWS (Segoni et al. 2018b). In many cases, thresholds undergo a very thorough validation procedure combining different approaches, including also ROC (receiving operator characteristics) analysis besides the aforementioned methods (e.g. Gariano et al. 2015; Galanti et al. 2018).

15.4 Discussion and Conclusion

The Italian scientific production on landslide analysis and prediction is wide (e.g. De Vita et al. 1998; Guzzetti et al. 2007, 2008; Wu et al. 2015; Gariano and Guzzetti 2016; Piciullo et al. 2018; Reichenbach et al. 2018; Segoni et al. 2018a, e), particularly on rainfall thresholds for landslide initiation, both in terms of quantity (39% of articles published on this topic worldwide in the 2008–2018 period) and quality (several methodological progresses were proposed in Italian case studies). The reason for such a large scientific production is probably related to the fact that in Italy rainfall thresholds represent a typical applied research topic aimed at addressing landslide hazard with twofold implications: first, Italy is undoubtedly highly exposed to landslide hazard and risk; second, to adequately manage the landslide risk, the implementation of a nationwide LEWS and a specific regional system in each of the 20 Italian administrative regions is dictated by the Italian body of laws on Civil Protection. This peculiarity clearly stands out from our literature review.

As a matter of fact, from the spatial distribution of the thresholds, it clearly stands out how the topic is important and widespread in the Italian research community. At least one case study is present in each region (in some cases some regions were considered together), achieving relevant outcomes. The areal extension of the case studies is very variable, highlighting that applications from small to large scales are considered of interest. The importance of rainfall thresholds is related to the high exposure of the Italian territory to high landside risk and, subsequently, to the urge of managing it with non-structural mitigation measures such as operative systems for landslide forecasting and early warning. To further substantiate this statement, it could be pointed out that 11 works describe operational or prototype LEWSs based on diverse thresholds that define four warning levels. This peculiarity is a clear attempt to fully meet the requirements of the Italian civil protection laws dictating that civil protection bodies are committed to issue warnings according to a standardized four levels scheme of increasing criticality (Dipartimento della Protezione Civile 2016) or probability of landslide occurrence.

In regional scale studies with abundance of input data, the strategy of partitioning the study area into several alert zones to be analyzed and monitored independently provides better results than using a single threshold for the whole region (e.g. Segoni et al. 2014a, 2015a). An alert zone can be defined according to homogeneous geomorphological and meteorological settings and it is more likely to have a dominant type of landslides. Therefore, as long as the number of data points is sufficient, a strong statistical correlation can be found. Moreover, the effectiveness of the warning system is higher because the spatial resolution is finer, as the spatial unit that can receive a warning is much smaller than the whole area covered by the warning system.

The need for increasing the spatial resolution of the threshold systems has recently led to exploring the possibility of combining rainfall thresholds and susceptibility maps. While the former can be used to accomplish a temporal forecasting, the latter can be used as a static base to differentiate the warning according to spatial units even smaller than the alert zones (Segoni et al. 2015b, 2018d; Tiranti et al. 2018).

Another issue arising from the literature review is the need of having landslide catalogues as complete, reliable, and updated as possible. To this end, a growing number of researchers are relying on internet news. While some of them use this source of information to integrate more traditional sources (Peruccacci et al. 2012, 2017), others focus entirely on internet news to build landslide datasets up to the national scale (Calvello and Pecoraro 2018) and some set up a semantic search engine capable of automatically building and updating in real time a landslide inventory for the whole Italian territory (Battistini et al. 2013, 2017).

Other relevant contributions of Italian works to the international literature on rainfall thresholds for landslide triggering aim at addressing some debated issues, including the following:

- The introduction of objective statistical techniques of analysis (Brunetti et al. 2010; Peruccacci et al. 2012; Rossi et al. 2017)
- The investigation of the role of environmental factors (e.g. lithology, soil type, land cover, climatic conditions, etc.) on the thresholds (Peruccacci et al. 2012, 2017; Vennari et al. 2014; Caracciolo et al. 2017; Palladino et al. 2018)
- The determination of automatic procedures to define the triggering rainfall conditions and to calculate the thresholds (Segoni et al. 2014a; Iadanza et al. 2016; Vessia et al. 2014, 2016; Rossi et al. 2017; Melillo et al. 2015, 2016, 2018)
- The analysis of several uncertainties (Gariano et al. 2015, Nikolopoulos et al. 2015; Marra et al. 2017; Rossi et al. 2017; Marra 2018; Peres et al. 2018)
- The application of quantitative validation procedures (Battistini et al. 2017; Gariano et al. 2015; Lagomarsino et al. 2015; Piciullo et al. 2017; Galanti et al. 2018)
- The need for considering diverse hydrological conditions of the hillslope system with more complex approaches (Ciabatta et al. 2016; Bogaard and Greco 2018; Segoni et al. 2018c)
- The adoption of rainfall data gathered from radars or satellites (Marra et al. 2016, 2017; Rossi et al. 2017; Brunetti et al. 2018)
- The evaluation of the effect of climatic and environmental changes on the thresholds (Ciabatta et al. 2016; Alvioli et al. 2018; Sangelantoni et al. 2018)
- The application of the thresholds into hazard management procedures and early warning systems (Rosi et al. 2015; Segoni et al. 2015a; Brigandì et al. 2017; Piciullo et al. 2017; Cremonini and Tiranti 2018; Devoli et al. 2018; Segoni et al. 2018d, e)

With regard to the latter, the implementation of rainfall thresholds into operational LEWS is not an easy and immediate task. A relevant number of works arise from a strict and long-lasting collaboration between the research community and the end users (Regional offices, Functional Centers or Civil Protection, National Department of Civil Protection), in which long-term applied research programs allowed the establishment of the thresholds, the setting up of warning systems, and their progressive and continuous amelioration over time (Peruccacci et al. 2017; Devoli et al. 2018; Segoni et al. 2018b; Tiranti et al. 2018).

Finally, a point that should be remarked is that the effectiveness and the reliability of a threshold depend both on the method used for its definition and on the quantity and quality of the input data. Therefore, we recommend that every paper describing the definition of a threshold should present clearly: the extent and features of the test site; the number, type and source of the analyzed landslides, the rainfall measurements' source, density, and temporal resolution; the period of analysis; the method adopted in the entire workflow. These details are also important for the reproducibility of the methods to other case studies. Regrettably, in our review, we found some works lacking relevant – and in some cases, simple – information about the study area extension, the data used, the methods of analysis, and even the threshold type.

Even with these recently established advances, both in the Italian and in the international scientific community, the definition of rainfall thresholds for landslide initiation is still a highly debated topic, with some unresolved issues. The efforts of scientists and technicians should aim to solve these issues, to enhance the threshold effectiveness, and to strengthen the societal perception of their reliability in operative systems for landslides prediction and early warning.

References

Alvioli M, Guzzetti F, Rossi M (2014) Scaling properties of rainfall induced landslides predicted by a physically based model. Geomorphology 213:38–47. https://doi.org/10.1016/j.geomorph.2013.12.039

Alvioli M, Melillo M, Guzzetti F, Rossi M, Palazzi E, von Hardenberg J, Brunetti MT, Peruccacci S (2018) Implications of climate change on landslide hazard in Central Italy. Sci Tot Environ 630:1528–1543. https://doi.org/10.1016/j.scitotenv.2018.02.315

Battistini A, Segoni S, Manzo G, Catani F, Casagli N (2013) Web data mining for automatic inventory of geohazards at national scale. Appl Geogr 43:147–158. https://doi.org/10.1016/j.apgeog.2013.06.012

Battistini A, Rosi A, Segoni S, Lagomarsino D, Catani F, Casagli N (2017) Validation of landslide hazard models using a semantic engine on online news. Appl Geogr 82:59–65. https://doi.org/10.1016/j.apgeog.2017.03.003

Berti M, Martina MLV, Franceschini S, Pignone S, Simoni A, Pizziolo M (2012) Probabilistic rainfall thresholds for landslide occurrence using a Bayesian approach. J Geophys Res 117:F04006. https://doi.org/10.1029/2012JF002367

Bogaard T, Greco R (2018) Invited perspectives: hydrological perspectives on precipitation intensity-duration thresholds for landslide initiation: proposing hydro-meteorological thresholds. Nat Hazards Earth Syst Sci 18:31–39. https://doi.org/10.5194/nhess-18-31-2018

Bovolo CI, Bathurst JC (2012) Modelling catchment-scale shallow landslide occurrence and sediment yield as a function of rainfall return period. Hydrol Proc 26(4):579–596. https://doi.org/10.1002/hyp.8158

Brigandì G, Aronica GT, Bonaccorso B, Gueli R, Basile G (2017) Flood and landslide warning based on rainfall thresholds and soil moisture indexes: the HEWS (Hydrohazards EarlyWarning System) for Sicily. Adv Geosci 44:79–88. https://doi.org/10.5194/adgeo-44-79-2017

Brunetti MT, Peruccacci S, Rossi M, Luciani S, Valigi D, Guzzetti F (2010) Rainfall thresholds for the possible occurrence of landslides in Italy. Nat Hazards Earth Syst Sci 10:447–458. https://doi.org/10.5194/nhess-10-447-2010

Brunetti MT, Melillo M, Peruccacci S, Ciabatta L, Brocca L (2018) How far are we from the use of satellite rainfall products in landslide forecasting? Remote Sens Environ 210:65–75. https://doi.org/10.1016/j.rse.2018.03.016

Caine N (1980) The rainfall intensity-duration control of shallow landslides and debris flows. Geografiska Annal 62A:23–27

Calvello M, Pecoraro G (2018) FraneItalia: a catalog of recent Italian landslides. Geoenviron Disasters 5:13. https://doi.org/10.1186/s40677-018-0105-5

Campbell RH (1975) Soil slips, debris flows, and rainstorms in the Santa Monica Mountains and vicinity, Southern California. US Geological Survey Professional Paper, 851, 51 pp

Caracciolo D, Arnone E, Lo Conti F, Noto LV (2017) Exploiting historical rainfall and landslide data in a spatial database for the derivation of critical rainfall thresholds. Environ Earth Sci 76:222. https://doi.org/10.1007/s12665-017-6545-5

Ciabatta L, Camici S, Brocca L, Ponziani F, Stelluti M, Berni N, Moramarco T (2016) Assessing the impact of climate-change scenarios on landslide occurrence in Umbria region, Italy. J Hydrol 541:285–295. https://doi.org/10.1016/j.jhydrol.2016.02.007

Ciervo F, Rianna G, Mercogliano P, Papa MN (2017) Effects of climate change on shallow landslides in a small coastal catchment in southern Italy. Landslides 14(3):1043–1055. https://doi.org/10.1007/s10346-016-0743-1

Cremonini R, Tiranti D (2018) The weather radar observations applied to shallow landslides prediction: a case study from North-Western Italy. Front Earth Sci 6:134. https://doi.org/10.3389/feart.2018.00134

De Luca DL, Versace P (2017a) A comprehensive framework for empirical modeling of landslides induced by rainfall: the Generalized FLaIR model (GFM). Landslides 14(3):1009–1030. https://doi.org/10.1007/s10346-016-0768-5

De Luca DL, Versace P (2017b) Diversity of rainfall thresholds for early warning of hydro-geological disasters. Adv Geosci 44:53–60. https://doi.org/10.5194/adgeo-44-53-2017

De Vita P, Reichenbach P, Bathurst JC, Borga M, Crozier GM, Glade T, Guzzetti F, Hansen A, Wasowski J (1998) Rainfall-triggered landslides: a reference list. Environ Geol 35:219–233

Devoli G, Tiranti D, Cremonini R, Sund M, Boje S (2018) Comparison of landslide forecasting services in Piedmont (Italy) and Norway, illustrated by events in late spring 2013. Nat Hazards Earth Syst Sci 18:1351–1372. https://doi.org/10.5194/nhess-18-1351-2018

Dipartimento della Protezione Civile (2016) Indicazioni per l'omogeneizzazione dei messaggi del Sistema di allertamento nazionale: livelli di criticità e di allerta e relativi scenari d'evento. http://www.protezionecivile.gov.it/documents/20182/823803/Allegato+1+livelli+di+criticit%C3%A0+e+di+allerta+e+relativi+scenari+d%27evento/108c1f84-c130-4fc1-bac9-d16596e83046. Accessed 10 July 2019 (in Italian)

Endo T (1969) Probable distribution of the amount of rainfall causing landslides, Annual Report 1968, Hokkaido Branch, For. Exp. Stn., Sapporo, pp 122–136

Floris M, Bozzano F (2008) Evaluation of landslide reactivation: a modified rainfall threshold model based on historical records of rainfall and landslides. Geomorphology 94(1):40–57. https://doi.org/10.1016/j.geomorph.2007.04.009

Floris M, D'Alpaos A, De Agostini A, Stevan G, Tessari G, Genevois R (2012) A process-based model for the definition of hydrological alert systems in landslide risk mitigation. Nat Hazards Earth Syst Sci 12:3343–3357. https://doi.org/10.5194/nhess-12-3343-2012

Frattini P, Crosta G, Sosio R (2009) Approaches for defining thresholds and return periods for rainfall-triggered shallow landslides. Hydrol Proc 23(10):1444–1460. https://doi.org/10.1002/hyp.7269

Froude MJ, Petley D (2018) Global fatal landslide occurrence from 2004 to 2016. Nat Hazards Earth Syst Sci 18:2161–2181. https://doi.org/10.5194/nhess-18-2161-2018

Galanti Y, Barsanti M, Cevasco A, D'Amato Avanzi G, Giannecchini R (2018) Comparison of statistical methods and multi-time validation for the determination of the shallow landslide rainfall thresholds. Landslides 15(5):937–952. https://doi.org/10.1007/s10346-017-0919-3

Gariano SL, Guzzetti F (2016) Landslides in a changing climate. Earth Sci Rev 62:227–252. https://doi.org/10.1016/j.earscirev.2016.08.011

Gariano SL, Brunetti MT, Iovine G, Melillo M, Peruccacci S, Terranova OG, Vennari C, Guzzetti F (2015) Calibration and validation of rainfall thresholds for shallow landslide forecasting in Sicily, southern Italy. Geomorphology 228:653–665. https://doi.org/10.1016/j.geomorph.2014.10.019

Giannecchini R, Galanti Y, D'Amato Avanzi G (2012) Critical rainfall thresholds for triggering shallow landslides in the Serchio River valley (Tuscany, Italy). Nat Hazards Earth Syst Sci 12:829–842. https://doi.org/10.5194/nhess-12-829-2012

Giannecchini R, Galanti Y, D'Amato Avanzi G, Barsanti M (2016) Probabilistic rainfall thresholds for triggering debris flows in a human-modified landscape. Geomorphology 257:94–107. http://dx.doi.org/10.1016/j.geomorph.2015.12.012

Gioia E, Carone T, Marincioni F (2015) Rainfall and land use empirically coupled to forecast landslides in the Esino river basin, Central Italy. Nat Hazards Earth Syst Sci 15(6):1289–1295. https://doi.org/10.5194/nhess-15-1289-2015

Greco R, Giorgio M, Capparelli G, Versace P (2013) Early warning of rainfall-induced landslides based on empirical mobility function predictor. Eng Geol 153:68–79. https://doi.org/10.1016/j.enggeo.2012.11.009

Guzzetti F, Peruccacci S, Rossi M, Stark CP (2007) Rainfall thresholds for the initiation of landslides in central and southern Europe. Meteorog Atmos Phys 98:239–267. https://doi.org/10.1007/s00703-007-0262-7

Guzzetti F, Peruccacci S, Rossi M, Stark CP (2008) The rainfall intensity–duration control of shallow landslides and debris flows: an update. Landslides 5(1):3–17. https://doi.org/10.1007/s10346-007-0112-1

Haque U, Blum P, Da Silva PF et al (2016) Fatal landslides in Europe. Landslides 13(6):1545–1554. https://doi.org/10.1007/s10346-016-0689-3

Haque U, da Silva PF, Devoli G, Pilz J, Zhao B, Khaloua A, Wilopo W, Andersen P, Lu P, Lee J, Yamamoto T, Keellings D, Wu J-H, Glass GE (2019) The human cost of global warming: deadly landslides and their triggers (1995–2014). Sci Tot Environ 682:673–684. https://doi.org/10.1016/j.scitotenv.2019.03.415

Iadanza C, Trigila A, Napolitano F (2016) Identification and characterization of rainfall events responsible for triggering of debris flows and shallow landslides. J Hydrol 541:230–245. https://doi.org/10.1016/j.jhydrol.2016.01.018

Lagomarsino D, Segoni S, Fanti R, Catani F (2013) Updating and tuning a regional-scale landslide early warning system. Landslides 10:91–97. https://doi.org/10.1007/s10346-012-0376-y

Lagomarsino D, Segoni S, Rosi A, Rossi G, Battistini A, Catani F, Casagli N (2015) Quantitative comparison between two different methodologies to define rainfall thresholds for landslide forecasting. Nat Hazards Earth Syst Sci 15:2413–2423. https://doi.org/10.5194/nhess-15-2413-2015

Marra F (2018) Rainfall thresholds for landslide occurrence: systematic underestimation using coarse temporal resolution data. Nat Hazards 95:1–8. https://doi.org/10.1007/s11069-018-3508-4

Marra F, Nikolopoulos EI, Creutin JD, Borga M (2016) Space–time organization of debris flows-triggering rainfall and its effect on the identification of the rainfall threshold relationship. J Hydrol 541:246–255. https://doi.org/10.1016/j.jhydrol.2015.10.010

Marra F, Destro E, Nikolopoulos EI, Zoccatelli D, Creutin JD, Guzzetti F, Borga M (2017) Impact of rainfall spatial aggregation on the identification of debris flow occurrence thresholds. Hydrol Earth Syst Sci 21:4525–4532. https://doi.org/10.5194/hess-21-4525-2017

Martelloni G, Segoni S, Fanti R, Catani F (2012) Rainfall thresholds for the forecasting of landslide occurrence at regional scale. Landslides 9(485–495):485. https://doi.org/10.1007/s10346-011-0308-2

Martelloni G, Segoni S, Lagomarsino D, Fanti R, Catani F (2013) Snow accumulation/melting model (SAMM) for integrated use in regional scale landslide early warning systems. Hydrol Earth Syst Sci 17:1229–1240. https://doi.org/10.5194/hess-17-1229-2013

Melillo M, Brunetti MT, Peruccacci S, Gariano SL, Guzzetti F (2015) An algorithm for the objective reconstruction of rainfall events responsible for landslides. Landslides 12 (2):311–320. https://doi.org/10.1007/s10346-014-0471-3

Melillo M, Brunetti MT, Peruccacci S, Gariano SL, Guzzetti F (2016) Rainfall thresholds for the possible landslide occurrence in Sicily (Southern Italy) based on the automatic reconstruction of rainfall events. Landslides 13(1):165–172. https://doi.org/10.1007/s10346-015-0630-1

Melillo M, Brunetti MT, Peruccacci S, Gariano SL, Roccati A, Guzzetti F (2018) A tool for the automatic calculation of rainfall thresholds for landslide occurrence. Environ Model Softw 105:230–243. https://doi.org/10.1016/j.envsoft.2018.03.024

Napolitano E, Fusco F, Baum RL, Godt JW, De Vita P (2016) Effect of antecedent hydrological conditions on rainfall triggering of debris flows in ash-fall pyroclastic mantled slopes of Campania (southern Italy). Landslides 13:967–983. https://doi.org/10.1007/s10346-015-0647-5

Nikolopoulos EI, Crema S, Marchi L, Marra F, Guzzetti F, Borga M (2014) Impact of uncertainty in rainfall estimation on the identification of rainfall thresholds for debris flow occurrence. Geomorphology 221:286–297. https://doi.org/10.1016/j.geomorph.2014.06.015

Nikolopoulos EI, Borga M, Creutin JD, Marra F (2015) Estimation of debris flow triggering rainfall: influence of rain gauge density and interpolation methods. Geomorphology 243:40–50. https://doi.org/10.1016/j.geomorph.2015.04.028

Nikolopoulos EI, Destro E, Maggioni V, Marra F, Borga M (2017) Satellite rainfall estimates for debris flow prediction: an evaluation based on rainfall accumulation–duration thresholds. J Hydrometeorol 18:2207–2214. https://doi.org/10.1175/JHM-D-17-0052.1

Onodera T, Yoshinaka R, Kazama H (1974) Slope failures caused by heavy rainfall in Japan. In: Proceedings 2nd international congress of the International Association Engineering Geology, San Paulo, 11, pp 1–10

Palladino MR, Viero A, Turconi L, Brunetti MT, Peruccacci S, Melillo M, Deganutti AM, Guzzetti F (2018) Rainfall thresholds for the activation of shallow landslides in the Italian Alps: the role of environmental conditioning factors. Geomorphology 303:53–67. https://doi.org/10.1016/j.geomorph.2017.11.009

Papa MN, Medina V, Ciervo F, Bateman A (2013) Derivation of critical rainfall thresholds for shallow landslides as a tool for debris flow early warning systems. Hydrol Earth Syst Sci 17 (10):4095–4107. https://doi.org/10.5194/hess-17-4095-2013

Peres DJ, Cancelliere A (2014) Derivation and evaluation of landslide-triggering thresholds by a Monte Carlo approach. Hydrol Earth Syst Sci 18:4913–4931. https://doi.org/10.5194/hess-18-4913-2014

Peres DJ, Cancelliere A, Greco R, Bogaard T (2018) Influence of uncertain identification of triggering rainfall on the assessment of landslide early warning thresholds. Nat Hazards Earth Syst Sci 18:633–646. https://doi.org/10.5194/nhess-18-633-2018

Peruccacci S, Brunetti MT, Luciani S, Vennari C, Guzzetti F (2012) Lithological and seasonal control of rainfall thresholds for the possible initiation of landslides in Central Italy. Geomorphology 139-140:79–90. https://doi.org/10.1016/j.geomorph.2011.10.005

Peruccacci S, Brunetti MT, Gariano SL, Melillo M, Rossi M, Guzzetti F (2017) Rainfall thresholds for possible landslide occurrence in Italy. Geomorphology 290:39–57. https://doi.org/10.1016/j.geomorph.2017.03.031

Piciullo L, Gariano SL, Melillo M, Brunetti MT, Peruccacci S, Guzzetti F, Calvello M (2017) Definition and performance of a threshold-based regional early warning model for rainfall-induced landslides. Landslides 14:995–1008. https://doi.org/10.1007/s10346-016-0750-2

Piciullo L, Calvello M, Cepeda JM (2018) Territorial early warning systems for rainfall-induced landslides. Earth Sci Rev 179:228–247. https://doi.org/10.1016/j.earscirev.2018.02.013

Ponziani F, Pandolfo C, Stelluti M, Berni N, Brocca L, Moramarco T (2012) Assessment of rainfall thresholds and soil moisture modeling for operational hydrogeological risk prevention in the

Umbria region (Central Italy). Landslides 9(2):229–237. https://doi.org/10.1007/s10346-011-0287-3

Reichenbach P, Rossi M, Malamud BD, Mihir M, Guzzetti F (2018) A review of statistically-based landslide susceptibility models. Earth Sci Rev 180:69–91. https://doi.org/10.1016/j.earscirev.2018.03.001

Roccati A, Faccini F, Luino F, Turconi L, Guzzetti F (2018) Rainfall events with shallow landslides in the Entella catchment, Liguria, northern Italy. Nat Hazards Earth Syst Sci 18:2367–2386. https://doi.org/10.5194/nhess-18-2367-2018

Rosi A, Segoni S, Catani F, Casagli N (2012) Statistical and environmental analyses for the definition of a regional rainfall threshold system for landslide triggering in Tuscany (Italy). J of Geogr Sci 22(4):617–629. https://doi.org/10.1007/s11442-012-0951-0

Rosi A, Lagomarsino D, Rossi G, Segoni S, Battistini A, Casagli N (2015) Updating EWS rainfall thresholds for the triggering of landslides. Nat Hazards 78:297–308. https://doi.org/10.1007/s11069-015-1717-7

Rossi M, Luciani S, Valigi D, Kirschbaum D, Brunetti MT, Peruccacci S, Guzzetti F (2017) Statistical approaches for the definition of landslide rainfall thresholds and their uncertainty using rain gauge and satellite data. Geomorphology 285:16–27. https://doi.org/10.1016/j.geomorph.2017.02.001

Salciarini D, Tamagnini C, Conversini P, Rapinesi S (2012) Spatially distributed rainfall thresholds for the initiation of shallow landslides. Nat Hazards 61:229–245. https://doi.org/10.1007/s11069-011-9739-2

Salvati P, Bianchi C, Fiorucci F, Giostrella P, Marchesini I, Guzzetti F (2014) Perception of flood and landslide risk in Italy: a preliminary analysis. Nat Hazards Earth Syst Sci 14:2589–2603. https://doi.org/10.5194/nhess-14-2589-2014

Sangelantoni L, Gioia E, Marincioni F (2018) Impact of climate change on landslides frequency: the Esino river basin case study (Central Italy). Nat Hazards 93:946–884. https://doi.org/10.1007/s11069-018-3328-6

Segoni S, Rossi G, Rosi A, Catani F (2014a) Landslides triggered by rainfall: a semiautomated procedure to define consistent intensity-duration thresholds. Comput Geosci 63:123–131. https://doi.org/10.1016/j.cageo.2013.10.009

Segoni S, Rosi A, Rossi G, Catani F, Casagni N (2014b) Analysing the relationship between rainfalls and landslides to define a mosaic of triggering thresholds for regional scale warning systems. Nat Hazards Earth Syst Sci 14:2637–2648. https://doi.org/10.5194/nhess-14-2637-2014

Segoni S, Battistini A, Rossi G, Rosi A, Lagomarsino D, Catani F, Moretti S, Casagli N (2015a) Technical note: an operational landslide early warning system at regional scale based on space–time-variable rainfall thresholds. Nat Hazards Earth Syst Sci 15:853–861. https://doi.org/10.5194/nhess-15-853-2015

Segoni S, Lagomarsino D, Fanti R, Moretti S, Casagli N (2015b) Integration of rainfall thresholds and susceptibility maps in the Emilia Romagna (Italy) regional-scale landslide warning system. Landslides 12:773–785. https://doi.org/10.1007/s10346-014-0502-0

Segoni S, Piciullo L, Gariano SL (2018a) A review of the recent literature on rainfall thresholds for landslide occurrence. Landslides 15:1483–1501. https://doi.org/10.1007/s10346-018-0966-4

Segoni S, Rosi A, Fanti R, Gallucci A, Monni A, Casagli N (2018b) A regional-scale landslide warning system based on 20 years of operational experience. Water 10(10):1297. https://doi.org/10.3390/w10101297

Segoni S, Rosi A, Lagomarsino D, Fanti R, Casagli N (2018c) Brief communication: using averaged soil moisture estimates to improve the performances of a regional-scale landslide early warning system. Nat Hazards Earth Syst Sci 18(3):807–812. https://doi.org/10.5194/nhess-18-807-2018

Segoni S, Tofani V, Rosi A, Catani F, Casagli N (2018d) Combination of rainfall thresholds and susceptibility maps for dynamic landslide hazard assessment at regional scale. Front Earth Sci 6:85. https://doi.org/10.3389/feart.2018.00085

Segoni S, Piciullo L, Gariano SL (2018e) Preface: landslide early warning systems: monitoring systems, rainfall thresholds, warning models, performance evaluation and risk perception. Nat Hazards Earth Syst Sci 18:3179–3186. https://doi.org/10.5194/nhess-18-3179-2018

Terranova OG, Gariano SL, Iaquinta P, Iovine G (2015) ^{GA}SAKe: forecasting landslide activations by a genetic-algorithms-based hydrological model. Geosci Model Dev 8:1955–1978. https://doi.org/10.5194/gmd-8-1955-2015

Terranova O, Gariano SL, Iaquinta P, Lupiano V, Rago V, Iovine G (2018) Examples of application of GASAKe for predicting the occurrence of rainfall-induced landslides in Southern Italy. Geosciences 8:78. https://doi.org/10.3390/geosciences8020078

Tiranti D, Rabuffetti D (2010) Estimation of rainfall thresholds triggering shallow implementation. Landslides 7:471–481. https://doi.org/10.1007/s10346-010-0198-8

Tiranti D, Nicolò G, Gaeta AR (2018) Shallow landslides predisposing and triggering factors in developing a regional early warning system. Landslides 16:235–251. https://doi.org/10.1007/s10346-018-1096-8

Vennari C, Gariano SL, Antronico L, Brunetti MT, Iovine G, Peruccacci S, Terranova O, Guzzetti F (2014) Rainfall thresholds for shallow landslide occurrence in Calabria, southern Italy. Nat Hazards Earth Syst Sci 14:317–330. https://doi.org/10.5194/nhess-14-317-2014

Vessia G, Parise M, Brunetti MT, Peruccacci S, Rossi M, Vennari C, Guzzetti F (2014) Automated reconstruction of rainfall events responsible for shallow landslides. Nat Hazards Earth Syst Sci 14(9):2399–2408. https://doi.org/10.5194/nhess-14-2399-2014

Vessia G, Pisano L, Vennari C, Rossi M, Parise M (2016) Mimic expert judgement through automated procedure for selecting rainfall events responsible for shallow landslide: a statistical approach to validation. Comput Geosci 86:146–153. https://doi.org/10.1016/j.cageo.2015.10.015

Wu X, Chen X, Zhan FB, Hong S (2015) Global research trends in landslides during 1991–2014: a bibliometric analysis. Landslides 12:1215–1226. https://doi.org/10.1007/s10346-015-0624-z

Chapter 16
Validation of a Shallow Landslide Susceptibility Analysis Through a Real Case Study: An Example of Application in Rome (Italy)

Geraud Poueme Djueyep, Carlo Esposito, Luca Schilirò, and Francesca Bozzano

Abstract In the last years, statistically based models (such as Logistic Regression) have been frequently used for evaluating the probability of landslide occurrence over large areas. In the case of Rome, over the years, more than 348 landslides have been recorded throughout the city. For this reason, in this study, we implemented and evaluated three main validation criteria of logistic regression to assess the rainfall-induced landslide susceptibility in a specific area of the city of Rome. Through the evaluation of the predictive performances, the best model has been identified and the results were also compared with those obtained in similar case studies.

Keywords Shallow landslide · Susceptibility · Statistical analyses · Rome

16.1 Introduction

Many rainfall-induced landslides all over the world are shallow-type, namely, the sliding surface is located at a depth from a few decimetres to some metres. They generally involve the so-called engineering soil (Varnes 1978; Cruden and Varnes 1996), which can be either weathered soil or colluvium. Also, such events generally lead to dramatic soil mass wasting in sloping areas, exposing the disturbed sites to further erosion. These phenomena can easily evolve into debris flows as they travel down steep slopes, especially along stream channels where they may mix with additional water and sediments.

However, considering the complexity of such phenomena, it has been always fastidious to precisely predict (both in spatial and in temporal terms) the occurrence of natural shallow landslides. For these reasons, over the years a great variety of

G. P. Djueyep (✉) · C. Esposito · L. Schilirò · F. Bozzano
Department of Earth Sciences, Sapienza University of Rome, Rome, Italy
e-mail: geraud.pouemedjueyep@uniroma1.it

M. De Maio, A. K. Tiwari (eds.), *Applied Geology*,
https://doi.org/10.1007/978-3-030-43953-8_16

approaches and methods have been proposed for landslide initiation susceptibility assessment. Specifically, a susceptibility analysis tends to define the spatial correlation between landslides and topography, geology, soil properties and land use characteristics (Guzzetti et al. 1999). The most largely used approaches to perform landslide susceptibility analysis are the statistical ones. Here, the influence or weight of predisposing factors is evaluated through conceptual models developed lying on the axiom that 'the past is the key for the future'. Bivariate and multivariate are the commonly used methods. Multivariate models (such as logistic regression (LR) see Ohlmacher and Davis 2003; Ayalew and Yamagishi 2005) evaluate the combined relationship between independent variables (landslide factors) and the dependent variable (either the presence or the absence of past landslides). As regards the LR technique, it was applied to different case studies all over the world, both in mountainous areas (such as Himalaya see Mathew et al. 2007; Chauhan et al. 2010; Devkota et al. 2013) and proximal ones (e.g. Mărgărint et al. 2013; Trigila et al. 2015), showing satisfactory results. The application of such a method in different morphoclimatic contexts and at different scales (Van Den Eeckhaut et al. 2012; Lin et al. 2017) implies a different risk exposure in relation to the eventual presence of urban settlements within the study area. Based on the above-mentioned issues, a statistical assessment of landslide susceptibility in the City of Rome has been implemented, with specific focus on the validation criteria of the performed approach. In fact, the city of Rome is characterized by the presence of hilly areas (culminating at around 140 m a.s.l.), but frequently affected by rainfall-induced landslides, which may cause relevant damage to buildings and infrastructure. The LR method used in this study implements a detailed landslide inventory referring to the landslide event that occurred in Rome between 31st January and 02nd February 2014. In this respect, the mitigation of rainfall-induced shallow landslide risk is a priority for such a vulnerable city due to its high population density, as well as to its precious archaeological and monumental heritage. Therefore, in a broader frame, this study will contribute to a better knowledge of the role of natural and anthropogenic factors, in order to address a suitable land use management and to support sustainable geo-hazard risk mitigation policies in similar contexts.

16.2 The Study Area and the 2014 Landslide Event

16.2.1 General Features of the Study Area

The study area lies in the North-western sector of the city of Rome which, in turn, is located in the central-western portion of the Italian Peninsula (Fig. 16.1), along the shores of the Tevere River (Fig. 16.1). This area is about 99 km^2 wide and it is characterized by a particularly high urban density (between 5001 and 12,225 inhabitants/km^2) and the presence of cultural goods and heritages estimated around 2.001–6.250 (source: ISTAT 2011; VIR; ISCR). The unbuilt areas of Rome

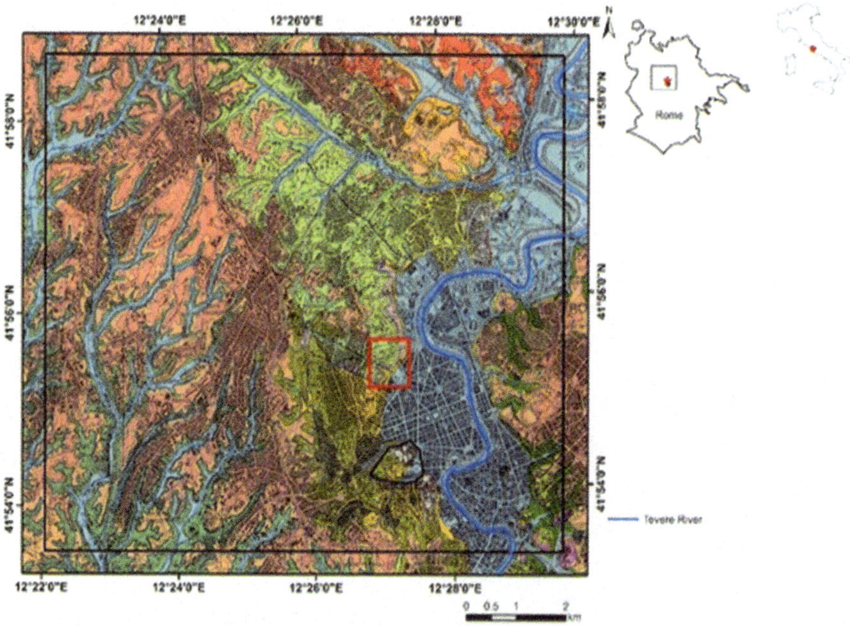

Fig. 16.1 Geological map of Rome comprising the study area and the Pilot site. (From Funiciello and Giordano 2008 mod.)

constitute 73% of the territory and are mainly occupied by farmland. In respect to this, the selection of the study area has been motivated considering the following criteria:

- the high landslide index (estimated to 2.5/km^2);
- the presence of 57 over the 71 (80%) inventoried triggering zones of shallow landslide occurred during the event of 2014 in Rome;
- the representativeness of lithological and geomorphological settings;

The study area envelops the main elevated areas of the city centre: Monte Mario and Monte Ciocci hills. In this respect, the pilot site chosen for this study is a part of Monte Mario hill, in which 18% of slope failures have been inventoried from the event of 2014.

16.2.2 The 2014 Landslide Event

The year 2014 was characterized, in many cases, by annual cumulative rainfall values among the highest in the historical series for many national meteorological stations in Italy. The main consequence of this fact concerned the number and extent of flood and landslide events, which have considerably exceeded what has occurred

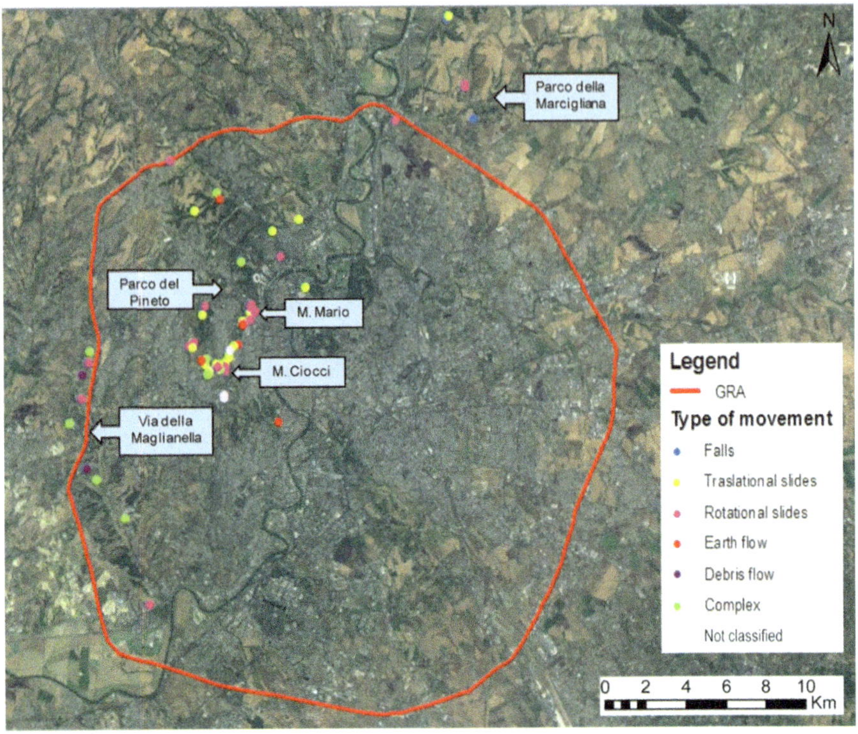

Fig. 16.2 Landslide inventory of the 2014 event in Rome. (Alessi et al. 2014)

in recent years. These exceptional rainfall values and associated hazards caused about 27 fatalities and economic losses estimated in many hundreds of millions of euro (Source: Technical Report ISPRA; February 2015). Particularly in Rome, the 2014 event caused substantial damages upon man-made built assets and cultural heritages. Specifically, a critical rainfall event led to the triggering of several landslides between January 31 and February 2 of 2014. During this event, some historical landslides have been reactivated such as those of Via Labriola and Viale Tiziano. According to the data from the rain gauge station located on Monte Mario hill, a 6-h cumulative rainfall of 146 mm (from 03:00 AM to 09:00 AM of January 31, 2014) was recorded, which was responsible for the occurrence of the majority of landslides. In this respect, 71 landslides have been inventoried all over the city (Fig. 16.2) by the CERI Research Centre in collaboration with Rome Municipality. Such landslides were mainly concentrated at the north-western sector (Alessi et al. 2014) of the city, where hilly areas are located. As a consequence of these phenomena, substantial damages to buildings and infrastructures have been recorded, along with vehicle traffic bans and inconveniences for citizens (Fig. 16.3). The involved materials were mainly deposits of sandy and sandy-silty nature deriving from Monte Mario, Ponte Galeria and Valle Giulia formations. Moreover, their prevalent clayey-

Fig. 16.3 Some damages encountered during the landslide event of 2014

silty or sandy nature (non-volcanic clastic deposits) generally makes them more susceptible to landslides induced by intense precipitation (Alessi et al. 2014; Del Monte et al. 2016). These deposits are particularly prone to instabilities also owing to their geomorphological setting (steep slopes) and to the strong anthropogenic pressure that they experienced in the second half of last century (Bozzano et al. 2006).

16.3 Material

16.3.1 Landslide Inventory Information

The inventory database is a pivotal tool when performing statistically based models of landslide susceptibility in a specific area. It is used first to train the models and second to test the resulting thematic maps and products. Landslide inventory can be either continuous in time or event-based related to a specific triggering phenomenon (Corominas et al. 2014).

In this study, the inventoried events have been collected, compiled and refined based on the availability of numerous datasets provided by the national, regional and local institutions and programs such as IFFI/ISPRA, Tevere Basin Authority/PAI, Rome Municipality Regulatory Plan, CERI – Sapienza University of Rome. Specific attention was given to the initiation zone (detachment area) regardless of the propagation pathway and the accumulation zone. The geometrical attributes of the resulting inventory were stored within a GIS platform and converted to a raster of

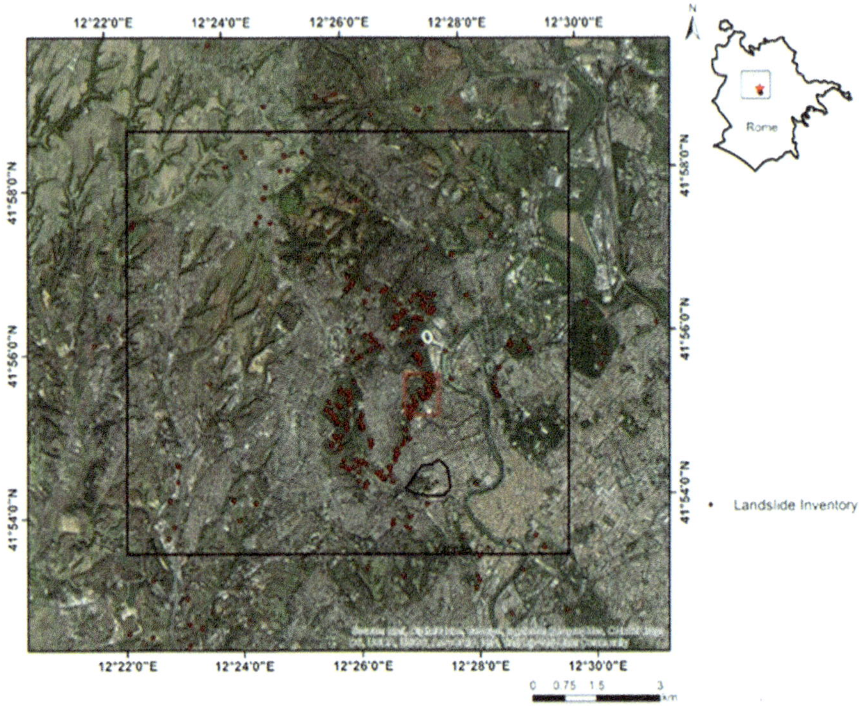

Fig. 16.4 Landslide inventory map of the study area

5 × 5 m resolution (corresponding to the raster resolution of the preparatory maps). After filtering and elaboration, 248 spots representing the most elevated points of landslide detachment areas have been considered as the inventory dataset within the 99 km^2 of the study area (Fig. 16.4).

16.3.2 Geo-Environmental Information Data

For performing a landslide statistical analysis, beside data regarding preceding landslides (dependent variable), information about the so-called geo-environmental predisposing factors (independent or explanatory variables) are also required (Carrara 1983; Guzzetti et al. 1999, 2006; Budimir et al. 2015). In this study, a set of 10 parameters has been selected, specifically:

1. *Morphometric or Terrain Parameters*

These thematic data derive from DEM processing. In this study, DEM has been elaborated through geostatistical models ('Topo to Raster' tool available in the GIS environment) from point elevation and surface contour lines. The shape file was provided by the 'Regione Lazio': 1/5.000 (http://www.regione.lazio.it/rl_sitr/?vw=contenutidettaglio&id=110).

From the DEM, morphometric parameters were extracted, such as: (i) – Elevation (ranging from 10 to 145 m a.s.l with an average value of 70 m a.s.l.); (ii) – Slope (ranging from 0 to 50° with average value of 4°); (iii) – Slope Aspect; (iv) – Plan, Profile and Total curvatures.

2. *Hydro-Morphometric Parameters*

Besides all those listed above, hydro-morphometric parameters were also extracted, such as: (i) – Topographic wetness index (TWI), which represents the spatial distribution and extent of zones of saturation. A high wetness index indicates regions characterized by low soil drainage and high-water availability whereas regions with low wetness index are susceptible to temporary soil drought. (ii) – Stream power index (SPI) that estimates the erosive power of the flowing water. It gives information on the ability of flowing water to detach and move particles downhill. The higher the value is, the more prone is the area to erosion.

$$\mathrm{TWI} = \mathrm{In}\left(A_g / \tan\beta\right) \tag{16.1}$$

$$\mathrm{SPI} = \ln\left(A_g * \tan\beta\right) \tag{16.2}$$

where A s is the specific catchment area expressed as m^2 per unit width orthogonal to the flow direction, and β is the slope angle expressed in radians.

3. *Geological Parameters*

The lithological map derives from the digitalization of the geology map of Rome territory (Ventriglia 2002). The study area displays 12 lithological classes of this thematic product: backfill; clays; Calcarenites; Litoid tuffs; Tuffs; Tufites; Silts/clays; Litoid tuffs/pozzolanes; gravels/Sands/clays; travertines; arenitic sands and Conglomeratic gravels.

4. *Anthropogenic Parameters*

Land use/cover has most of the time been considered as a static factor. By the ongoing anthropogenic activities, land cover has been progressively modified, then causing a different proneness to landslide and other related hazards. However, there is an increasing number of studies that have analysed the effect of land-use changes in landslide susceptibility assessment (e.g. Glade 2003, Schilirò et al. 2018). Land use maps are made on a routine basis medium-resolution satellite imagery. The thematic land cover product is delivered by the European program 'COoRdination de l'INformation sur l'Environnement' (2012 CORINE Land Cover, release available on the website http://www.pcn.minambiente.it/geoportal/catalog/search/resource/details.page?uuid=%7BA0.

16.4 Methodology

Mapping unit is an important pre-requisite depending on the data availability, the extension of the study area and the aim of statistically based landslide susceptibility modelling (Guzzetti et al. 1999, 2006). In this study, we used the pixel as the mapping unit.

16.4.1 *The Logistic Regression*

Logistic Regression (LR) is one of the mostly used statistically based methods for landslide susceptibility assessment (Guzzetti et al. 1999): precisely 18.5% according to Reichenbach et al. (2018). Logistic Regression analysis is more complex as data need to be converted from GIS to a statistical software program (Park et al. 2013). However, it allows to evaluate the continuous independent variables in addition to discrete forms (Wang et al. 2013).

The LR can be expressed in the following Eqs. (16.1) and (16.2):

$$P = \frac{1}{1 + e^{-z}} \tag{16.3}$$

$$Z = \alpha_0 + \alpha_1 X_1 + \alpha_2 X_2 + \alpha_3 X_3 + \cdots + \alpha_n X_n \tag{16.4}$$

where α_0 is the intercept of the model, n is the number of variables, $\alpha_{1,}$ $\alpha_{2, \ldots,}$ α_{n} are the values associated with each of the independent variables, P is the probability which varies between 0 and 1.

16.4.2 *Susceptibility Model Evaluation*

To evaluate the performances of susceptibility models deriving from the statically based methods, the following scheme has been followed: (a) specific model validation criteria of the study area, (b) model fitting performance, and (c) model prediction performance.

(a) *Specific Model Validation Criteria*

Statistically based models for susceptibility zonation build relationships between predisposing factors and landslide occurrence through training sub-datasets and then verify these relationships using testing sub-datasets (Guzzetti et al. 2006).

In this work, random validation, multi-temporal validation and multi-spatial validation of Logistic Regression have been implemented following Ayalew and Yamagishi (2005):

- For the training of the statistical model through the random validation, there is not only the need of considering landslide-presence pixel, but also landslide-absence pixel in order to estimate the Landslide Susceptibility Index from the combination of independent factors leading to landslide (value 1 of the dependent variable). Two hundred and fifty landslide-absence pixels' dataset have been randomly generated for evaluating also the combination of independent factors tending to have no landslide (value 0 of dependent variable) within the model. Therefore, the output landslide dataset combining both landslide-presence pixels and landslide-absence pixels has been randomly split into two sub-datasets, i.e. 80% for model training and 20% for the model test.
- For the multi-temporal validation, the training sub-dataset was made from the landslide inventory before 2014 and the testing sub-dataset consisted on the 2014 landslide inventory.
- For the multi-spatial validation, a specific area representing 1/20 of the entire study area was selected, in which models have been calibrated (59 landslides from inventory dataset) and then the resulting coefficients have been applied on the whole area in order to be tested. This validation approach was also used to evaluate the representativeness of the specific portion in terms of predisposing factors of the whole study area.

(b) *Model Fitting Performance*

To assess the reliability of the susceptibility model built with the inventory training subset, the Success Rate Curve (SRC) is frequently used (Chung and Fabbri 1999; Carrara et al. 2008; Frattini et al. 2010). This is a graphical representation which allows to estimate the performance and the accuracy of the used model (model fitting). The x axis is the cumulated proportion of areas predicted as susceptible (%) and the y axis is the cumulative proportion of landslide occurrence (%) (Chung and Fabbri 2003).

(c) *Model Prediction Performance*

For estimating the model prediction performance, the Receiver Operating Characteristics (ROC) was used, while the confusion matrix represents the most crucial metric commonly used to evaluate classification models (Chung and Fabbri 2003).

16.5 Results and Discussion

The resulting coefficients from random and multi-temporal validation (Table 16.1) indicate a slightly similar response in terms of main contributing factors, such as in the case of slope. On the contrary, the low contribution of this specific factor in the multi-spatial can be explained by its low variability within the area where this validation criterion has been trained. In this sense, the relatively high ongoing human activities of this portion explains the high influence of the anthropogenic factor with respect to the other two validation criteria. Figures 16.5, 16.6 and 16.7

Table 16.1 Coefficients of predisposing factors' logistic regression validation criteria

Variables	Coeff (LRrandom)	Coeff (LRmulti-temporal)	Coeff (LRmulti-spatial)
Lithology	−0.050	−0.054	−0.311
CLC	0.255	0.248	0.289
Slope	0.177	0.284	0.039
Plan curvature	−0.209	−0.116	−0.368
Profile curvature	−0.252	−0.343	0.392
Aspect	−0.002	−0.001	−0.003
TWI	−0.292	−0.013	−0.965
SPI	0.198	0.004	0.140
Elevation	0.006	0.008	0.022
Constant	−0.142	−3.008	6.940

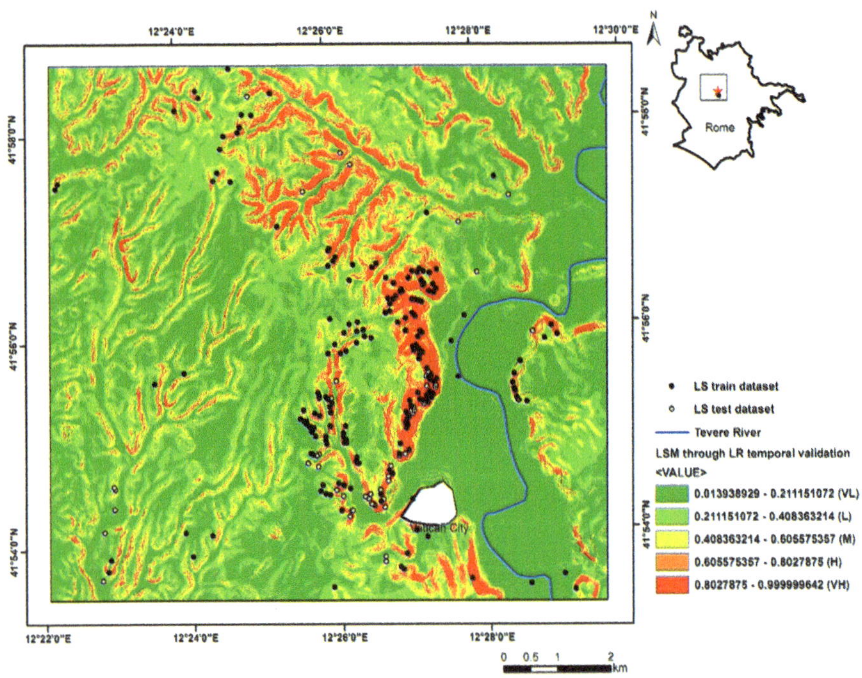

Fig. 16.5 Landslide susceptibility map of the study area from random validation Logistic regression

are, respectively, the susceptibility maps of random, multi-temporal and multi-spatial validations. Each map has been classified into five zones of landslide probability according to the equal intervals criterion: Very low (VL), Low (L), Medium (M), High (H) and Very high (VH).

The fitting performance obtained through the calculation of the area under the curve of the success rate curve (SRC/AUC) has been estimated as 86.08% for the

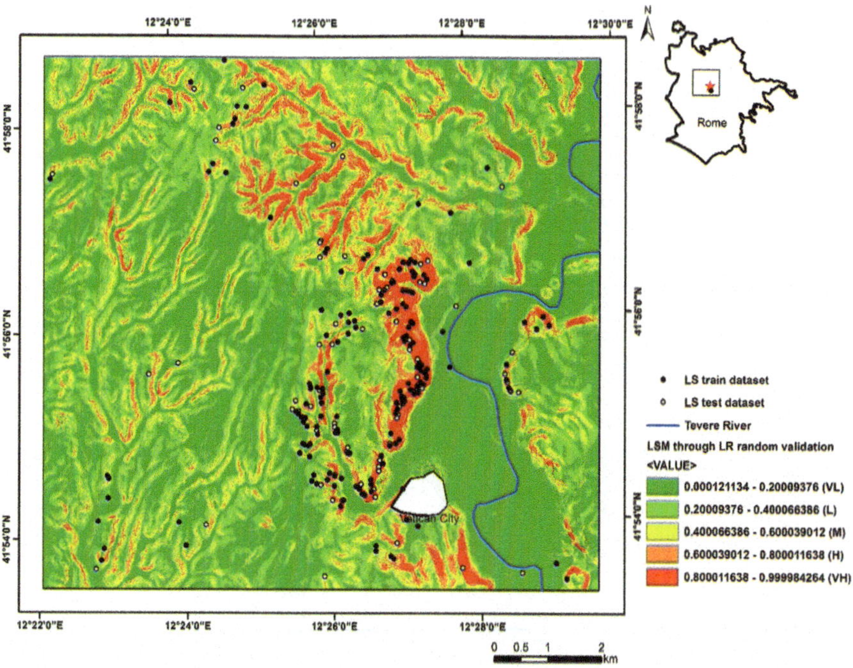

Fig. 16.6 Landslide susceptibility map of the study area from multi-temporal validation Logistic regression

random validation, 84.62% for the multi-temporal validation and 85.24% for the multi-spatial validation (Fig. 16.8).

The values of the three validation criteria are slightly similar. However, the slightly higher performance observed in the model fitting accuracy in random validation might be justified by the fact of the highest number of sub-dataset used for the training phase.

The predictive performance derived through the calculation of the area under the curve of receiver operating characteristics (ROC/AUC) ranges over 80.90%, 84.14% and 73.69% for random and multi-temporal and multi-spatial validations, respectively (Fig. 16.9).

The multi-temporal validation criterion displays the best predictive accuracy. This is likely related to the concise inventory carried out after the event of 2014. It was suitable to test the logistic regression statistical analysis with temporal event-based data. The lowest value of predictive accuracy has been displayed by the multi-spatial validation criterion. This was implemented in order to test the representativeness of the entire study area.

The predictive performances of the random validation, multi-temporal and multi-spatial validation logistic regression model have been analysed through contingency tables (Table 16.2). These are suitable to assess predictive robustness of rightly modelled unstable areas (TPR) and wrongly modelled unstable areas FPR).

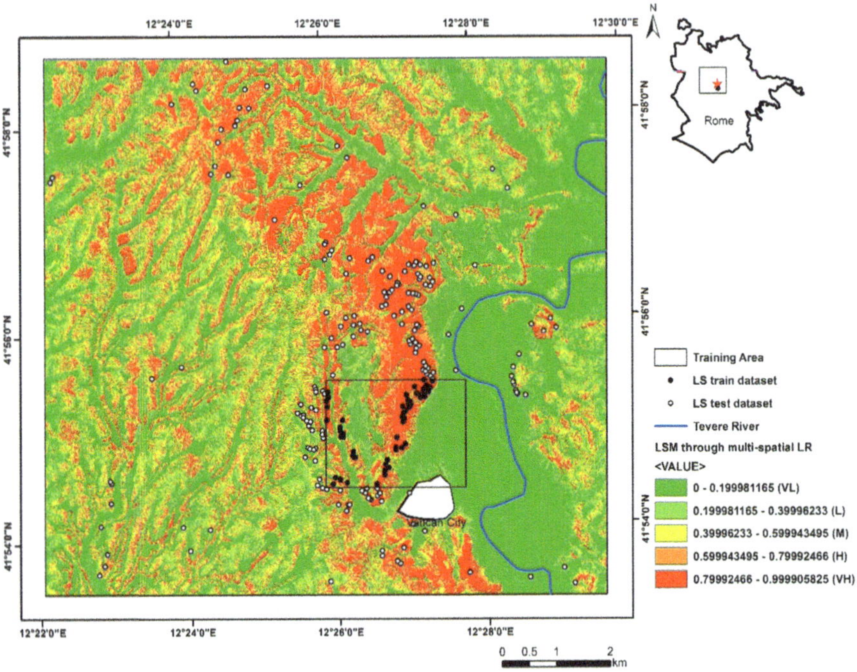

Fig. 16.7 Landslide susceptibility map of the study area from multi-spatial validation Logistic regression

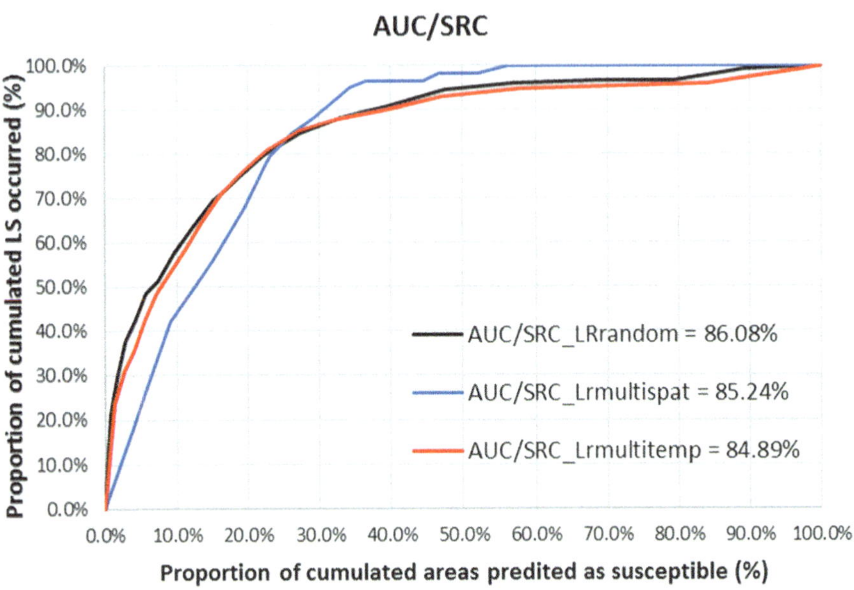

Fig. 16.8 ARC/SRC of landslide susceptibility analyses

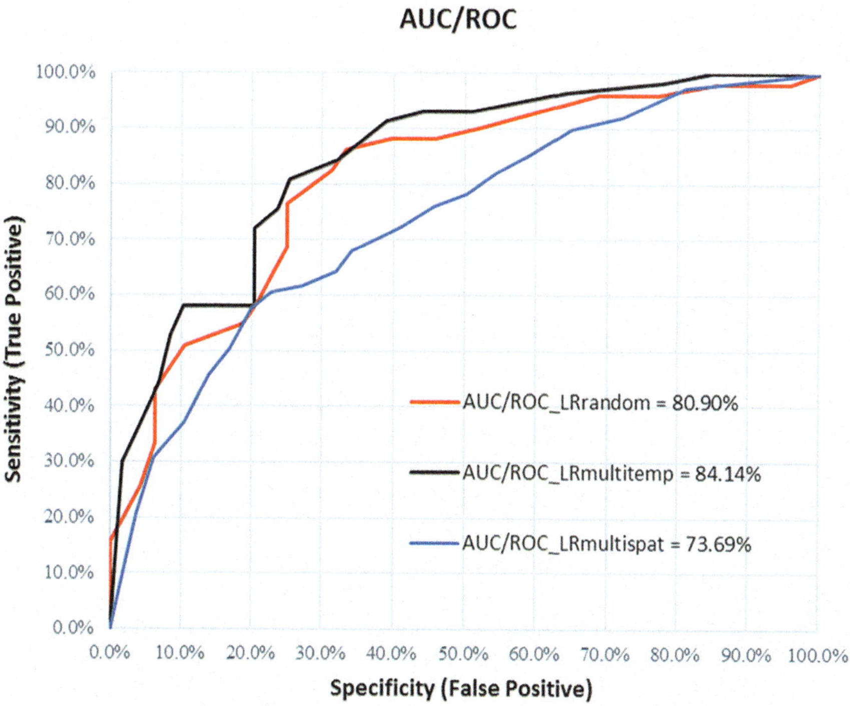

Fig. 16.9 ARC/ROC of landslide susceptibility analyses

Table 16.2 Contingency table of the predictive performance of logistic regression validation criteria

	Random validation LR model		Multi-temporal validation LR model		Multi-spatial validation LR model	
Susceptible area percentage	True positive rate (%)	False positive rate (%)	True positive rate (%)	False positive rate (%)	True positive rate (%)	False positive rate (%)
0	0.00	0.00	0.00	0.00	0.00	0.00
15	33.33	6.25	52.63	8.47	37.23	10.40
30	50.98	10.42	57.89	20.34	57.45	19.96
50	76.47	25.00	80.70	25.42	68.09	34.10
70	88.24	45.83	92.98	44.07	78.19	50.31
85	96.08	77.08	98.25	77.97	89.89	65.07
100	100.00	100.00	100.00	100.00	100.00	100.00

As can be noted, the best compromise is observed in the multi-temporal validation model. It correctly predicts 80.70% of the observed landsides at 50% of susceptibility probability. In parallel, it wrongly and over-predicts 25.42% at the same susceptibility probability threshold.

Table 16.3 Landslide distribution in predicted landslide susceptibility zones for multi-temporal validation

landslide susceptibility zones	% area of predicted zones (a)	% area of observed landslides per class (b)	landslide density (b/a)
VLS	43	1.8	0.0408
LS	29.7	14	0.47
MS	15.2	15.8	1.04
HS	8	12.3	1.53
VHS	4.1	56.1	13.7

The results show that 68.4% of landslides are found in 12.1% of the very high and high susceptibility areas: these data agree with those obtained from a recent study performed in Chamoli region of the Himalayas (Chauhan et al. 2010), where 71.13% of landslides are located in 21.96% of the territory predicted as of very high and high susceptibility. The same model applied to a similar case study in southern Italy (Trigila et al. 2015) has given congruent results of the ROC prediction model. Also, in another case study in Europe, more specifically in Romania, the landslide susceptibility analysis performed through the logistic regression resulted in AUC/ROC values ranging between 85.1% and 94%. (Mărgărint et al. 2013).

The comparison with different case studies shows how the performed methods can be suitable for different morphoclimatic environments. In this respect, the AUC of the ROC curve equal to 84.14% for the multi-temporal validation (the best validation model of this study) as seen in the Table 16.3 is comparable with the values of 88.27% and 86.29% obtained in similar studies respectively from Mathew et al. (2007) in India and Devkota et al. (2013) in Nepal.

16.6 Conclusions

In this study, three main validation criteria of logistic regression have been used to assess the rainfall-induced shallow landslide susceptibility in a specific area of the city of Rome. The spatial likelihood has also been assessed through a specific event that occurred in 2014. From the estimation of the predictors through the application of logistic regression, it resulted that land cover and slope are the most influencing factors, specifically for random and multi-temporal validation criteria. The multi-temporal validation model displayed the best compromise of landslide susceptibility and may be used to identify critical susceptible areas in which a specific focus should be addressed while implementing land use management plans as well as hazard risk mitigation policies of such a vulnerable city.

References

Alessi D, Bozzano F, Di Lisa A, Esposito C, Fantini A, Loffredo A, Martino S, Mele F, Moretto S, Noviello A, Prestininzi A, Sarandrea P, Scarascia Mugnozza G, Schiliro L, Varone C (2014) Geological risk in large cities: the landslides triggered in the city of Rome (Italy) by the rainfall of 31 January–2 February. Italia J Eng Geol Environ:15–34. https://doi.org/10.4408/IJEGE.2014-01-0-02

Ayalew L, Yamagishi H (2005) The application of GIS-based logistic regression for landslide susceptibility mapping in the Kakuda-Yahiko Mountains, Central Japan. Geomorphology 65:15–31

Bozzano F, Salvatore M, Priori M (2006) Natural and man-made induced stress evolution of slope: the Monte Mario hill of Rome. Environ Geol 50:505–524. https://doi.org/10.1007/s00254-006-0228-y

Budimir MEA, Atkinson PM, Lewis HG (2015) A systematic review of landslide probability mapping using logistic regression. Landslides 12:419–436. https://doi.org/10.1007/s10346-014-0550-5

Carrara A (1983) Multivariate models for landslide hazard evaluation. J Int Assoc Math Geol 15 (3):403–426

Carrara A, Crosta G, Frattini P (2008) Comparing models of debris-flow susceptibility in the alpine environment. Geomorphology 94:353–378. https://doi.org/10.1016/j.geomorph.2006.10.033

Chauhan S, Sharma M, Arora MK (2010) Landslide susceptibility zonation of the Chamoli region, Garhwal Himalayas, using logistic regression model. Landslides 7:411–423. https://doi.org/10.1007/s10346-010-0202-3

Chung CF, Fabbri AG (1999) Probabilistic prediction models for landslide hazard mapping. Photogramm Eng Remote Sens 65–12:1389–1399

Chung CF, Fabbri AG (2003) Validation of spatial prediction models for landslide hazard mapping. Nat Hazards 30(3):451–472

Corominas J, van Westen CJ, Frattini P, Cascini L, Malet J-P, Fotopoulou S, Catani S, Van Den Eeckhaut M, Mavrouli O, Agliardi F, Pitilakis K, Winter MG, Pastor M, Ferlisi S, Tofani V, Hervas J, Smith JT (2014) Recommendations for the quantitative analysis of landslide risk. Bull Eng Geol Environ 73:209–263. https://doi.org/10.1007/s10064-013-0538-8

Cruden DM, Varnes DJ (1996) Landslide Types and Processes. In: Turner AK, Shuster RL (eds) Landslide: investigation and Mitigation, Special Report, Transportation Research Board, National Academy of Sciences, 247. National Academy Press, Washington, DC, pp 36–75

Del Monte M, D'orefice M, Luberti GM, Marini R, Pica A, Vergari F (2016) Geomorphological classification of urban landscapes: the case study of Rome (Italy). J Maps 12(Suppl 1):178–189. https://doi.org/10.1080/17445647.2016.1187977

Devkota CK, Regmi DA, Pourghasemi HR, Yoshida K, Pradhan B, Ryu IC, Dhital MR, Althuwaynee OF (2013) Landslide susceptibility mapping using certainty factor, index of entropy and logistic regression models in GIS and their comparison at Mugling–Narayanghat road section in Nepal Himalaya. Nat Hazards 65:135–165. https://doi.org/10.1007/s11069-012-0347-6

Frattini P, Crosta GB, Carrara A (2010) Techniques for evaluating the performance of landslide susceptibility models. Eng Geol 111(1–4):62–72. https://doi.org/10.1016/j.enggeo.2009.12.004

Funiciello R, Giordano G (2008) La nuova Carta Geologica di Roma: litostratigrafia e organizzazione stratigrafica. Mem Descr Carta Geol D'It 80(1):39–85, 22 figg., Firenze

Glade T (2003) Landslide occurrence as a response to land use change: review of evidence from New Zealand. Catena 50(3–41):297–314

Guzzetti F, Carrara A, Cardinali M, Reichenbach P (1999) Landslide hazard evaluation: a review of current techniques and their application in a multi-scale study, Central Italy. Geomorphology 31:181–216

Guzzetti F, Reichenbach P, Cardinali M, Galli M, Ardizzone F (2006) Probability landslide hazard assessment at the basin scale. Geomorphology 81:272–299

ISTAT (2011) Istituto Nazionale di Statistica (eng. National Institute of Statistics) of Italy

Lin L, Lin Q, Wang Y (2017) Landslide susceptibility mapping on a global scale using the method of logistic regression. Nat Hazards Earth Syst Sci 17:1411–1424. https://doi.org/10.5194/nhess-17-1411-2017

Mărgărint MC, Grozavu A, Patriche CV (2013) Assessing the spatial variability of coefficients of landslide predictors in different regions of Romania using logistic regression. Nat Hazards Earth Syst Sci 13:3339–3355. https://doi.org/10.5194/nhess-13-3339-2013

Mathew J, Jha VK, Rawat GS (2007) Application of binary logistic regression analysis and its validation for landslide susceptibility mapping in part of Garhwal Himalaya, India. Int J Remote Sens 28(10):2257–2275. https://doi.org/10.1080/01431160600928583

Ohlmacher GC, Davis JC (2003) Using multiple logistic regression and GIS technology to predict landslide hazard in Northeast Kansas, USA. J Eng Geol:331–343. https://doi.org/10.1016/S0013-7952(03)00069-3

Park DW, Nikhil NV, Lee SR (2013) Landslide and debris flow susceptibility zonation using TRIGRS for 2011 Seoul landslide event. Nat Hazards Earth Syst Sci 13:2833–2849. https://doi.org/10.5194/nhess-13-2833-2013

Reichenbach P, Rossi M, Malamud BD, Mihir M, Guzzetti F (2018) A review of statistically-based landslide susceptibility models. Earth Sci Rev 180:60–91

Schilirò L, Cevasco A, Esposito C, Scarascia MG (2018) Shallow landslide initiation on terraced slopes: inferences from a physically based approach. Geomatics Nat Hazards Risk 9 (1):295–324. https://doi.org/10.1080/19475705.2018.1430066

Trigila A, Iadanza C, Esposito C, Scarascia MG (2015) Comparison of logistic regression and random forests techniques for shallow landslide susceptibility assessment in Giampilieri (NE Sicily, Italy). Geomorphology 249:119–136. https://doi.org/10.1016/j.geomorph.2015.06.001

Van Den Eeckhaut M, Hervás J, Jaedicke C, Malet JP, Montanarella M, Nadim F (2012) Statistical modelling of Europe-wide landslide susceptibility using limited landslide inventory data. Landslides 9:357–369. https://doi.org/10.1007/s10346-011-0299-z

Varnes DJ (1978) Slope movement types and processes. In: Schuster RL, Krizek RJ (eds) Landslides: analysis and control, special report 176. Transportation research Board, national research council, Washington, DC, pp 11–33

Ventriglia U (2002) Geologia del territorio Comune di Roma. A cura del Servizio Geologico, Difesa del Suolo – Provincia di Roma, 809 pagg., 13 tavole fuori testo.

Wang L-J, Sawada K, Moriguchi S (2013) Landslide susceptibility analysis with logistic regression model based on FCM sampling strategy. Comput Geosci 57:81–92. https://doi.org/10.1016/j.cageo.2013.04.006

Part VI
Landslide: Control

Chapter 17
Application of a Generalized Criterion: Time-of-Failure Forecast and Alert Thresholds Assessment for Landslides

Alessandro Valletta, Andrea Segalini, and Andrea Carri

Abstract The forecast of a landslide time-of-failure and alert thresholds assessment are fundamental aspects of landslide risk prevention and mitigation. The problem has been often considered with a 'site-specific' approach, allowing to describe the single event with great accuracy, while hindering its automatic application to other cases featuring different properties. A procedure aimed to define a common behaviour between different cases was developed with the purpose of improving the approach to this topic. Starting from displacement data referring to landslides documented in scientific literature, a series of normalized velocity vs time trends was defined. These curves highlighted a common trend in landslides' evolution towards failure, thus allowing the assessment of general alert thresholds. In this paper, the generalized criterion is applied to other landslides not included within the initial database, in order to validate the proposed methodology and investigate its effectiveness. The process has been applied to these new cases, studying and describing their evolution from the beginning of the monitoring activity to the final collapse of the landslide, thus simulating real-time data collection and elaboration in order to compare the results with the developed criterion.

Keywords Landslides · Monitoring · Displacements · Inverse Velocity Method · Time of failure · Early warning system

A. Valletta (✉) · A. Segalini
Università di Parma, DIA, Parma, Italy
e-mail: alessandro.valletta@unipr.it

A. Carri
ASE – Advanced Slope Engineering S.r.l., Parma, Italy

M. De Maio, A. K. Tiwari (eds.), *Applied Geology*,
https://doi.org/10.1007/978-3-030-43953-8_17

17.1 Introduction

Nowadays, the development and application of Early Warning Systems (EWSs) plays a key role in the framework of landslide risk reduction and mitigation. This task involves not merely technical and scientific concepts, but also regards social aspects, since the main goal of EWSs is to alert people of imminent hazards and allow them to get to safety (Stähli et al. 2015). Given the major importance of alert systems applied to natural events, several studies (e.g. UN-ISDR 2006; Intrieri et al. 2013; Michoud et al. 2013; Thiebes and Glade 2016) provided a selection of common elements and actions that should be included in a EWS design process. These components can be summarized as follows:

- Knowledge about the risk threatening a community, including different risk scenarios and design criteria assessment;
- Monitoring system, able to record data related to the phenomenon and transmit them to the elaboration centre;
- Analysis and forecasting method, which describes the landslide evolution, predicts its most probable behaviour and identifies critical events on the basis of alert thresholds;
- Warnings and dissemination of alert messages, to notify the community about the ongoing critical event in order to activate emergency plans and measures.

Among these features, failure forecasting and thresholds assessment represent probably the most problematic components both from a conceptual and social point of view. In fact, an inaccurate model application could lead to missed warnings if the procedure underestimates the phenomenon severity, or it can cause false alarms when an ordinary event is misinterpreted as a critical occurrence (Paté-Cornell 1986).

Starting from concepts previously exposed, several warning systems including collapse predictions and/or threshold assessment have been developed and applied to different case studies over the years. Intrieri et al. (2012) designed a site-specific EWS featuring three different velocity threshold levels, based on the analysis and elaboration of the most critical time periods of the Torgiovannetto landslide entire dataset. Manconi and Giordan (2015) presented a procedure where alert thresholds are updated on the basis of failure forecasting results reliability, thanks to a system able to acquire data in near-real time. According to the authors, this approach should contribute to solve the problem arising from the use of pre-defined warning levels, which return no additional information after their overcoming. Crosta and Agliardi (2002) developed a methodology to define 'characteristic velocity curves' starting from the accelerating creep theory by Voight (1988), giving a theoretical description of the future behaviour of a landslide and allowing the definition of several warning levels. In their study, the authors provided an application of the proposed method on the Ruinon landslide, defining a different curve for each monitored sub-area of the slope. Ju et al. (2015) designed a EWS specifically for the Longjingwan landslide, Southwest China, based on four alert velocity thresholds. In this particular case, warning levels were assessed thanks to expert suggestions and further phases of experts' judgement were included to avoid false alarms as far as possible.

More recently, Segalini et al. (2018) presented a new generalized criterion aimed at providing a single graph to be used as a general reference for alert threshold definition. The study was intended to partially overcome the 'site-specific' concept linked to failure forecasting methods, thanks to a procedure that allowed the identification of similarities in the behaviour of landslides characterized by different features. In this paper, the generalized criterion is applied in order to test its reliability and efficiency as a EWS component.

17.2 Materials and Methods

17.2.1 Time-of-Failure Forecasting

Starting from the 1950s, when Terzaghi (1950) first studied the ground movements that precede a landslide, several authors approached the challenging problem of developing a reliable method to predict the date and time of a landslide collapse. In particular, the so-called 'empirical approaches', based on creep theory and closely related to the observational method, gained considerable relevance (Mazzanti et al. 2012). These methods rely on the hypothesis that the landslide collapse is anticipated by an accelerating stage, corresponding to the 'tertiary creep' phase as described in the model proposed by Varnes (1978).

One of the most commonly used forecasting models, called Inverse Velocity Method (IVM), was theorized in the '80s by Fukuzono (1985), which studied and improved the creep curve method previously introduced by Saito (1965, 1969). Starting from small-scale laboratory tests, the author derived that the logarithm of acceleration is proportional to the logarithm of displacement rate, according to the following power law equation:

$$\frac{d^2x}{dt^2} = A\left(\frac{dx}{dt}\right)^{\alpha} \tag{17.1}$$

where x represents surface displacements, t is time, A and α are dimensionless parameters. While the value of A coefficient displays significant variations, α usually ranges between 1.5 and 2.2 (Fukuzono 1985, 1990) when considering natural slopes. Moreover, on the basis of the relationship previously presented, the following formulation was proposed as a tool to predict the time-of-failure:

$$\frac{1}{v} = (A(\alpha - 1)(t_f - t))^{\frac{1}{1-\alpha}} \tag{17.2}$$

where v is the displacement rate and t_f represents the time-of-failure. This equation works under the theoretical assumption that a landslide would display increasing values of displacement rate as it gets closer to collapse, therefore approaching a vertical asymptote at $v \rightarrow \infty$. Consequently, the time-of-failure t_f can be derived as the point where the inverse of velocity vs time trend reaches the x-axis, representing $\frac{1}{v} = 0$.

After its formulation, Fukuzono's theory has been validated by Voight (1988, 1989) and it has been applied to a large number of different case studies, displaying satisfying results (e.g. Petley 2004; Rose and Hungr 2007; Gigli et al. 2011; Mazzanti et al. 2017; Carlà et al. 2017). Moreover, Intrieri and Gigli (2016) provided a comparison between different failure forecasting methods, obtaining good performances compared to other prediction models. These applications showed also that the value of α parameter is frequently near to 2, representing a linear trend in the inverse-velocity vs time plot.

Nonetheless, it should be specified that IVM application must be carefully evaluated depending on the case. In fact, the forecasted time-of-failure should never be considered as an exact date & time prediction, since the method generally indicates that the failure is likely in proximity of the intersection point (Carlà et al. 2018b). Moreover, the following considerations concerning the IVM implementation should be taken into account:

- Given the inherent complexity of these phenomena, failure forecasting methods should not be considered as tools able to return an exhaustive description of the landslide behaviour.
- The forecasting quality is strongly influenced by the quality of monitoring data, which can be elaborated with post-processing and statistical tools to improve their meaningfulness.
- Monitoring systems able to guarantee high sampling frequencies are recommended in order to accurately represent the observed phenomenon. This is particularly relevant for landslides featuring brittle failure mechanisms, which are more difficult to represent by using IVM due to their extremely rapid evolution.
- The monitoring activity should continue as long as possible during the critical phase of the landslide, allowing the identification of possible trend changes in displacement evolution that could heavily influence the prediction's reliability.

17.2.2 Generalized Criterion

As previously mentioned, Segalini et al. (2018) presented a new generalized criterion based on Fukuzono's results, aiming to provide a standard procedure to partially overcome the 'site-specific' nature of alert thresholds assessment. The criterion development followed a series of steps summarized in Fig. 17.1 and described below:

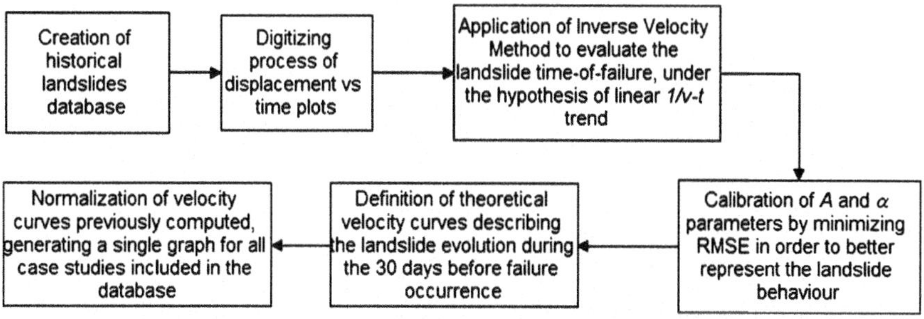

Fig. 17.1 Flow chart summarizing the procedure followed to develop the generalized criterion

- Creation of a database including 26 different landslides reported in scientific literature, consisting of displacement and/or velocity datasets recorded by different monitoring systems. Case studies included in the database present relevant differences in terms of volume, failure mechanism, triggering factors, monitoring typology and duration.
- Application of a digitizing software to generate a series of numerical coordinates, representing the displacement vs time trend referred to each case study included in the database.
- Application of the Inverse Velocity Method to evaluate the time-of-failure t_f for each landslide. In this phase, the hypothesis of the linear relationship between inverse of velocity and time is considered; hence, the collapse date is evaluated with a linear regression in the $1/v - t$ plot, keeping the α parameter constant and equal to 2.0.
- Calibration of A and α parameters of the IVM in order to improve the model ability to represent the studied dataset. It should be specified that this step involves only the parameters' calibration, while the time-of-failure value evaluated at the previous step is still considered valid. The variation in parameters aims to minimize the Root Mean Square Error (RMSE) resulting from a comparison between data recorded by the monitoring system and evaluated by the forecasting model.
- Definition of a velocity vs time curve for each single case, representing the landslide theoretical behaviour during the 30 days before the collapse. These curves are computed by introducing the parameters previously calibrated in the Fukuzono-Voight equation relating displacement velocity v_{FV} and time t:

$$v_{FV} = \left(A(\alpha - 1)(t_f - t)\right)^{\frac{1}{\alpha - 1}} \tag{17.3}$$

- Normalization of velocity curves previously computed, in order to identify a series of generalized velocity vs time-of-failure curves to be represented in a

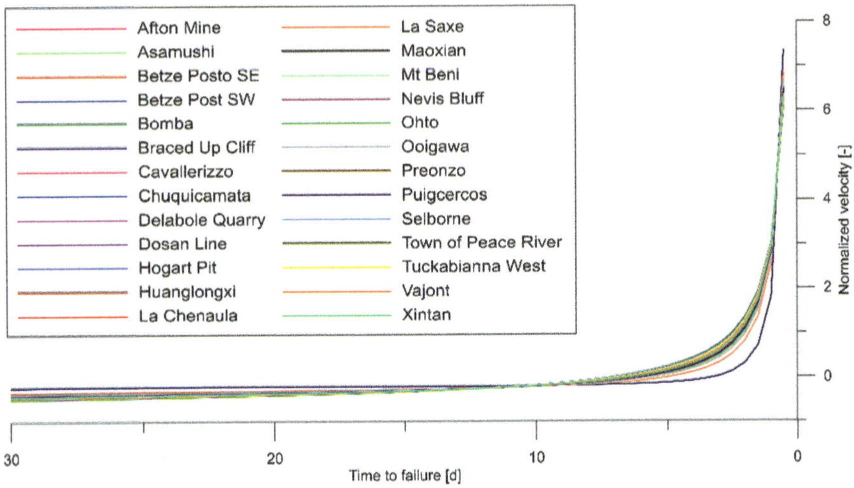

Fig. 17.2 Normalized velocity vs time plot, reporting all the different case studies included in the historical landslides database (Segalini et al. 2018)

single graph, thus allowing its application to case studies with different features. The normalization equation is reported below:

$$v_n = \frac{v_{FV} - \mu_{v_{FN}}}{\sigma_{v_{FN}}} \tag{17.4}$$

where $\mu_{v_{FN}}$ and $\sigma_{v_{FN}}$ are the mean value and standard deviation of v_{FN} values computed at the previous step.

Figure 17.2 reports the single graph obtained from the normalization procedure, allowing the identification of a common trend between different case studies. In fact, the proposed process successfully generated a dimensionless velocity parameter that could be taken as a general reference to assess alert thresholds for different case studies. It should be noted that parameter A does not influence the curves shape and concavity, which depend only on parameter α. In particular, by taking as a reference the 'linear' $\alpha = 2$ curve, a value of $\alpha < 2$ leads to a delayed increasing of normalized velocity. Instead, curves featuring $\alpha > 2$ are less concave since the flex point appears earlier (Segalini et al. 2018).

A major feature of the generalized approach is its broad applicability. In fact, as mentioned before, the landslides database used for the model development included phenomena presenting significant differences in terms of features like volume, material, triggering factors and event duration. Ultimately, the generalization procedure allowed to find a common trend for landslides with different features. For this reason, from a theoretical point of view the proposed method could be applied to all those phenomena displaying a behaviour that satisfies the IVM hypotheses, while also sharing some limitations of this method (e.g. difficulty to predict a fast-moving landslide collapse).

The procedure was followed to develop the generalized criterion and to generate the plot reported in Fig. 17.2, and involved the full dataset for each single landslide analysed. Because of this, the approach adopted should not be considered as representative of a real-time case study, where displacement data are acquired and become available as the monitoring activity progresses.

This paper presents a series of applications where a real-time sampling and elaboration is simulated, in order to test the generalized criterion in a more realistic scenario. These case studies were not included in the historical database created to develop the generalized criterion. Moreover, they are analysed by simulating progressive acquisition of monitoring data, according to the following procedure:

I. After the identification of the first point of the accelerating phase, the following three displacement data are included to generate the first dataset to be analysed.
II. Linear Inverse Velocity Method is applied to the selected dataset in order to evaluate the time-of-failure value.
III. Equation (17.2) is applied to evaluate the theoretical displacement velocity, by using the A and α obtained from the application of IVM under the hypothesis of linearity in the $1/v - t$ trend.
IV. Model parameters are calibrated by minimizing the RMSE value, computed by comparing theoretical and monitoring data.
V. The theoretical velocity curve for the specific case study is evaluated by implementing into Eq. (17.3) the parameters evaluated at step II (time-of-failure t_f) and step IV (A and α coefficients).
VI. The last data recorded by the monitoring tool are normalized by applying Eq. (17.4), where mean and standard deviation values derive from the velocity curve determined in the previous step. As a result, it should be possible to compare the normalized velocity value with one of the theoretical curves generated by the generalized criterion, chosen according to the computed value of the α parameter.
VII. If new monitoring data are available, they are added to the dataset and the procedure is repeated starting from step II.

17.3 Results and Discussion

17.3.1 New Tredegar Colliery

The first case study analysed in this paper regards the New Tredegar site, located in the Rhymney Valley in Caerphilly County Borough, South Wales. As reported by several authors (Bentley and Siddle 2000; Siddle et al. 2007; Carey 2011), the construction of a railway crossing the area in 1856 first evidenced the presence of

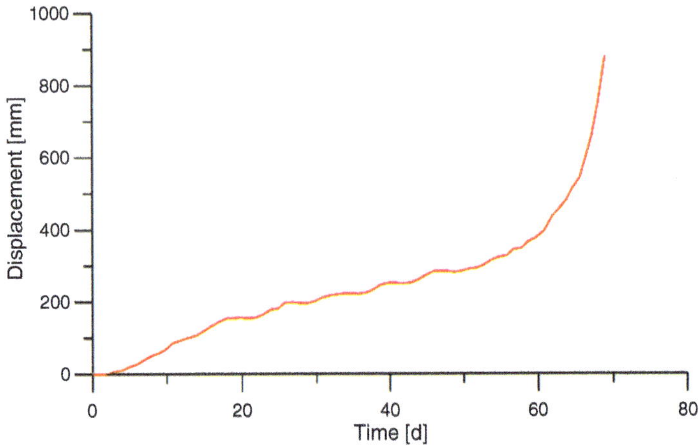

Fig. 17.3 Cumulative displacements at the ground level recorded at New Tredegar, date of collapse $t_f = 69$ days (Digitized after Siddle et al. 2007)

ground displacements affecting the colliery. In March 1905, extreme rainfalls activated a large landslide that damaged the railway line and displaced the road laterally up to 50 feet (Knox 1927).

After this event, monitoring activities started in the site area. Moreover, early movements in the first part of 1930 led to the detection of fissures in the upper part of the slope that caused significant concern. For this reason, the colliery commissioned daily checks of ground movements. Finally, a catastrophic landslide occurred on April 12, 1930 completely destroying the colliery and overwhelming the road (Bentley and Siddle 2000). Figure 17.3 represents recorded surface displacements up to the collapse date, corresponding to $t = 69$ days in the displacement vs time plot.

In this case, onset of acceleration was identified approximately at $t = 56$ days with respect to the zero-reference. It should be noted that, since the monitoring activity started after the identification of previous movements in the area, available data do not provide clear information about the primary creep phase (Carey 2011).

Table 17.1 reports the results obtained from the application of the generalized criterion. In particular, values resulting from Inverse Velocity Method application, parameters' calibration and normalization procedure are presented. Additionally, it is also indicated the theoretical warning time represented by the difference between estimated time-of-failure and acquisition date of the last monitoring data. It can be noted that time-of-failure forecasting returns a t_f value that varies along the monitoring activity duration, getting closer to the actual collapse date as new data are acquired. Calibration of α parameter shows a similar trend, starting from values close to the linear reference $\alpha = 2.0$ and reaching a final value of 1.94 in correspondence with the last dataset analysed for this case study.

Table 17.1 Results obtained from the application of the generalized criterion for New Tredegar dataset

Time [d]	Predicted time-of-failure [d]	Warning time [d]	Calibrated A parameter [−]	Calibrated α parameter [−]	Normalized velocity [−]
59.0	66.6	7.6	0.0096	2.01	−0.0674
60.0	66.6	6.6	0.0099	2.00	−0.0110
61.0	65.5	4.5	0.0111	2.02	0.3022
62.0	66.2	4.2	0.0107	1.99	0.2171
63.0	66.7	3.7	0.0104	1.98	0.2874
64.0	67.4	3.4	0.0101	1.97	0.2943
65.0	68.2	3.2	0.0099	1.95	0.2275
66.0	68.4	2.4	0.0096	1.95	0.7190
67.0	68.4	1.4	0.0097	1.94	1.6706

Table 17.2 Warning levels for New Tredegar landslide

Warning level	Theoretical distance from the collapse day [d]	Normalized velocity [−]
1	10	−0.1989
2	6	0.0082
3	3	0.5474

The calibration procedure is a key component of alert threshold assessment. In fact, the generalized curve to take into account as a theoretical reference for normalized velocity values should be chosen according to results deriving from this procedure. In this case, a curve featuring $\alpha = 1.95$ was selected to be compared to case study results and to assess different warning levels, represented by normalized velocity values.

Table 17.2 reports the alert threshold values selected for this case study. Warning levels here presented were chosen by the authors for demonstration purposes, in order to illustrate possible implementations of the generalized criterion. In real case applications, it is advised that an expert of the specific case study, such as the monitoring responsible holding adequate knowledge on the monitored phenomenon, should address the definition of these thresholds.

Figure 17.4 presents the comparison between the theoretical curve featuring $\alpha = 1.95$ and monitoring data after their elaboration and normalization performed by following the procedure described in the previous paragraph. It can be noted that, despite minor inaccuracies, the generalized criterion provides an effective tool to describe the landslide evolution towards collapse. Moreover, Fig. 17.4 also presents a graphical depiction of generalized warning levels assessed for this case study. It can be observed how the first monitoring point is located already above the first threshold, while the second level is consistently exceeded starting from the third normalized velocity value. Finally, the last alert threshold is overcome by two monitoring data, as the landslide gets closer to the collapse day. Table 17.3 reports the numerical comparison between normalized velocity values and warning levels.

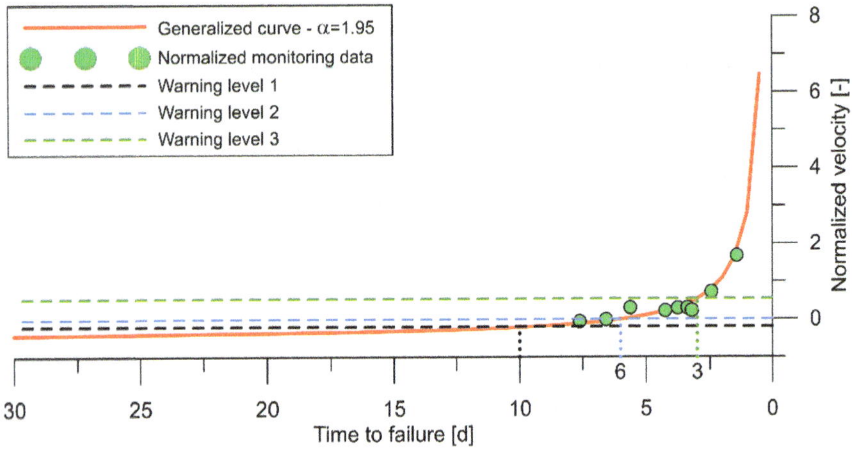

Fig. 17.4 Graphical representation of normalized monitoring data and generalized curve, compared with three warning levels – New Tredegar

Table 17.3 Warning levels' activation at different times during monitoring activity – New Tredegar

Time [d]	Normalized velocity [−]	Alert thresholds overcoming		
		Warning Level 1	Warning Level 2	Warning Level 3
59.0	−0.0674	x		
60.0	−0.0110	x		
61.0	0.3022	x	x	
62.0	0.2171	x	x	
63.0	0.2874	x	x	
64.0	0.2943	x	x	
65.0	0.2275	x	x	
66.0	0.7190	x	x	x
67.0	1.6706	x	x	x

17.3.2 Letlhakane Mine

The second case study presented in this paper regards a collapse observed at the Letlhakane diamond mine in Botswana, active since 1975. The landslide occurred on July 14, 2005 and involved about 233′000 m^3 of material, with a spatial development of 183 metres along the slope direction (Kayesa 2006). A geodetic Monitoring System (*Geomos*), composed of two theodolites located in concrete shelters, was installed on-site. The system featured an automatic procedure to acquire the position of retroreflectors placed in several locations on the mine wall, aiming to detect slope displacements.

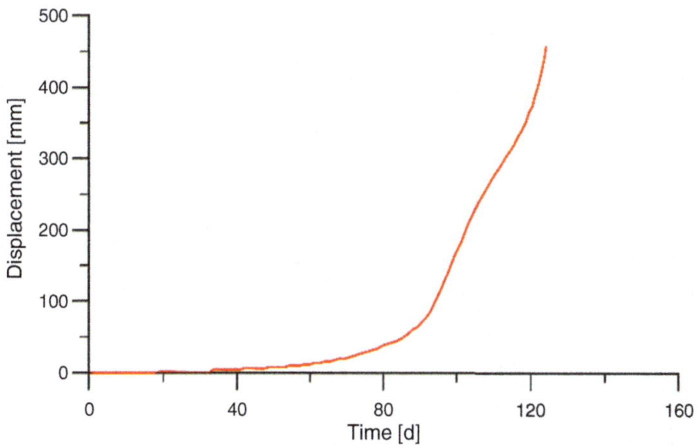

Fig. 17.5 Cumulative displacements at the ground level recorded at Letlhakane mine, date of collapse $t_f = 127$ days (Digitized after Kayesa 2006)

The monitoring system allowed the analysis of the landslide evolution, which can be observed in Fig. 17.5. First instability signals were recorded at the end of May 2005, where tension cracks started to appear in some parts of the monitored slope (Kayesa 2006). Since June 13, 2005 the instrumentation detected a significant increment of displacement rate, which remained constant for the following week. Starting from July 7, after a temporary reduction of velocity values, monitored displacements showed a clear accelerating trend that ultimately led to the slope collapse on July 14, 2005 corresponding to $t = 127$ days in Fig. 17.5.

As previously mentioned, the first dataset comprises the point representing the onset of acceleration of slope displacement, together with the following three points. For this case study, the starting point was identified approximately at $t = 114$ days.

Results obtained from IVM application, calibration of model parameters and normalization procedure are reported in Table 17.4, including also the warning time represented by the difference between estimated time-of-failure and acquisition date of the monitoring data. In accordance with previous hypothesis, it can be observed that the accuracy of the forecasting model significantly increases with the progressing of monitoring activity. In fact, values of t_f initially evidence a relevant fluctuation around the collapse date $t = 127$ days, which is correctly identified in the last part of the dataset analysis.

Moreover, a similar behaviour can be observed in the calibration procedure that concerns α parameter, reaching a stable value of 1.99 in the last three data analysed. However, it should be noted that these are minor differences that have quite less significance with respect to the collapse forecasting variations. On the basis of results obtained in the calibration phase, in the following step it was decided to consider the curve featuring $\alpha = 1.99$. Table 17.5 reports the hypothetical warning levels considered in the present analysis for the Letlhakane landslide.

Table 17.4 Results obtained from the application of the generalized criterion for Letlhakane dataset

Time [d]	Predicted time-of-failure [d]	Warning time [d]	Calibrated *A* parameter [–]	Calibrated *α* parameter [–]	Normalized velocity [–]
117.6	128.5	10.9	0.0107	1.96	–0.2559
118.6	128.4	9.8	0.0100	1.98	–0.1897
119.4	125.1	5.8	0.0122	2.01	0.1860
120.4	128.5	8.1	0.0100	1.97	–0.1959
121.0	126.4	5.4	0.0110	2.00	0.3019
121.8	127.0	5.2	0.0108	1.99	0.0463
122.4	126.8	4.4	0.0108	2.00	0.2575
123.0	127.0	4.0	0.0107	1.99	0.2359
123.5	127.1	3.6	0.0106	1.99	0.3512
123.9	127.0	3.1	0.0106	1.99	0.7473

Table 17.5 Warning levels for the Letlhakane landslide

Warning level	Theoretical distance from the collapse day [d]	Normalized velocity [–]
1	14	–0.2914
2	7	–0.0453
3	3	0.6111

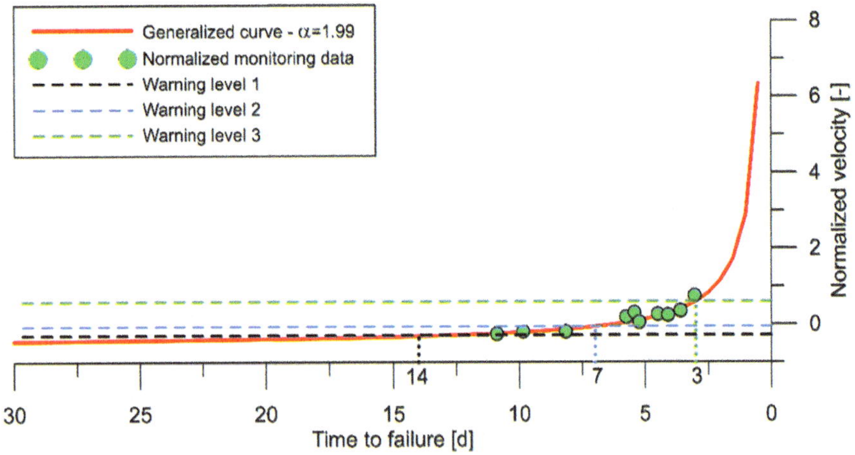

Fig. 17.6 Graphical representation of normalized monitoring data and generalized curve, compared with three warning levels – Letlhakane

Figure 17.6 presents the comparison between elaborated monitoring data and theoretical curve from the generalized criterion, displaying also a graphical representation of alert thresholds previously assessed. First, it can be observed that the velocity curve featuring $\alpha = 1.99$ properly represents the trend displayed by

Table 17.6 Warning levels' activation at different times during monitoring activity – Letlhakane Mine

		Alert thresholds overcoming		
Time [d]	Normalized velocity [–]	Warning Level 1	Warning Level 2	Warning Level 3
117.6	–0.2559	x		
118.6	–0.1897	x		
119.4	0.1860	x	x	
120.4	–0.1959	x		
121.0	0.3019	x	x	
121.8	0.0463	x	x	
122.4	0.2575	x	x	
123.0	0.2359	x	x	
123.5	0.3512	x	x	
123.9	0.7473	x	x	x

normalized monitoring data. For what concern warning levels, it can be noted how the first data included in the analysis already overcome the first alert threshold. A significant number of normalized velocity data exceeds the second warning level, highlighting the accelerating trend of the phenomenon. At the end, the last available monitoring data overcomes the third warning level, confirming that the landslide is getting closer to the collapse date. Table 17.6 summarizes results obtained in this step.

Numerical values here reported show also an unusual behaviour, where the fourth point features a velocity value lower than the previous one, leading to a downgrading from second to first warning level. This singularity could be due to a temporary decreasing of displacement rates, causing the forecasting model to calculate a collapse date further in time (125.1 days for the third point and 128.5 for the fourth point respectively, as previously reported in Table 17.4).

17.4 Conclusions

This paper presents the practical application of a new method for failure forecasting and alert threshold assessment, developed by the authors and described in detail in Segalini et al. (2018). The method aims to overcome the 'site-specific' approach that characterizes the definition of warning levels for landslides, providing a general reference to be used as a tool for Early Warning purposes. The generalized criterion is applied to two case studies by simulating a progressive acquisition of monitoring data. The implemented procedure includes the time-of-failure evaluation with the Inverse Velocity Method (Fukuzono 1985), model parameters calibration, theoretical velocity curve selection and alert thresholds assessment.

Results obtained for both case studies highlighted the ability of the proposed method to provide a reliable tool to describe the accelerating trend of a landslide. It can also be noted that the failure forecasting procedure increased its reliability as the monitoring activity progressed, obtaining predictions that are more accurate as the landslide approaches the collapse date. Following the time-of-failure evaluation, the calibration procedure allowed to determine the α parameter value in order to identify the appropriate velocity curve to be used as a theoretical reference. In particular, for the New Tredegar case study a curve featuring $\alpha = 1.95$ was selected, while dataset from Letlhakane Mine was compared to a curve with $\alpha = 1.99$. After the normalization procedure, monitoring data were in good agreement with generalized velocity curves, allowing to use them as a reference to assess warning levels. A graphical and numerical comparison with elaborated monitoring data permitted to identify a threshold overcoming, indicating that the landslide is progressively approaching a critical state. It should be remembered that, in a real case application, the overcoming of these thresholds should lead to specific actions aimed to ensure the safety of persons and properties.

While the proposed method provided positive results, it should be always taken into account that several factors influence the model efficiency and accuracy. In particular, adequate quality of monitoring data is strongly advised when dealing with failure forecasting procedures and, more in general, in case of Early Warning applications. On this topic, positive results could derive from the implementation of data processing operations such as smoothing algorithms, filters and de-spiking procedures (Dick et al. 2015; Carlà et al. 2018a).

Moreover, the criterion proved to be an effective tool when applied to case studies displaying a relatively long accelerating phase (i.e. days, weeks). An interesting evolution could be the adjustment of the proposed methodology to faster displacements, which at this stage of the criterion development are not interpreted accurately by the generalized approach.

References

Bentley SP, Siddle HJ (2000) New Tredegar landslide, Rhymney Valley. In: Landslides and landslide management in South Wales. National Museums & Galleries of Wales, Cardiff

Carey JM (2011) The progressive development and post-failure behaviour of deep-seated landslide complexes. PhD thesis, Durham University. Available at Durham E-Theses Online: http://etheses.dur.ac.uk/3186/

Carlà T, Farina P, Intrieri E, Botsialas K, Casagli N (2017) On the monitoring and early-warning of brittle slope failures in hard rock masses: examples from an open-pit mine. Eng Geol 228:71–81. https://doi.org/10.1016/j.enggeo.2017.08.007

Carlà T, Intrieri E, Di Traglia F, Gigli G, Casagli N (2018a) Methods to improve the reliability of time of slope failure predictions and to setup alarm levels based on the inverse velocity method. In: Sassa K et al (eds) Landslide dynamics: ISDR-ICL landslide interactive teaching tools. https://doi.org/10.1007/978-3-319-57774-6_39

Carlà T, Macciotta R, Hendry M, Martin D, Edwards T, Evans T, Farina P, Intrieri E, Casagli N (2018b) Displacement of a landslides retaining wall and application of an enhanced failure forecasting approach. Landslides 15:489–505. https://doi.org/10.1007/s10346-017-0887-7

Crosta GB, Agliardi F (2002) How to obtain alert velocity threshold for large rockslides. Phys Chem Earth 27(2002):1557–1565. https://doi.org/10.1016/S1474-7065(02)00177-8
Dick GJ, Eberhardt E, Cabrejo-Liévano AG, Stead D, Rose ND (2015) Development of an early-warning time-of-failure analysis methodology for open-pit mine slopes utilizing ground-based slope stability radar monitoring data. Can Geotech J 52:515–529. https://doi.org/10.1139/cgj-2014-0028
Fukuzono T (1985) A new method for predicting the failure time of a slope. In: Proceedings of the fourth international 51 conference and field workshop on landslides (Tokyo; 1985). Tokyo University Press 1985:145–150
Fukuzono T (1990) Recent studies on time prediction of slope failure. Landslide News 4:9–12
Gigli G, Fanti R, Canuti P, Casagli N (2011) Integration of advanced monitoring and numerical modelling techniques for the complete risk scenario analysis of rockslides the case of Mt. Beni (Florence, Italy). Eng Geol 120:48–59. https://doi.org/10.1016/j.enggeo.2011.03.017
Intrieri E, Gigli G (2016) Landslide forecasting and factors influencing predictability. Nat Hazards Earth Syst Sci 16:2501–2510. https://doi.org/10.5194/nhess-16-2501-2016
Intrieri E, Gigli G, Mugnai F, Fanti R, Casagli N (2012) Design and implementation of a landslide early warning system. Eng Geol 147-148(2012):124–136. https://doi.org/10.1016/j.enggeo.2012.07.017
Intrieri E, Gigli G, Casagli N, Nadim F (2013) Landslide early warning system: toolbox and general concepts. Nat Hazards Earth Syst Sci 13(85–90):2013. https://doi.org/10.5194/nhess-13-85-2013
Ju NP, Huang J, Huang RQ, He CY, Li YR (2015) A real-time monitoring and early warning system for landslides in Southwest China. J Mt Sci 12(5):1219. https://doi.org/10.1007/s11629-014-3307-7
Kayesa G (2006) Prediction of slope failure at Letlhakane mine with the Geomos slope monitoring system. In: Proceeding of the international symposium on stability of rock slopes in open pit mining and civil engineering, Cape Town, South Africa, 3–6 April 2006. South African Institute of Mining and Metallurgy, Johannesburg, pp 605–622
Knox G (1927) Landslides in South Wales Valleys. In: Proceedings of the South Wales Institute of Engineers, 43
Manconi A, Giordan D (2015) Landslide early warning based on failure forecast models: the example of the Mt. de la Saxe rockslide, Northern Italy. Nat Hazards Earth Syst Sci 15:1639–1644. https://doi.org/10.5194/nhess-15-1639-2015
Mazzanti P, Rocca A, Bozzano F, Cossu R, Floris M (2012) Landslides forecasting analysis by displacement time series derived from satellite InSAR data: preliminary results. In: Proceedings 'Fringe 2011 Workshop', Frascati, Italy, 19-23 September 2011 (ESA SP-697, January 2012)
Mazzanti P, Bozzano F, Cipriani I, Prestininzi A (2017) New insights into the temporal prediction of landslides by a terrestrial SAR interferometry case study. Landslides 1:55–68. https://doi.org/10.1007/s10346-014-0469-x
Michoud C, Bazin S, Blikra LH, Derron MH, Jaboyedoff M (2013) Experiences from site specific landslide early warning systems. Nat Hazards Earth Syst Sci 2013(13):2659–2673. https://doi.org/10.5194/nhess-13-2659-2013
Paté-Cornell ME (1986) Warning system in risk management. Risk Anal 6(2):223–234
Petley DN (2004) The evolution of slope failures: mechanisms of rupture propagation. Nat Hazards Earth Syst Sci 4(1):147–152, Copernicus Publications on behalf of the European Geosciences Union. https://doi.org/10.5194/nhess-4-147-2004
Rose ND, Hungr O (2007) Forecasting potential rock slope failure in open pit mines using the inverse-velocity method. Int J Rock Mech Min Sci 44(2):308–320. https://doi.org/10.1016/j.ijrmms.2006.07.014
Saito M (1965) Forecasting the time of occurrence of a slope failure. In: Proceedings of the 6th international conference on soil mechanics and foundation engineering, 2, pp 537–539
Saito M (1969) Forecasting time of slope failure by tertiary creep. In: Proceedings of the seventh international conference on soil mechanics foundation engineering, 1, pp 315–318 (Paris)

Segalini A, Valletta A, Carri A (2018) Landslide time-of-failure forecast and alert threshold assessment: a generalized criterion. Eng Geol 245(1):72–80. https://doi.org/10.1016/j.enggeo.2018.08.003

Siddle HJ, Moore R, Carey JM, Petley DN (2007) Pre-failure behavior of slope materials and their significance in the progressive failure of landslides. In: Mathie E, McInnes R, Fairbank H, Jakeways J (eds) Landslides and climate change: challenges and solutions. Proceedings of the international conference of landslides and climate change, Ventnor, Isle of Wight. Taylor & Francis, UK, pp 21–24

Stähli M, Sättele M, Huggel C, McArdell BW, Lehmann P, Van Herwijnen A, Berne A, Schleiss M, Ferrari A, Kos A, Or D, Springman SM (2015) Monitoring and prediction in early warning systems for rapid mass movements. Nat Hazards Earth Syst Sci 15:905–917. https://doi.org/10.5194/nhess-15-905-2015

Terzaghi K (1950) The mechanism of landslides. In: Paige S (ed) Application of geology to engineering practice, Berkey volume. Geological Society of America, Boulder, pp 83–123

Thiebes B, Glade D (2016) Landslide early warning systems – fundamental concepts and innovative applications. In: Aversa S, Cascini L, Picarelli L, Scavia C (eds) Proceedings of the 12th international symposium on landslides, June 12–19, 2016, 3. CRC Press, Napoli, pp 1903–1911. https://doi.org/10.1201/b21520-238

UN-ISDR (2006) Developing early warning systems, a checklist: third international conference on early warning (EWC III), 27–29 March 2006, Bonn, Germany – UNISDR Available online: https://www.unisdr.org/we/inform/publications/608. Accessed on 21 Jan 2019

Varnes DJ (1978) Slope movement types and processes. In: Schuster RL, Krizek RJ (eds) Special report 176: landslides: analysis and control. Transportation and Road Research Board, National Academy of Science, Washington, DC, pp 11–33

Voight B (1988) A method for prediction of volcanic eruptions. Nature 332:125–130. https://doi.org/10.1038/332125a0

Voight B (1989) A relation to describe rate-dependent material failure. Science 243:200–203. https://doi.org/10.1126/science.243.4888.200

Chapter 18
Evaluation of prediction capability of the MaxEnt and Frequency Ratio methods for landslide susceptibility in the Vernazza catchment (Cinque Terre, Italy)

Emanuele Raso, Diego Di Martire, Andrea Cevasco, Domenico Calcaterra, Patrizio Scarpellini, and Marco Firpo

Abstract The research is focused on the Vernazza catchment, an area of 5,75 km^2 belonging to the Vernazza municipality in the Cinque Terre National Park, Italy; here, landslide susceptibility maps are produced using two different statistical methods by analyzing several intrinsic factors controlling landslides. It is also intended to evaluate the maps to determine the comparison between the coverage of high susceptibility areas obtained through different methods.

The first statistically based method is a presence–absence (Frequency Ratio) method, while the second one is a presence-only (MaxEnt) method; the acquisition and preparation of the predisposition factors are also described, as well as their sensitivity and hierarchy regarding the landslide susceptibility models. Furthermore, in order to understand the effective improvement brought by the performance of the models, a validation using the receiving operator characteristics (ROC) and the area under curve (AUC) techniques has been carried out.

The role played by variables such as land use and FAS is well visible: the outputs generated through both methods show a uniform distribution of very high

E. Raso (✉)
Department of Earth Sciences, Environment and Resources (DISTAR), Federico II University of Napoli, Complesso Universitario di Monte Sant'Angelo, Napoli, Italy

Ente Parco Nazionale delle Cinque Terre, Manarola, Riomaggiore, Italy
e-mail: emanuele.raso@unina.it

D. Di Martire · D. Calcaterra
Department of Earth, Environment and Resources Sciences, Federico II University of Naples, Complesso Universitario Monte Sant'Angelo, Naples, Italy

A. Cevasco · M. Firpo
Department of Earth, Environmental and Life Sciences, University of Genova, Genoa, Italy

P. Scarpellini
Ente Parco Nazionale delle Cinque Terre, Manarola, Riomaggiore, Italy

M. De Maio, A. K. Tiwari (eds.), *Applied Geology*,
https://doi.org/10.1007/978-3-030-43953-8_18

susceptibility values on the medium-lower right portion of the catchment, and also the "aspect" variable, in which the value of each cell in a dataset indicates the direction of the cell's slope faces, strongly influences the results since the west–south west-facing cells are considered as prone to generate landslides.

Results obtained for assessing landslide susceptibility show good prediction rate curves for both the tested methodologies, with higher values for the frequency ratio susceptibility model. However, for the MaxEnt susceptibility models, these values are lower, though, without ever decreasing below 0.60. In both cases, future developments of the adopted methods could involve a further distinction of landslide type to evaluate the potential of model prediction specifically for each landslide category.

Keywords Cinque Terre national park · Landslide susceptibility · Statistical methods · Maxent · Frequency ratio · Vernazza catchment · Terraced slopes

18.1 Introduction

Landslides have been historically considered as the most frequent geomorphological hazard in Italy and furthermore, in the last 20 years, a higher frequency of events has been registered (Cruden and Varnes 1996); during the last decades, mainly due to rise of geomorphological risk, research on landslides has increased in Italy and worldwide (Guzzetti 2000; Crozier and Glade 2006; Corominas et al., 2014). The Geotechnical Society Commission, on behalf of the United Nations and UNESCO, proposed a catalogue of the world's landslides in 1993 and established guidelines for the standardized description of landslides and classification criteria. A recent important contribution in terms of landslides classification is provided by Hungr et al. (2014).

Liguria Region and, more specifically, the Cinque Terre area present geological, geomorphological, hydrological, climatic, and anthropogenic features predisposing to landslides: the inventory of Italian landslides (IFFI) highlights about 7500 landslides, approximately 8% of the whole territory. Most of the surveyed landslides in IFFI catalogue are complex type, followed by rotational/translational sliding (Bottero et al., 2004). In this region, many landslides were triggered by heavy rainfall events, whose number is rising year after year (Cevasco et al., 2013, 2014; Faccini et al., 2015; Galve et al., 2015; Raso et al., 2017, 2019).

The Cinque Terre climate is mostly influenced by the Ligurian Sea and the Northern Apennine chain, which provide relief from Italy's north winds, turning the climate of the Ligurian coast to the mild side; the average temperature during the summer is around 24°C, while the mean temperature values during the winter are observed around 9°C (Raso et al., 2017). The rainfall data belonging to the weather stations in the Cinque Terre area show remarkable average precipitation amounts, with heavy and frequent rains mostly occurring during spring (mean rainfall values: 80,38 mm/month) and fall (mean rainfall values: 118,56 mm/month) as shown by the tragic events of October 25th, 2011 during which flash floods and landslides badly hit the Cinque Terre villages: the rainfall data related to that dramatic event were exceptional, with a precipitation climax of 382 mm/24 h in Monterosso al Mare (Cevasco et al., 2013).

Within the context of slope instability, landslides can affect human settlements and influence their activity (Glade and Crozier 2005): several landslide susceptibility mapping methods such as direct geomorphological mapping, analysis of landslide inventories, heuristic methods, neural networks (Choi et al., 2012), logistic regression (Tsangaratos and Ilia 2016) fuzzy logic, expert systems, and process-based models have been used to assess the vulnerability of different parts of an area to landslide (Nourani et al., 2014).

Thus, landslides susceptibility mapping, especially in vulnerable areas such as Liguria, is essential for management of several social activities and to help decision-makers to be more effective; the main goal of this research contribution is therefore to reduce the shortcoming in knowledge of the causes of landslides by providing different landslide susceptibility maps through the implementation of statistically based methods.

18.2 Study Area and Landslide Inventory

The Vernazza catchment is located between the Ligurian Sea and the Northern Apennines chain. It sums up characteristic morphological features of small coastal catchments such as (1) a few km^2 wide area (5.75 km^2); (2) steep to sharp slopes due to their closeness to the sea, and (3) small, narrow rivers characterized by remarkable erosive power and capacity to transfer sediment, thanks to their sharp profiles influenced by a tectonically controlled stream network (Abbate 1969; Cevasco 2007). Mt. Malpertuso (815 m a.s.l.) is the highest mountain in the area, and it is located in the middle portion of the National Park between the Cinque Terre and Vara valley.

The bedrock, geographically belonging to the Northern Apennine chain sector, is characterized by sedimentary rocks belonging to three different tectonic units (from the older to the younger): Tuscan Nappe, Marra unit, and Canetolo complex (Giammarino and Giglia 1990). The Marra Unit and the Canetolo complex mainly crop out along a NW-SE oriented direction, occupying the central portion of the catchment. The first one is prevalently composed of marls and siltstones (Pignone Marls Fm.) while the second one includes shales with a lower content of limestones and silty sandstones (Canetolo Shales and Limestones Fm.), marly limestones and calcarenitic turbidites (Groppo del Vescovo Limestones Fm.), and fine-grained sandstone turbidites (Ponte Bratica Sandstones Fm.) (Pepe et al., 2019). On the other hand, the Tuscan Nappe crops out in the majority of the basin area and chiefly consists of a turbiditic sandstone–siltstone flysch, with coarse and medium-grained sandstone beds and thin interbedded siltstones (Macigno Fm).

The Tuscan Nappe and the Canetolo Units are included in a wide overturned southwest-verging antiform fold whose axis strikes 150 N. These units are bounded to the Northeast by a major normal fault (the La Spezia Fault), beyond which the Ligurian Units outcrop (Federici et al., 2001).

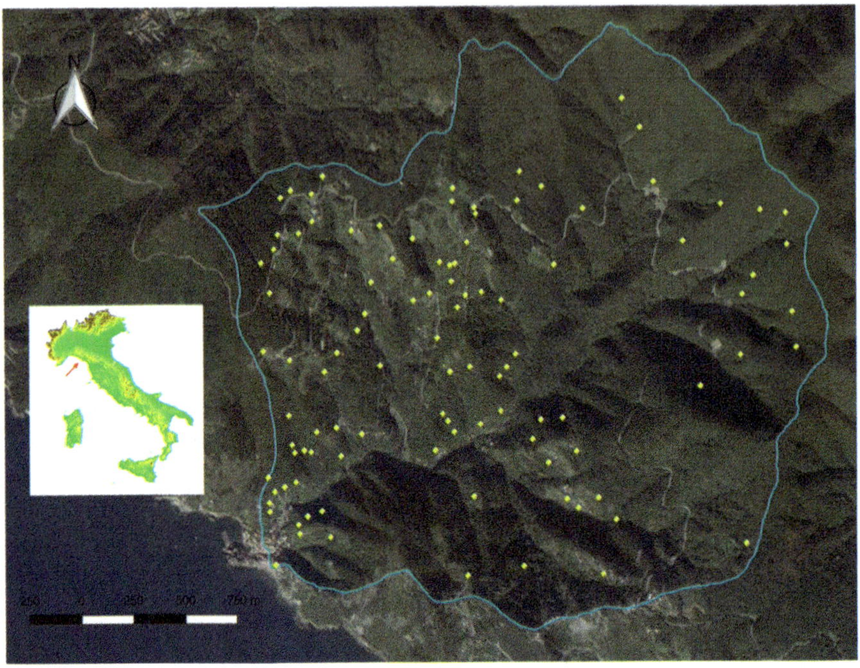

Fig. 18.1 Distribution of 106 landslide source spots (referred to landslides showing an area higher than 25 m^2) in the Vernazza catchment. (After Cevasco et al., 2013)

Peculiar land-use conditions characterize the Cinque Terre and, in particular, the slopes within the Vernazza catchment. These slopes have been almost completely terraced with vineyards and olive groves during the past millennium, reaching their apex at the end of nineteenth century. Following the exodus of farmers during the last century, terraced slopes have been progressively abandoned and covered by Mediterranean scrub and pine, leading to increasing geomorphological instability (Brandolini 2010; Terranova et al., 2002, 2006). In general, even in currently cultivated terraced areas, the state of preservation of dry stonewalls is poor due to limited maintenance.

Here, a total of 106 landslides (Fig. 18.1) with areal extension >25 m^2 belonging to different typologies such as rotational/translational slides, debris flows and debris/gravel slides were considered for a susceptibility analysis.

18.3 Materials and Methods

Two different approaches have been chosen for predicting the spatial distribution of landslides. The first one (Frequency Ratio – FR (Bonham-Carter 1994) tries to reveal the main mechanisms that the environmental variables exert on a determinate phenomenon. The second (MaxEnt – Phillips et al., 2004) applies a mathematical function, based on various classification algorithms, to correlate the areal distribution of the landslides with the selected variables (Brenning 2005). In scientific literature, several statistical methods have been increasingly used in landslide susceptibility mapping (Moosavi and Niazi 2016), comprehending the two methods used in this contribution.

Statistical methods are applied to cartographic units defined a priori, as raster units, units of unique condition, geological-geomorphological units, morpho-hydrographical units, or administrative units (Guzzetti et al. 2005). As it was already mentioned, raster (matrix) units with 5 m pixel are used in the present work.

The statistical methods are bivariated or multivariated: in this study, a bivariated statistical method (e.g. the Frequency ratio method) is used (Constantin et al., 2011; Chen et al., 2018), to model landslide susceptibility. In addition, a multivariated method, the Maximum Entropy (MaxEnt), is adopted in order to perform, in addition to landslide susceptibility modelling, also an expeditious sensitivity analysis of each variable used as the predisposing factor.

18.3.1 Variables Used as Predictors

The choice for the set of predisposing factors, which generate slope instability, is an important task for landslide susceptibility assessed statistically. Accordingly, it is expected a discriminatory capacity from these factors with the aim of rebuilding the constraints leading to slope mass movements during the previous decades (Havenith et al., 2015). These predisposing factors, inherent to the terrain, are often static (e.g., aspect, lithology) and influence the relationship with the different types of landslide and inventories and the determination of their sensitivity and hierarchy.

Another distinction has to be stressed out regarding the variables values: some of them are continuous (e.g., aspect and slope, even if subsequently discretized in categories), some are discrete (lithology, road distance, stream distance, land use), and some others (FAS, soil thickness) are derived from interpolation of punctual values (Table 18.1).

18.3.2 Frequency Ratio

The frequency ratio method (FR – Bonham-Carter 1994) shows a quantitative approach because the calculation of this parameter derives from the percentage ratio between the pixel number of the area hit by landslides, related to a determined parameter (e.g., slope, exposure, land use, etc.), often divided into subcategories (slope range, etc.) and the pixel number of the class in which this phenomenon occurs.

Table 18.1 Continue and discrete independent variables used as predictors in susceptibility analysis

 Lithology	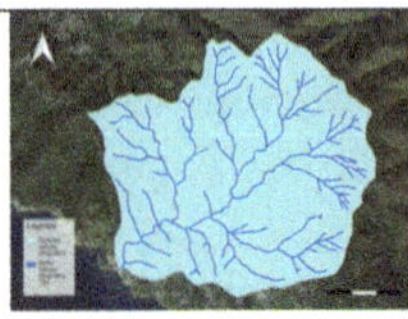 Stream distance	 Road distance	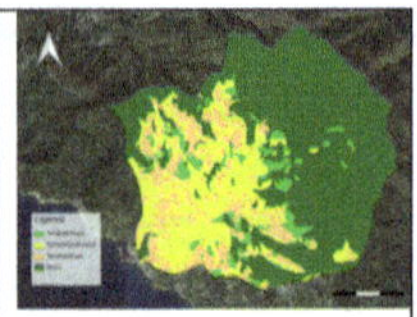 Land use
 Slope	 Aspect	 FAS	 Soil thickness

The FR value is then calculated and extended to each class “*I*” of a “*p*” parameter.

$$FR_{pi} = \frac{A_{Lpi}/A_L}{A_{pi}/A}$$

where

FR_{pi} = *frequency ratio* value related to a class *i* of a *p* parameter.
A_{Lpi} = landslide area inside the class *i* of a *p* parameter.
A_L = total landslide area
A_{pi} = total area of class *i* of a *p* parameter
A = total analyzed area.

FR values higher than 1 imply a positive correlation between landslide occurrences and landslide-causing variable (high susceptibility), while FR values lower than 1 identify layers with low landslide susceptibility (Carratù et al., 2015); In fact, the percentage of the area interested by mass movements, falling in a class *i* of a *p* parameter is greater than the percentage between the area of the class of a *P* parameter and the total area.

Subsequently, it is possible to obtain the landslide susceptibility index (LSI) summing the FR values extended to all classes and defined for each factor:

$$\text{LSI} = \sum_{p=p_1}^{p_j} \sum_{i=1}^{n_k} FR_{pi}$$

Once this parameter has been obtained and extended to each raster cell composing the grid, a classification of 5 different susceptibility levels is performed (I1: Low Susceptibility; I2: Moderate Susceptibility; I3: Medium Susceptibility; I4: High Susceptibility; I5: Very High Susceptibility), diving the total range of LSI values in 5 classes.

The entire procedure has been implemented in a GIS environment, specifically QGIS and GrassGIS – free and open source software. In fact, every single thematic query (*p* parameter) can be intersected (overlay operation) with the "landslides" parameter, with the aim to obtain quantitative areal distributions in terms of individual *i* class parameters.

18.3.3 *MaxEnt*

The MaxEnt (Maximum Entropy) GIS application (Phillips et al., 2004) weights each variable with a constant value. The probability output corresponds to the weighted sum of each variable divided by a scaling constant to make sure that the scale of output values goes from 0 to 1. The model starts with a steady distribution of the probability and iteratively changes the weight at each time, to increase the probability to achieve the best likelihood distribution. The algorithm is always convergent and hence the outputs are deterministic.

$$H = \sum_{i=1}^{n} p_i \ln\left(\frac{1}{p_i}\right)$$

$$H_{\min} = 1 \cdot \ln\left(\frac{1}{1}\right) = 0$$

$$H_{\max} = \sum_{i=1}^{n} \frac{1}{x} \ln(x) = x\left(\frac{1}{x} \ln x\right) = \ln x$$

This model attempts, after a set of iterations, to obtain the distribution of probability values of the highest entropy based on limitations imposed by the data distribution and the constraints of the entire study area. This algorithm also contemplates the possibility to incorporate a nonabsence data for the dependent variable, which are obtained through the generation of pseudo-random absences (Henriques, 2014).

In the landslide environment, Maxent characterizes the distributions of a casual landslide occurrence with an equal probability to fall into any set of predisposing factors. When behaving in a nonrandom way, e.g., when landslides occur on specific conjugations of predisposing factors, the distribution output is irregular and entropy is lower, therefore MaxEnt automatically invert the values, converting 0 (absence of entropy) to 1 (maximum probability of landslide occurrence).

18.4 Results and Discussion

18.4.1 Variables Used as Predictors

In order to depict the spatial susceptibility of the detected mass movements, we adopted both a presence–absence and a presence-only method, together with a set of morphometric, geotechnical, geologic, and land use predictors.

The predictors were chosen with reference to previous scientific studies on shallow landslide activations: the lithology of the area comprises two different formations (Macigno formation (MAC) and Argille and Calcari di Canetolo (ACC)) and many studies worldwide have shown that landslides are often conditioned by the bedrock composition (Luzi and Pergalani 1999; Duman et al., 2006; Yalcin et al., 2011).

The river network map was employed to calculate an attribute that reflects the proximity of cells to streams, a buffer distance to channel network fixed on 10 m, since drainage networks have negative impacts on landslide susceptibility due to the erosion of the slope base and saturation of the underwater section of the material forming the slope (Vijith and Madhu 2008).

Distance to roads is considered as an important factor for triggering landslide occurrences, and has been accepted as one of the most important anthropogenic factors by many investigators (Yilmaz 2009): here, a buffer zone with a 10 m width on both sides considering the centerline was generated.

Land-use classes that were visually identified and used in the analysis were four: cultivated terraces, abandoned terrace with poor cover (bush), abandoned terrace with dense cover (wood), scrubland and wood. Terraces with poor cover are those that have been abandoned for a short time and have more natural herbaceous cover, whereas abandoned terraces with dense cover have been abandoned for a longer period of time and are mainly occupied by forest tree species or scrub (Cevasco et al., 2013). Scrubland and wood areas usually show a soil thickness reduction as slope angle increases, until they assume a lenticular shape with high slope angle values (De Vita and Celico 2006). These groupings were chosen in order to aggregate similar land-use zones that may also have a similar effect on erosion processes.

The slope angle variable was calculated using a second-degree polynomial method (Zevenbergen and Thorne 1987). Since the aspect of slopes is not a linear variable (its lower and higher values have the same meaning), it was classified into five categories.

Long recognized as an important topographic variable, this aspect affects the amount and daily cycle of solar radiation received at different times of the year and has a strong influence on the microclimate, especially air temperature, humidity, and soil moisture (Rosenberg et al., 1983). Since the aspect of slopes is not a linear variable (its lower and higher values have the same meaning), it was classified into eight categories, each one representing 45 degrees.

Soil thickness is, however, widely recognized as a controlling factor in numerous surface and subsurface processes, e.g., landscape evolution, sediment budgets, and landslides (Catani et al., 2010). Its values, classified into nine classes, were obtained through interpolation of punctual thickness values along the Vernazza catchment – a GIS tool was used to accomplish the purpose, in particular r.spline (GrassGIS), using a cubic equation for points weighting on neighborhood cells.

Through the FAS (Friction Angle minus Slope), the combination of the friction angles with the slope angle is based on the relationship established in the Mohr–Coulomb failure criterion (Roopnarine et al., 2013), which indicates that the soil friction angle is a component of shear strength such that:

$$\mathrm{T}f = c + \sigma(\tan \phi)$$

where

$\mathrm{T}f$ is the shear strength; σ is the applied normal stress; ϕ soil friction angle and c is the apparent cohesion. Classification of FAS was based on the resultant value of rasterized analysis of Friction angle minus Slope, where both the slope pixel values and the friction angle ones were derived from a 5×5 m resolution Digital Terrain Model.

18.4.2 Results – Frequency Ratio

Results obtained from the bivariate, presence–absence Frequency Ratio method highlight some interesting deductions: first, the high control on landslide susceptibility Index expressed by variables with linear, bidimensional development (e.g., River Network (stream distance) and Road network (Road distance)); secondly, the influence of some discrete variables such as Land Use, in which the higher FR value has been registered (Table 18.2) by the TB (terraced – bush covered) class, meaning that the terraced area abandoned for a short time is more prone to fail over intense rainfall.

Continuous variables such as FAS (Table 18.3) also show interesting results, in particular an expected decrease of FR value with the increase of FAS value, which is physically reasonable; low FR values for the poorest class (−15;5) could be explained by the fact that these FAS values cover a small area (0,15% of the total catchment area) compared to the other ones (Fig. 18.2).

The resulting landslide susceptibility map made through Frequency Ratio method highlights a strong influence by linear networks on landslide triggering: very high susceptibility areas mainly follow the previously described buffer zones; other clear factors influencing the dynamics of the Vernazza catchment slopes are land use (determining high susceptibility values along the slopes belonging to the lower right portion of the catchment where terraces characterized by recent abandonment are more common to find) and soil thickness, increasing considerably the FR values

Table 18.2 Summary of FR values calculated for each I class related to a P parameter

	Class	% of the studied area (a)	% of landslide area per category (b)	Frequency ratio (b/a)
Slope	1 (0°–10°)	0,70%	0,27%	0.39
	2 (11°–20°)	6,38%	7,26%	1,14
	3 (21°–30°)	28,83%	29,73%	1,03
	4 (31°–40°)	54,76%	52,00%	0,95
	5 (>41°)	9,33%	10,75%	1,15
Aspect	0°–45°	4,72%	3,45%	0,73
	46°–90°	3,56%	4,80%	1,35
	91°–135°	13,82%	19,43%	1,41
	136°–180°	13,91%	13,38%	0,96
	181°–225°	18,42%	17,51%	0,95
	226°–270°	19,46%	22,52%	1,16
	271°–315°	16,72%	13,69%	0,82
	316°–360°	9,39%	5,23%	0,65
Road distance	0–20 m	12,69%	17,71%	1,40
	> 20 m	87,31%	82,29%	0,94
Lithology	MAC	66,65%	64,01%	0,96
	ACC	33,35%	35,99%	1,08
PFAS	1 (−15;−5)	0,15%	0,02%	0,12
	2 (−4;5)	6,22%	7,81%	1,26
	3 (6;15)	66,34%	73,82%	1,11
	4 (16;25)	26,18%	18,71%	0,71
	5 (26;35)	0,99%	0,04%	0,04
	6 (36;45)	0,12%	0,00%	0,00
Stream distance	0–10 m	10,44%	18,13%	1,74
	> 10 m	89,56%	81,87%	0,91
Land use	Terraced – Wood	5,92%	5,34%	0,90
	Terraced – Cultivated	21,12%	31,93%	1,51
	Terraced – Brush	13,92%	22,72%	1,63
	Wood	59,04%	40,01%	0,68
Soil thickness	1 (0–50)	0,54%	1,84%	3,41
	2 (51–100)	1,94%	1,66%	0,86
	3 (101–150)	12,05%	7,75%	0,64
	4 (151–200)	42,83%	34,48%	0,80
	5 (201–250)	32,84%	36,90%	1,12
	6 (251–300)	4,78%	8,29%	1,73
	7 (301–350)	3,22%	5,65%	1,76
	8 (351–400)	1,06%	3,52%	3,31
	9 (401–450)	0,57%	2,27%	3,99
	10 (451–500)	0,17%	1,19%	6,87

Table 18.3 Comparison between FR values extended to different class on land use (in green) and FAS (in red)

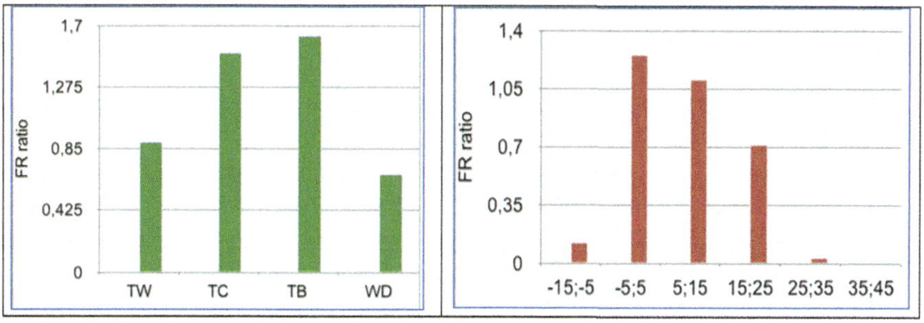

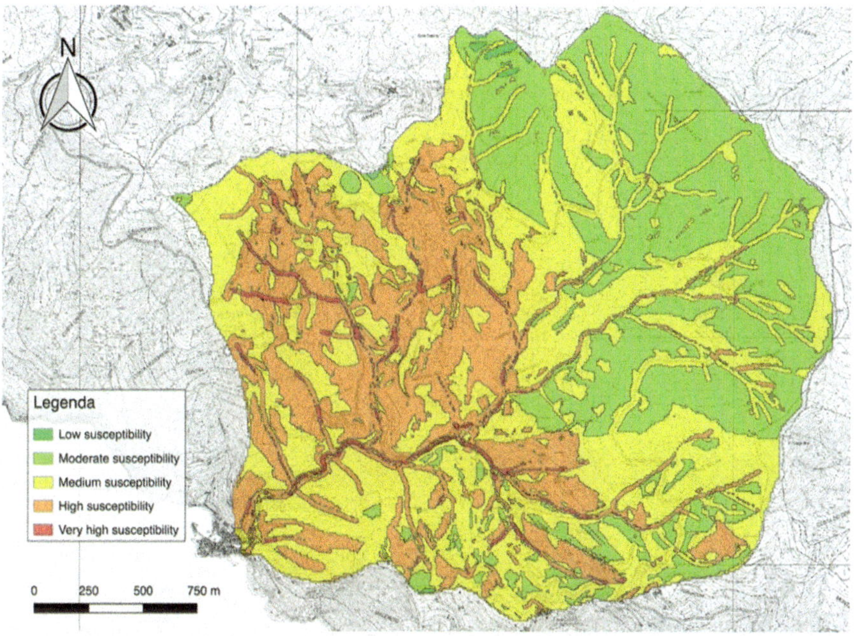

Fig. 18.2 Map showing landslide susceptibility values obtained through frequency ratio method

when high amounts of soil covers are detected; here, as previously anticipated with reference to FAS values, it is evident how poorly represented classes could overestimate or underestimate the consequent FR value (see 0–50 cm class of soil thickness and, as stated before, −15;−5 class of FAS).

18.4.3 Results – MaxEnt

The software is composed of a jar file (maxent.jar), usable on any calculator running an updated Java version. Maxent and some related literature and scientific papers can be obtained as a free software from the website www.cs.princeton.edu/~schapire/maxent (Fig. 18.3).

Through the Maxent elaboration, the probability distribution is defined mathematically and therefore the formulation of the model: this presence-only methodology is relatively transparent and examines interconnections between the chosen variables. Moreover, it could estimate the impact of each variable in the final output.

The role played by variables such as land use and FAS is well visible: there is a uniform distribution of very high susceptibility values on the medium-lower right portion of the catchment, and also this aspect heavily influences the results since the west–south west-facing slope is considered as prone to generate landslides.

The potential disadvantages of this method are to be found in a lack of selection in procedure variables, the extreme importance of computational procedures and possible weaknesses in dealing with biased samples, which gives importance to the dataset goodness.

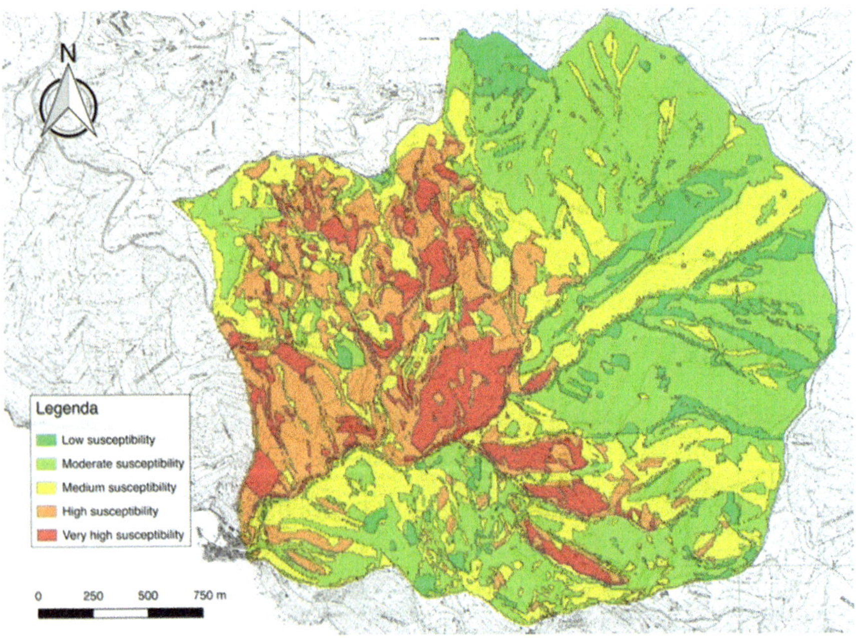

Fig. 18.3 Map showing landslide susceptibility values obtained through MaxEnt calculation

18.4.4 Results Validation – ROC Curves

ROC analysis originally came out in order to establish the performance of radar receivers in the defense field; since then, it has been used in several research areas (Adams and Hand 1999; Provost and Fawcett 2001). The area comprised below the ROC curve (area under curve – AUC) can be considered as a benchmark to determine the global model prediction capacity (Hanley and Mcneil 1982): the wider the surface under the curve, the higher the model predictive attitude over the whole range of possible outputs. The dots on the ROC curve (FP, TP) are couples of values obtained from different contingency tables created through the insertion of distinct cutoffs (Fig. 18.4). Points located near the upper right portion mean lower cutoff values. More in general, an ROC curve has a higher prediction rate if it is closer to the upper left corner.

If the performance of a model is assessed through the use of a dataset not involved in the development of the model and both the ROC curves for the evaluation and production of datasets are located close to each other in the ROC graph and AUC

	frana	non frana
pericoloso	tp	fp
non pericoloso	fn	tn

Sensitività =	tp/tp+fp
Specificità =	tn/fn+tn

1 - specificità	sensitività
0	0
0,053	0,296
0,059	0,274
0,119	0,672
1	1

FR		
i4 + i5	Positive	Negative
Test Positive	17895	47495
Test Negative	9671	153795
Sensitività	0,274	
Specificità	0,941	
Correz. %		
Hit Ratio		

FR		
i5	Positive	Negative
Test Positive	1166	568
Test Negative	26400	195170
Sensitività	0,672	
Specificità	0,881	
Correz. %		
Hit Ratio		

FR		
i4	Positive	Negative
Test Positive	18969	45056
Test Negative	8567	151795
Sensitività	0,296	
Specificità	0,947	
Correz. %		
Hit Ratio		

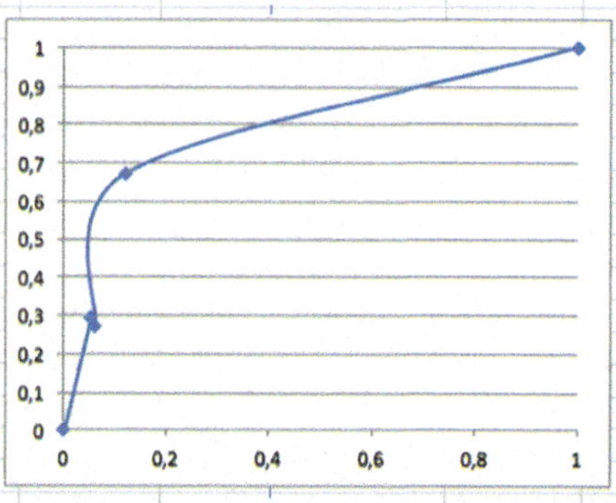

Fig. 18.4 ROC curves results obtained from the FR model

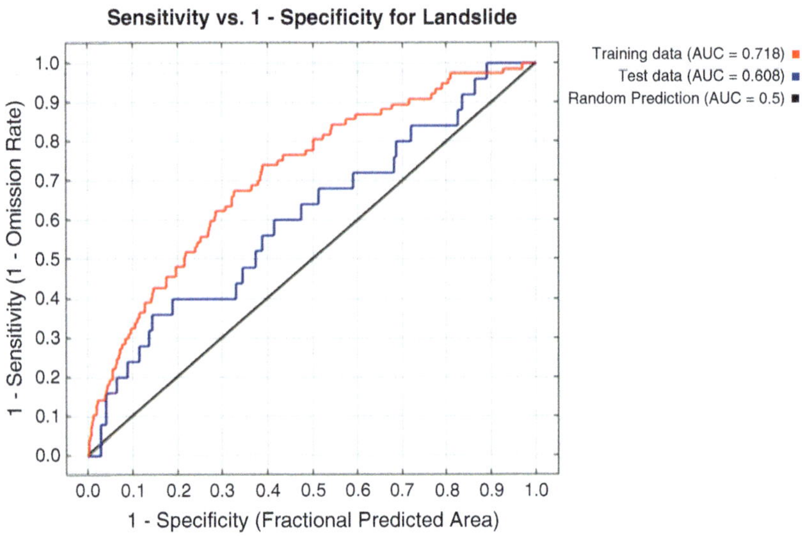

Fig. 18.5 ROC curve representing training and test data obtained from Maxent model

values fall in the range between 0.6 (middle accuracy) and 0.8, then the model accuracy could be evaluated as high (highly accurate; Swets 1988).

Here, the results obtained from validation of the FR model are shown: only the two most relevant susceptibility categories are considered (I4 – high susceptibility, and I5 – very high susceptibility). Since the model is based on a presence–absence (dichotomic) scheme, the ROC curve will consider as sensitivity (Y) the ratio between the landslide cell (positive) in a determinate class and the sum between the landslide cell (positive) and no-landslide cells (negative) for the same class; specificity (X), qualitatively speaking, is the total (1) minus the ratio between no-landslide cells (negative) and the sum of landslide cells (positive) and no-landslide cells (negative).

First, results of ROC curves analysis are shown for a run conducted with only training data: this is a typical ROC curve scenario in which, differently from what previously assessed for FR method, the sensitivity is based on omission rates (since MaxEnt is not a method based on the presence and absence of the dichotomic variable but a presence-only one) and specificity is based on the fraction of progressive predicted area, value changing while the model is running through a convergence.

If the data are used for testing and training (respectively blue and red lines in Fig. 18.5), then the curve's path will be the same. If data are divided into two groups, one for testing and the other for training, it is normal for the blue (testing) line to show a lower AUC than the red (training) line. The blue (testing) line shows the fit of the model to the testing data, and is the real test of the model predictive power. The red (training) line indicates the "fit" of the model to the training dataset. The black line shows the path that would be expected if the model would have a random behavior. If the blue line (the test line) goes under the black line, then this means that the model output shows worse results than a random model would. The more toward the top left of the graph the blue line tends to be, the better the model is at predicting the presence contained in the test sample of the dataset (Fielding and Bell 1997).

Since there are only occurrence data and no absence data, the "fractional predicted area" (the portion of the total study area predicted at a determinate time) is applied instead of the fraction of predicted absences predicted, considered a more standard parameter. It is important to note that AUC values tend to be higher for data with narrow distribution ranges, relative to the research area in which the environmental data are distributed. This does not necessarily mean that the models are better, since this behavior is considered as an artifact of the AUC statistic.

18.5 Conclusions

During last years, landslide susceptibility research has increased considerably, as well as its role in urban planning and management: when a landslide susceptibility model is applied in practice, the classification of land according to spatial probability of landslide triggering results in social and economic consequences: indeed, the misclassification of terrain (stable and unstable areas) in a model produces an increase in economic costs and risk exposure. Hence, the evaluation of the models performance and their suitability to different contests are fundamental actions to be carried out.

Quantitative susceptibility assessment of the Vernazza catchment in a GIS environment was obtained through different statistical methods: for the present study, and in order to understand the possible advantages and disadvantages of using different statistically based methods, a simple, presence-absence, statistical bivariated method (Frequency Ratio) was selected, as well as a presence-only, computationally elaborated methodology (MaxEnt) for landslide susceptibility assessment. The obtained prediction rate curves for both the tested methodologies seem to be acceptable (according to the values proposed by Guzzetti et al. (2005)), with slightly higher values for the Frequency Ratio susceptibility model. However, for the MaxEnt susceptibility models these values are lower, though, without ever decreasing below 0.60. In both cases, future developments of the adopted methods could involve a further distinction of landslide type to evaluate the potential of model prediction specifically for each landslide category, as well as the volume calculation and the intensity of the catalogued phenomena.

In terms of spatial distribution of different susceptibility classes, the resulting landslide susceptibility map made through the FR method highlights the role of linear networks to landslide triggering: very high susceptibility areas mainly follow the previously described buffer zones; other clear factors influencing the dynamics of the Vernazza catchment slopes are land use (determining high susceptibility values along the slopes belonging to the lower right portion of the catchment were terraces characterized by recent abandonment are more common to be found) and soil thickness, increasing considerably the FR values when high amounts of soil covers are detected.

On the other hand, the map obtained through the MaxEnt model highlights the role played by variables such as Land Use and FAS: there is a uniform distribution of very high susceptibility values on the medium-lower right portion of the catchment, and also this aspect heavily influences the results since the West-South West-facing slope is considered as prone to generate landslides.

Quantitative susceptibility analysis has contributed to detect the main landslide issues, which could affect a small, coastal municipality whose geomorphological and demographic characteristics are found to be frequent along the North-Western Italian coastline: more specifically, in the Vernazza municipality, a considerable amount of economic resources should be addresses to both upgrade the current emergency plans and provide effective engineering solutions with the aim of decreasing landslide risk; in particular, hazard mitigation techniques should be applied to debris flows and shallow landslides damage mitigation, two landslide types that have considerably increased the destructive power of flash floods triggered by heavy rainfall events during the last years.

References

Abbate E (1969) Geologia delle Cinque Terre e dell'entroterra di Levanto (Liguria Orientale). Mem Soc Geol Ital 8:923–1014

Adams NM, Hand DJ (1999) Comparing classifiers when the misallocation costs are uncertain. Pattern Recogn 32:1139–1147

Bonham-Carter GF (1994) Geographic information Systems for Geoscientists: modeling with GIS. Pergamon Press, Oxford, pp 398–403

Bottero D, Cavallo C, De Stefanis E, Gorziglia G, Poggi F (2004) Progetto Iffi: primi risultati in Liguria. Regione Liguria – Settore politiche dell'assetto del territorio, Genova

Brandolini P (2010) I versanti terrazzati delle Cinque Terre: un patrimonio ambientale e paesaggistico a rischio. Rivista SLM, Ente Italiano della Montagna 36:54–56

Brenning A (2005) Climatic and geomorphological controls of rock glaciers in the Andes of Central Chile: combining statistical modelling and field mapping, Dissertation, Mathematisch-Naturwissenschaftliche Fakultät II, Humboldt-Universität zu Berlin, Berlin, urn:nbn:de: kobv:11–10049648, pp, 412–423

Carratù MT, Di Martire D, Finicelli GF, Calcaterra D (2015) Comparison of bivariate and multivariate analyses for landslide susceptibility mapping in the Phlegraean district: the case study of Camaldoli hill (Napoli, Italy). Rend Online Soc Geol It 35:50–53

Catani F, Segoni S, Falorni G (2010) An empirical geomorphology based approach to the spatial prediction of soil thickness at catchment scale. Water Resour Res 46:W05508

Cevasco A (2007) I fenomeni di instabilità nell'evoluzione della costa alta delle Cinque Terre (Liguria Orientale). Studi costieri 13:93–109

Cevasco A, Brandolini P, Scopesi C, Rellini I (2013) Relationships between geo-hydrological processes induced by heavy rainfalls and land use: the case of 25 October 2011 in the vernazza catchment (Cinque Terre, nw Italy). J Maps 9:289. https://doi.org/10.1080/17445647.2013.780188

Cevasco A, Pepe G, Brandolini P (2014) The influences of geological and land use settings on shallow landslides triggered by an intense rainfall event in a coastal terraced environment. Bull Eng Geol Environ 73:859

Chen W, Peng J, Hong H, Shahabi H, Pradhan B, Liu J, Zhu X, Pei X, Duan Z (2018) Landslide susceptibility modelling using GIS-based machine learning techniques for Chongren County, Jiangxi Province, China. Sci Total Environ 626:1121–1135

Choi J, Oh HJ, Lee HJ, Lee C, Lee S (2012) Combining landslide susceptibility maps obtained from frequency ratio, logistic regression, and artificial neural network models using ASTER images and GIS. Eng Geol 124:12–23

Constantin M, Bednarik M, Jurchescu MC, Vlaicu M (2011) Landslide susceptibility assessment using the bivariate statistical analysis and the index of entropy in the Sibiciu Basin (Romania). Environ Earth Sci 63(2):397–406

Corominas J, Van Westen C, Frattini P, Cascini L, Malet JP, Fotopoulou S, Pitilakis K (2014) Recommendations for the quantitative analysis of landslide risk. Bull Eng Geol Environ 73 (2):209–263

Crozier MJ, Glade T (2006) Landslide hazard and risk: issues, concepts and approach, Landslide hazard and risk. Wiley, West Sussex, pp 1–40

Cruden DM, Varnes DJ (1996) Landslide types and processes. In: Turner AK, Shuster RL (eds) Landslides: investigation and mitigation, Transportation Research Board Special Report 247. National Academy Press, Washington, DC, pp 36–75

De Vita P, Celico P (2006) Distribuzione delle coltri piroclastiche sui versanti carbonatici perivesuviani e suscettibilità a franare. Giornale di Geologia Applicata 3:145–151

Duman TY, Can T, Gokceoglu C, Nefeslioglu HA, Sonmez H (2006) Application of logistic regression for landslide susceptibility zoning of Cekmece Area, Istanbul, Turkey. Environ Geol 51(2):241–256

Faccini F, Raso E, Malgarotto C, Antonielli G (2015) Rockfall risk assessment and management along the "Via Dell'amore" (Lovers' lane) in the Cinque Terre National Park (Italy). In: Engineering Geology for Society and Territory – Volume 2: Landslide Processes. Springer, Cham, pp 1979–1983

Federici PR, Baldacci F, Petresi A, Serani A (2001) Atlante dei Centri Abitati Instabili della Liguria. I Provincia della Spezia. In: Università degli Studi di Pisa. Genova, Italy

Fielding AH, Bell JF (1997) A review of methods for the assessment of prediction errors in conservation presence/absence models. Environ Conserv 24(1):38–49

Galve JP, Cevasco A, Brandolini P, Soldati M (2015) Assessment of shallow landslide risk mitigation measures based on land use planning through probabilistic modelling. Landslides 12:101–114

Giammarino S, Giglia G (1990) Gli elementi strutturali della piega di La Spezia nel contesto geodinamico dell'Appennino Settentrionale. Boll Soc Geol Ital 109:683–692

Glade T, Crozier M (2005) The nature of landslide hazard impact. In: Gladem T, Andersonm MM, Crozierm M (eds) Landslide hazard and risk. Wiley, Chichester, pp 43–74

Guzzetti F (2000) Landslides fatalities and the evaluation of landslide risk in Italy. Eng Geol 58:89–107

Guzzetti F, Stark CP, Salvati P (2005) Evaluation of flood and landslide risk to the population of Italy. Environ Manag 36:15–36

Guzzetti F, Reichenbach P, Cardinali M, Galli M, Ardizzone F (2005) Probabilistic landslide hazard assessment at the basin scale. Geomorphology 72(272):299

Hanley JA, Mcneil BJ (1982) The meaning and use of the area under a receiver operating characteristic (ROC) curve. Radiology 143(1):29–36

Havenith HB, Torgoev A, Schlögel R, Braun A, Torgoev I, Ischuk A (2015) Tien Shan Geohazards database: landslide susceptibility analysis. Geomorphology 249:32–43

Henriques CS (2014) Landslide susceptibility evaluation and validation at a regional scale. Doutoramento em Geografia Física, Universidade de Lisboa – Instituto de Geografia e Ordenamento do Território

Hungr O, Leroueil S, Picarelli L (2014) The Varnes classification of landslide types, an update. Landslides 11:167–194
Luzi L, Pergalani F (1999) Slope instability in static and dynamic conditions for urban planning: the "Oltre Po Pavese" case history (Regione Lombardia Italy). Nat Hazards 20(1):57–82
Moosavi V, Niazi Y (2016) Development of hybrid wavelet packet-statistical models (WP-SM) for landslide susceptibility mapping. Landslides 13:97–114
Nourani V, Pradhan B, Ghaffari H, Sharifi SS (2014) Landslide susceptibility mapping at Zonouz Plain, Iran using genetic programming and comparison with frequency ratio, logistic regression, and artificial neural network models. Nat Hazards 71:523–547. https://doi.org/10.1007/s11069-013-0932-3
Pepe M, Raso S, Brandolini C (2019) Investigation on Farmland Abandonment of Terraced Slopes Using Multitemporal Data Sources Comparison and Its Implication on Hydro-Geomorphological Processes. Water 11(8):1552
Phillips SJ, Dudík ME, Schapire RE (2004) A maximum entropy approach to species distribution modeling. In: Proceedings of the twenty first international conference on machine learning, pp 655–662
Provost F, Fawcett T (2001) Robust classification for imprecise environments. Mach Learn 42 (3):203–231
Raso E, Brandolini P, Faccini F, Realini E, Caldera S, Firpo M (2017) Geomorphological evolution and monitoring of San Bernardino – Guvano coastal landslide (eastern Liguria, Italy). Geogr Fis Din Quat 40:197–210. https://doi.org/10.4461/GFDQ2017.40.12
Raso E, Mandarino A, Pepe G, Di Martire D, Cevasco A, Calcaterra D Firpo M (2019) Landslide inventory of the Cinque Terre National Park, Italy. IAEG/AEG annual meeting proceedings, San Francisco, California, 2018—Volume 1, https://doi.org/10.1007/978-3-319-93124-1_24
Roopnarine R, Eudoxie G, Opadeyi J (2013) Soil friction angle as an instability factor in landslide susceptibility modeling. J Earth Sci Geotech Eng 3(1):55–71
Rosenberg R, Arntz WE, De Flores EC, Flores LA, Carbajal G, Finger I, Tarazona J (1983) Benthos biomass and oxygen deficiency in the upwelling system off Peru. J Mar Res 41(2):263–279
Swets JA (1988) Measuring the accuracy of diagnostic systems. Science 240(4857):1285–1293
Terranova R, Brandolini P, Spotorno M, Rota MP, Montanari C, Galassi D, Nicchia P, Leale S, Bruzzo R, Renzi L, Seronello G, Dagorne A, Bally E, Castex JM, Reynes Trias A, Alomar Canyelles G, Ferrer Garcia I, Rodriguez Gomila R, Grimalt Gelabert M, Mus Amezquita M (2002) Patrimoni de marjades a la Mediterrania Occidental. Una proposta de catalogaciò. 1–243, Commissio Europea DGX, Programa Raphael, Palma Di Mallorca: Fodesma
Terranova R, Zanzucchi G, Bernini M, Brandolini P, Campobasso S, Faccini F, Zanzucchi F (2006) Geologia, geomorfologia e vini del Parco Nazionale delle Cinque Terre (Liguria, Italia). Bollettino della Società Geologica Italiana, Spec 6:115–128
Tsangaratos P, Ilia I (2016) Comparison of a logistic regression and Naïve Bayes classifier in landslide susceptibility assessments: the influence of models complexity and training dataset size. Catena 145:164–179
Vijith H, Madhu G (2008) Estimating potential landslide sites of anupland subwatershed in Western Ghat's of Kerala (India) through frequency ratio and GIS. Environ Geol 55:1397–1405
Yalcin A, Reis S, Aydinoglu AC, Yomralioglu T (2011) A GIS-based comparative study of frequency ratio, analytical hierarchy process, bivariate statistics and logistics regression methods for landslide susceptibility mapping in Trabzon, NE Turkey. Catena 85:274–287
Yilmaz I (2009) Landslide susceptibility mapping using frequency ratio, logistic regression, artificial neural networks and their comparison: a case study from Kat landslides (Tokat—Turkey). Comput Geosci 35:1125–1138
Zevenbergen LW, Thorne CR (1987) Quantitative analysis of land surface topography. Earth Surf Process Land 12:47–56

The manufacturer's authorised representative in the EU is Springer Nature Customer Service Centre GmbH, Europaplatz 3, 69115 Heidelberg, Germany. If you have any concerns regarding our products, please contact ProductSafety@springernature.com

Printed and bound by CPI Group (UK) Ltd, Croydon, CR0 4YY
28/09/2026
02224085-0008